Lost Crops
of the
Incas

Little-Known Plants
of the Andes with
Promise for Worldwide Cultivation

Report of an Ad Hoc Panel of the
Advisory Committee on Technology Innovation
Board on Science and Technology for International Development
National Research Council

National Academy Press
Washington. D.C. 1989

F
3429.3
,A4
L67
1989

The Board on Science and Technology for International Development (BOSTID) of the Office of International Affairs addresses a range of issues arising from the ways in which science and technology in developing countries can stimulate and complement the complex processes of social and economic development. It oversees a broad program of bilateral workshops with scientific organizations in developing countries and conducts special studies. BOSTID's Advisory Committee on Technology Innovation publishes topical reviews of technical processes and biological resources of potential importance to developing countries.

This report has been prepared by an ad hoc advisory panel of the Advisory Committee on Technology Innovation, Board on Science and Technology for International Development, Office of International Affairs, National Research Council. Program costs were provided by the Jesse Smith Noyes Foundation. Staff support was funded by the Office of the Science Advisor, Agency for International Development, under Grant No. DAN 5538-G-SS-1023-00.

Library of Congress Catalog Card Number: 89-42566
ISBN 0-309-04264-X

First Printing, July 1989
Second Printing, June 1990

20397360

2/22/91 M

Panel on Lost Crops of the Incas

HUGH POPENOE, International Program in Agriculture, University of Florida, Gainesville, *Chairman*

STEVEN R. KING, Latin America Science Program, The Nature Conservancy, Rosslyn, Virginia

JORGE LEÓN, Plant Genetics Resource Center, Centro Agronómica Tropical de Investigación y Enseñanza (CATIE) (retired), San José, Costa Rica

LUIS SUMAR KALINOWSKI, Centro de Investigaciones de Cultivos Andinos, Universidad Nacional Técnica del Altiplano, Cuzco, Peru

NOEL D. VIETMEYER, *Senior Program Officer, Inca Crops Study Director* and *Scientific Editor*

MARK DAFFORN, *Staff Associate*

National Research Council Staff
F. R. RUSKIN, *BOSTID Editor*
MARY JANE ENGQUIST, *Staff Associate*
ELIZABETH MOUZON, *Senior Secretary*

Front cover: pepino (Turner and Growers)

Back cover: quinoa (S. King)

This book is dedicated to the memory of *Martín Cárdenas Hermosa (1899–1973), outstanding botanist, Quechua scholar, historian, and a pioneer advocate of Andean crops. He was born of Indian parents and became Rector (President) of the Universidad Autónoma de Cochabamba, Bolivia.*

Cárdenas was friend and advisor to innumerable horticulturists and botanists from many countries. He enhanced gardens around the world with his generous samples of the rich and varied flora of the Andes. A wise and tireless researcher and traveler whom all remember with affection, he published a large number of articles and two books. His 1969 Manual de Plantas Económicas de Bolivia—*now unfortunately out of print—is the most comprehensive review of useful Andean plants ever made. Cárdenas left his herbarium containing about 6,500 Bolivian specimens—many originally described by him—to the Universidad de Tucumán, and his home and library in Cochabamba to the local university for the use of researchers.*

Preface

The primary purpose here is to draw attention to overlooked food crops of the Andes. The crops are not truly lost; indeed, most are well known in many areas of the Andes, especially among Indian groups. It is to the mainstream of international science and to people outside the Andes that they are "lost." Moreover, most of these crops were developed by ancient Indian tribes and were established foods long before the beginning of the Inca Empire about 1400 AD. For all that, however, it was the Incas who, by the time of the Spanish Conquest, had brought these plants to their highest state of development and, in many cases, had spread them throughout the Andean region.

It should be understood that we are not the first to appreciate the potential of these crops. Several agronomists and ethnobotanists—many of them working in the Andes—have begun preserving what remains of these traditional Indian foods. Indeed, a handful of dedicated Andean researchers have studied these plants intensively, and have struggled for decades to promote them in the face of deeply ingrained prejudices in favor of European food. Moreover, their efforts have sparked interest outside the region. Some of the plants are already showing promise in exploratory trials in other tropical highlands as well as in more temperate zones. For instance, cultivation of quinoa (a grain) has begun in the United States, oca (a tuber) is an increasingly popular food with New Zealanders, tarwi (a grain legume) is stimulating attention in Eastern Europe, and cherimoya (a fruit) has long been an important crop in Spain.

This study had its origins in a 1984 seminar held at the National Research Council.[1] Subsequently, the staff mailed questionnaires to about 200 plant scientists worldwide requesting nominations of under-exploited Andean crop species for inclusion in this report. The result was a flood of suggestions and information that was fashioned into a first draft of this book. Each of the first-draft chapters was then mailed back to the original contributors as well as to other experts identified by the staff. As a result, several thousand suggestions for corrections

---

[1] The participants included G.J. Anderson, T. Brown, F. Caplan, D. Cusack, D.W. Gade, C.B. Heiser, T. Johns, C. Kauffman, S. King, D. Plucknett, H. Popenoe, C. Rick, R.E. Schultes, L.L. Schulze, C.R. Sperling, D. Ugent, and K. Zimmerer.

and improvements were received, and each was evaluated and integrated into the second draft. At the same time, a dozen or so additional species were included. Subsequently, all chapters were mailed out at least once more, and several thousand more comments were received and incorporated into what, after editing, became the current text. Finally, the panel and staff visited the Andes in mid–1988 to assess the accuracy and balance of each chapter.

All told, more than 600 people from 56 countries (see Research Contacts) have directly contributed to this book. A few species described—capuli cherry and zambo squash, for example—are not Andean natives but are included because the Andean types have much to offer the rest of the world.

This report has been written for dissemination to administrators, entrepreneurs, and researchers in developing countries as well as in North America, Europe, and Australasia. It is not a handbook or scientific monograph: references are provided for readers seeking additional information. Its purpose is to provide a brief introduction to the plants selected, and it is intended as a tool for economic development rather than a textbook or survey of Andean botany or agriculture. The ultimate aim is to raise nutritional levels and create economic opportunities, particularly in the Andes. The report, however, deliberately describes the promise of these plants for markets in industrialized nations. It is in these countries (where a concentration of research facilities and discretionary research funds may be found) that many important research contributions are likely to be made.

Because the book is written for audiences of both laymen and researchers, each chapter is organized in increasing levels of detail. The lead paragraphs and prospects sections are intended primarily for nonspecialists. The later sections contain background information from which specialists can better assess a plant's potential for their regions or research programs. Finally, appendixes provide the addresses of researchers who know the individual plants, information on potential sources of germplasm, and lists of carefully selected papers that provide more detail than can be presented here. Because these plants are so little studied, the literature on them is often old, difficult to find, or available only locally. This is unfortunate, and we hope that this book will stimulate monographs on each of the species.

This book has been produced under the auspices of the Advisory Committee on Technology Innovation (ACTI) of the Board on Science and Technology for International Development, National Research Council. ACTI is mandated to assess innovative scientific and technological advances, particularly emphasizing those appropriate for developing countries. Since its founding in 1971, it has produced almost 40 reports identifying unconventional scientific subjects of promise for developing countries. These have covered subjects as diverse as the uses of water buffalo, butterfly farming, fast-growing trees, and techniques to provide more water for arid lands (see list of current titles at the back of this report). Publications dealing with underexploited food plants are:

- *Tropical Legumes: Resources for the Future* (1979)
- *The Winged Bean: A High Protein Crop for the Tropics* (Second Edition, 1981)
- *Amaranth: Modern Prospects for an Ancient Crop* (1983)
- *Quality-Protein Maize* (1988)
- *Triticale: A Promising Addition to the World's Cereal Grains* (1989).

The panel wishes to thank those who contributed to the success of its South American visit—in particular, Luis Sumar Kalinowski, Jorge Soria, Modesto Soria, Jaime Pacheco, Vincente Callaunapa, and the staff of the Instituto Nacional de Investigaciones Agropecuarias (INIAP).

This activity was stimulated by a Program Initiation Funding grant from the National Research Council. Program costs for the study were provided by the Jesse Smith Noyes Foundation. Staff support costs were provided by the Office of the Science Advisor, Agency for International Development.

How to cite this report:
National Research Council. 1989. *Lost Crops of the Incas: Little-Known Plants of the Andes with Promise for Worldwide Cultivation.* National Academy Press, Washington, D.C.

Contents

Art Credits

The small drawings that appear at the end of some chapters are redrawn from ancient pottery and artwork dug up in the Andean region.

The face of the country was shagged over with forests of gigantic growth, and occasionally traversed by ridges of barren land, that seemed like shoots of the adjacent Andes, breaking up the region into little sequestered valleys of singular loveliness. The soil, though rarely watered by the rains of heaven, was naturally rich, and wherever it was refreshed with moisture, as on the margins of the streams, it was enamelled with the brightest verdure. The industry of the inhabitants, moreover, had turned these streams to the best account, and canals and aqueducts were seen crossing the low lands in all directions, and spreading over the country, like a vast network, diffusing fertility and beauty around them. The air was scented with the sweet odors of flowers, and everywhere the eye was refreshed by the sight of orchards laden with unknown fruits, and of fields waving with yellow grain, and rich in luscious vegetables of every description that teem in the sunny clime of the equator.

W.H. Prescott
The Conquest of Peru

Introduction

At the time of the Spanish conquest, the Incas cultivated almost as many species of plants as the farmers of all Asia or Europe.[1] On mountainsides up to four kilometers high along the spine of a whole continent and in climates varying from tropical to polar, they grew a wealth of roots, grains, legumes, vegetables, fruits, and nuts.

Without money, iron, wheels, or work animals for plowing, the Indians terraced and irrigated and produced abundant food for fifteen million or more people—roughly as many as inhabit the highlands today. Throughout the vast Inca Empire, sprawling from southern Colombia to central Chile—an area as great as that governed by Rome at its zenith[2]—storehouses overflowed with grains and dried tubers. Because of the Inca's productive agriculture and remarkable public organization, it was usual to have 3–7 years' supply of food in storage.

But Pizarro and most of the later Spaniards who conquered Peru repressed the Indians, suppressed their traditions, and destroyed much of the intricate agricultural system. They considered the natives to be backward and uncreative. Both Crown and Church prized silver and souls—not plants. Crops that had held honored positions in Indian society for thousands of years were deliberately replaced by European species (notably wheat, barley, carrots, and broad beans) that the conquerors demanded be grown.

Forced into obscurity were at least a dozen native root crops, three grains, three legumes, and more than a dozen fruits. Domesticated plants such as oca, maca, tarwi, nuñas, and lucuma have remained in the highlands during the almost 500 years since Pizarro's conquest. Lacking a modern constituency, they have received little scientific respect, research, or commercial advancement. Yet they include some widely adaptable, extremely nutritious, and remarkably tasty foods.

This botanical colonialism closed off from the rest of the world a major center of crop diversity. Food plants of Asia, Mexico, and

---

[1] It has been estimated that Andean Indians domesticated as many as 70 separate crop species. Cook, 1925.
[2] Britain to Persia.

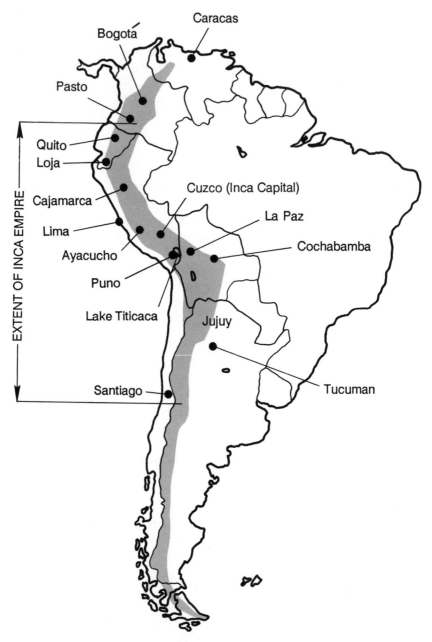

The Inca Empire measured more than 4,000 km from end to end. Superimposed on a map of modern South America, it would begin on Colombia's southern frontier, stretch southward along the coast and highlands of Ecuador and Peru, sprawl across highland Bolivia into northwestern Argentina, and reach down into central Chile to just below Santiago. This vast territory was probably the largest ever formed anywhere based on a "Bronze Age" level of technology.

especially of Europe became prominent; those of the Andes were largely lost to the outside world.[3]

However, it is not too late to rescue these foods from oblivion. Although most have been hidden from outsiders, they did not become extinct. Today in the high Andes, the ancient influences still persist with rural peasants, who are largely pure-blooded Indian and continue to grow the crops of their forebears. During the centuries, they have maintained the Inca's food crops in the face of neglect, and even scorn, by much of the society around them. In local markets, women in distinctive hats and homespun jackets (many incorporating vivid designs inspired by plant forms and prescribed by the Incas more than 500 years ago) sit behind sacks of glowing grains, baskets of beans of every color, and bowls containing luscious fruits. At their feet are piles of strangely shaped tubers—red, yellow, purple, even candy striped; some as round and bright as billiard balls, others long and thin and wrinkled. These are the "lost crops of the Incas."

That these traditional native crops have a possible role in future food production is indicated by the success of the few that escaped the colonial confines. Among the Inca's wealth of root crops, the domesticated potato, an ancient staple previously unknown outside the Andes, proved a convenient food for slaves in the Spanish silver mines and sailors on the Spanish galleons. Almost inadvertently, it was introduced to Spain, where, over several centuries, it spread out across Europe and was genetically transformed. Eventually, the new form rose to become the fourth largest crop on earth. Other Andean crops that reached the outside world and enjoyed spectacular success were lima beans, peppers, and the tomato.[4]

In light of this, it is surprising that more than 30 promising Inca staples remain largely restricted to their native lands and unappreciated elsewhere. Given research, these, too, could become important new contributors to the modern world's food supply.

ANDEAN ENVIRONMENT

The Andean region became an important center for domestication of crop species because of its striking geographical contrasts. Along its western margin stretches a narrow coastal desert that is all but uninhabitable except where some forty small, fertile river valleys cross

---

[3] Most New World crops that rose to global prominence came from outside the Andes: cassava and sweet potato (Caribbean); corn, beans, most squashes, pineapple, vanilla, and chocolate (Mexico); and peanuts (Brazil), for example.

[4] See pages 163, 191, and 195. Although the tomato species had its origins in the Andes, it was domesticated as a food plant elsewhere, and is therefore perhaps not strictly an Andean crop. Potatoes are the only exclusively Andean plant to have gone worldwide.

it. Behind this mostly barren plain towers the world's second-highest mountain range, the Andes, reaching an average of over 3,000 meters elevation. Its glacial heights are also uninhabitable, but intermontane valleys and basins are well suited to human occupation, and these became the home of the Inca rulers. Beyond the mountain valleys, on the eastern face of the Andes, are found subtropical cloud forests gently sloping into the Amazon jungle.

The Andean region was quite unlike the other regions where clusters of crops were domesticated. Here were no vast, unending plains of uniformly fertile, well-watered land as in Asia, Europe, or the Middle East. Instead, there was an almost total lack of flat, fertile, well-watered soil. Andean peoples grew their crops on millions of tiny plots scattered over a length of thousands of kilometers and perched one above another up mountainsides rising thousands of meters.

This complicated ecological mosaic created countless microclimates—including some of the driest and wettest, coldest and hottest, and lowest and highest found anywhere in the world. Perhaps no other contiguous region has such a broad range of environments as in the ancient Inca Empire. And the region is so fragmented that rainfall, frost, sunlight, and soil type can vary over distances as short as a few meters. For instance, a valley floor may have thick soils, abundant sunshine in the daytime, and severe frost at night, whereas immediately adjacent slopes may be thin soiled, shaded, and frost free.

To protect themselves against crop failure, ancient Andean farmers utilized all the microenvironments they could. Conditions causing poor harvests in one could produce bumper crops at another. Farmers deliberately maintained fields at different elevations, and this vertically diversified farming fostered the development of a cornucopia of crop varieties, each with slightly different tolerances to soil type, moisture, temperature, insolation, and other factors.

The resulting diversity of crops served as a form of farm insurance, but the differing growth cycles of different elevations also permitted work to be staggered and therefore more area to be cultivated.

INCA AGRICULTURE

Western South America's dramatic stage—coast, valleys, highlands, and cloud forest—formed the setting for the evolution of Andean civilization, which emerged some 4,500 years ago. On the semiarid

Opposite: Laraos (Department of Lima), Peru. The highlands of Peru contain more than 600,000 hectares of terraces, most constructed in prehistoric times. These staircase farms, built up steep mountain slopes with stone retaining walls, contributed vast amounts of food to the Incas. They provided tillable land, controlled erosion, and protected crops during freezing nights. Many were irrigated with water carried long distances through stone canals. Today, as in the distant past, the chief crops on these terraces are native tubers, such as potatoes, oca, and ulluco. (A. Cardich)

coast, up the precipitous slopes, across the high plateaus, and down into the subtropical jungles of the eastern face of the Andes, dozens of cultures flourished and faded before the rise of the Incas in about 1400 A.D.

The Incas inherited and built upon the products of thousands of years of organized human endeavor. It was they who, through military and diplomatic genius, first united a vast realm running the length of the Andes. Employing an inspired, if rigid, administration, they promulgated a social uniformity from their capital, Cuzco. The entire empire was a single nation, governed by the same laws, privileges, and customs.

The union within the Inca Empire was surprising because the various lands it covered were so vastly different: seared desert, saline flats, vertical valley walls, windswept barrens, triple-canopy jungle, glacial sands, floodplains, saline crusts, perpetual snow, and equatorial heat. This diversity is reflected in the Incas' own name for their empire: Tahuantinsuyu—Kingdom of the Four Corners—coast, plateau, mountain, and jungle. Yet the Incas learned to manage the desolation and the variety of these most demanding habitats, and they made these regions bloom.

This success was owing to several factors.

First, the Incas were master agriculturalists. They borrowed seeds and roots from their conquered neighbors and forcibly spread a wealth of food crops throughout their empire, even into regions where they were previously unknown. To enhance the chances of success, the Incas purposefully transplanted the plants with their farmers, thereby spreading both the species and the knowledge of how to cultivate them.

Second, the Incas created a vast infrastructure to support (or perhaps to enforce) the empire's agriculture. For example, they modified and conserved steeply sloping erodible terrain by constructing terraces and irrigation works, and by fostering the use of farming systems that attenuated the extremes of temperature and water. These included, for example, ridged fields and planting in small pits. In some areas, Inca terraces and irrigation systems covered thousands of hectares. Many are still in use.

Third, contributing to the infrastructure were roads and footpaths that provided an extensive system for transporting products to all corners of the realm. As a result, massive amounts of food could be moved on the backs of llamas and humans—for example, corn into the highlands, quinoa to the lowlands, and tropical fruits from the eastern jungles to the heights of Cuzco. To implement this superb organization without paper or a written language, an accounting system was developed that used knots tied in strands of yarn (*quipu*). The code of the knots has never been solved, so today they cannot be "read," but it resembles the binary system of computers, and could maintain highly elaborate and complex accounts.

Further, the roads and footpaths made possible the exchange of information. Instructions and advice were carried quickly throughout the empire by an organized corps of runners. In this way, Inca sages sent predictions of the weather for the upcoming cropping season to and from all regions. The predictions were based on natural indicators such as the behavior of animals, the flowering of certain plants, and the patterns of the clouds and rainfall. The Incas were familiar, for example, with the phenomenon known as "El Niño" that periodically changes the ocean currents off the coasts of Peru and Ecuador.

Also, the Incas developed methods for preserving their harvests for years, when necessary. It is estimated that in the central highlands of Peru alone there were tens of thousands of large, rock-walled silos and warehouses. Such stores were filled each year with dried and salted meat (this was called "*charqui*," which is the source of the English word "jerky"). They also contained roots preserved by freeze-drying. When potatoes, for example, had been harvested at the highest altitudes, they were spread out and left overnight in the freezing air. The next day, men, women, and children walked over the partly withered tubers, squeezing out the moisture that had been released by the freezing. The same process was repeated over several nights and days, after which the potatoes were completely dehydrated and could be stored safely (see page 10).

THE INCAS' DESCENDANTS

For all its size and splendor, the Inca Empire endured for only a century, and it was brought down by fewer than 200 Spanish adventurers. Today, the region of the empire—the highlands from Colombia through Chile—is one of the world's most depressed areas. The infant mortality rate is one of the highest on the South American continent—more than one-fourth of the children die before their first birthday, a rate more that twice that of Latin America at large and about 50 times that of Sweden. Only 1 in 7 homes has potable water, and only 1 in 40 has indoor plumbing. Add to this the disruption caused by guerrillas, who have launched an armed campaign of terror in the Peruvian highlands, and it is no surprise that massive migration from the countryside to the cities is occurring.

Exacerbating the highlands' difficulties are cultural and ethnic divisions. The Indians, who make up about half of the population, live a life apart from the modern sector. Most still speak Quechua, the

RECREATING PREHISTORIC ABUNDANCE

About 3,000 years ago, an ingenious form of agriculture was devised on the high plains of the Peruvian Andes. It employed platforms of soil surrounded by ditches filled with water. For centuries this method flourished because it produced bumper crops in the face of floods, droughts, and the killing frosts of those 3,800-m altitudes.

Around Lake Titicaca, remnants of over 80,000 hectares of these raised fields (*waru waru*) can still be found. Many date back at least 2,000 years.

Now, in a dramatic resurrection, modern-day Peruvians working with archeologists have reconstructed some of the ancient farms, and the results have been amazing. They have found, for instance, that this method can triple the yield of potatoes. In at least one experiment, potato yields outstripped those from nearby fields that were chemically fertilized. As a result of such observations, local farmers have begun restoring the ancient *waru waru* on their own. Government-sponsored restoration projects are also under way.

The combination of raised beds and canals has proved to have remarkably sophisticated environmental effects. For one thing, it reduces the impacts of extremes of moisture. During droughts, moisture from the

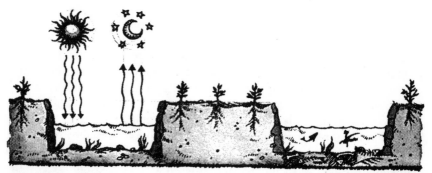

Water in the canals absorbs the sun's heat by day and radiates it back by night, helping protect crops against frost. The more fields cultivated this way, the bigger the effect on the microenvironment.

The platforms are generally 13 to 33 feet wide, 33 to 330 feet long, and about 3 feet high, built with soil dug from canals of similar size and depth.

Sediment in the canals, nitrogen-rich algae, and plant and animal remains provide fertilizer for crops. In an experiment, potato yields outstripped those from chemically fertilized fields.

canals slowly ascends to the roots by capillarity, and during floods, the furrows drain away excess runoff.

For another, it reduces the impact of temperature extremes. Water in the canals absorbs the sun's heat by day and radiates it back by night, thereby keeping the air warm and helping protect crops against frost. On the raised beds, nighttime temperatures can be several degrees higher than in the surrounding region.

For a third, it maintains fertility in the soil. In the canals, silt, sediment, nitrogen-rich algae, and plant and animal remains decay into a nutrient-rich muck. Seasonal accumulation can be dug out of the furrows and added to the raised beds, providing nutrients to the plants.

The prehistoric technology has proved so productive and inexpensive that it is seen as a possible alternative for much of the Third World where scarce resources and harsh local conditions have frustrated the advance of modern agriculture. It requires no modern tools or fertilizers; the main expense is for labor to dig canals and build up the platforms with dirt held in by blocks of sod on the sides.

Illustrations by Narda Lebo

CHUÑO

In addition to possessing ingenious farming systems and outstanding public works, the Incas and their forebears had remarkable ways to preserve food.

One technique was to freeze-dry root crops. In the Andean uplands, the nights are so cold and the days are so dry that tubers left out in the open for a few nights and days become freeze-dried. Usually, the people help the process along by covering the tubers at night to keep off dew and by trampling on the tubers during the day to squeeze out the water released by the previous night's freezing.

The resulting product, made mostly from potatoes and known as *chuño*, was vital to the Incas' ability to carry out their conquests and maintain command of the empire. For instance, it enabled the millions of inhabitants to withstand natural disasters, it supplied passing armies, and it was a long-term insurance against crop failure (a constant threat in this frost-prone region). The Incas planned so well that conquistador Hernando de Soto was moved to say: "There was never hunger known in their realm." The conquistadores quickly recognized *chuño's* virtues. Indeed, some Spaniards made fortunes shipping *chuño* by llama train to the barren heights of Potosí (in today's Bolivia), where it was the main food for slaves working in the silver mines.

Chuño can be kept for years without refrigeration or special care, and it is still widely made. Even today, it can comprise up to 80 percent of the diet of the highland Indians in times of crop failure. The natural freeze-drying process is also extensively used with some bitter species of potatoes (see page 97). The bitter glycoalkaloids are water soluble and they get squeezed out with the water, eliminating most of the bitterness.

The figure shows modern samples of *chuño* and related products from the market in Ayacucho, Peru. Clockwise from top: *chuño* and related products made from oca (*kaya*), from fresh potatoes (true *chuño*), from ulluco (*llingli*), and from boiled potatoes (*papa seca*). (S. King)

---

lingua franca of the Incas; a few around Lake Titicaca on the Peru-Bolivia border speak Aymara, an even older language. The Indians' rural lives have not changed appreciably for generations.

On the other hand, the whites and mestizos (persons of mixed European and Indian ancestry), who make up the other half of the population, speak Spanish and live in a modern urban world that is undergoing rapid change.

The classes, therefore, are separate and unequal. And a concomitant notion is that their food plants are separate and unequal as well. It may seem irrational, but crops the world over are stigmatized by the prejudices held against the peoples who use them most.[5]

Over the centuries, the Spanish view that native crops are inferior to European crops such as wheat, barley, and broad beans has persisted. Indian foods are still equated with lower status. The conquistadores would undoubtedly be amazed to see potatoes, tomatoes, peppers, and limas contributing significantly to modern Spain's cuisine. But they would see that their prejudices against oca, tarwi, quinoa, and dozens of other Inca foods are still largely in place in South America.

---

[5] The English refused to eat potatoes for two centuries, in part because the Irish ate them; northern Europeans ignored tomatoes even longer, in part because Italians ate them; and even today in the United States, collard greens are unacceptable to many people who consider them "poor folks food."

RECENT INTEREST IN INCA CROPS

The urgency for exploring the Indians' native crops has been heightened in recent years as more and more highland Indians have forsaken their indigenous crops. In part, this abandonment has been caused by migration to lowland cities. Also, in the 1960s—when imported wheat became cheap—products such as bread or noodles were widely promoted, and traditional dietary habits were abandoned.

Nonetheless, the Indians' agricultural techniques are now regaining the attention of Latin American governments and scientists. For example, an experimental project in Puno, Peru, has rejuvenated raised fields (built up to enhance drainage) built long before the time of the Incas, and found that they produced triple the yield of fertilized potatoes in adjacent fields. The results were so spectacular that local people began restoring raised fields on their own (see pages 8–9). This caught the attention of Peru's president, who hopes to use the knowledge to help highland villagers grow more food and thereby halt migration to the already overcrowded cities.

The traditional crops are also gaining more modern respect. Over the past 25 years, Andean researchers of all backgrounds have begun focusing on native food plants. In 1964, the Instituto Interamericano de Ciencias Agrícolas (IICA) published a classic work on food crops of the Andes.[6] In 1968, an international convention on quinoa and kaniwa was held in Puno.[7] In 1976, IICA helped organize a second international convention in Potosí, Bolivia, on these two traditional grain crops. By then, the number of Andean scientists working on indigenous crops had increased markedly, and it was decided that an international congress would be held every few years and would be expanded to cover the full complement of Andean crops. At about this time, also, the International Board of Plant Genetic Resources (IBPGR) began providing support for maintaining germplasm collections, and IICA and the Canadian International Development Research Centre (IDRC) began providing resources for horticultural research on Andean crops.

All this was heartening, although it was still a tiny part of the region's overall agricultural research effort. A handful of dedicated Andean researchers had recognized the value of their agricultural heritage but were receiving little local encouragement. Nonetheless, substantial progress was made. In 1977, for example, the First International Congress on Andean Crops was held in Ayacucho, Peru. In 1979 a

---

[6] León, 1964.
[7] This was initiated by researchers concerned about what seemed to be a bleak future for these high-protein grain crops. The researchers included Martín Cárdenas and Humberto Gandarillas of Bolivia and Mario Tapia of Peru.

major collaborative work was published on quinoa and kaniwa.[8] And shortly thereafter, major collections of quinoa were made and modern machinery for harvesting and processing the grain was developed.

The early 1980s brought yet more advances. From 1980 to 1985, the Peruvian university system undertook research on Andean crops and mountain agricultural systems. Peru's Programa Nacional de Sistemas Andinos de Producción Agropecuarias funded more than 50 agronomic research projects, involving native crops and researchers from universities in Huancayo, Ayacucho, Cuzco, and Puno. Also, a massive program for the selection and improvement of quinoa was undertaken.[9]

Now, in the late 1980s, as a result of all these efforts, Andean crops are increasingly being viewed as national resources in Colombia, Ecuador, Peru, and Bolivia. Peru, for instance, is encouraging farmers to plant kiwicha, a nutritious Andean grain, and has supported programs that use kiwicha, quinoa, and tarwi in childhood nutrition projects. Ecuador has established urban-nutrition education programs, with the goal of reintroducing people to the centuries-old, nutritious crops of the region.

Even with all these efforts, there still exists a vast lack of understanding of the crops of the Incas, and many people still retain the prejudice that the plants are second-rate. However, the cultural barriers that once kept the plants suppressed are starting to crumble. Behind them, researchers now can glimpse the promise of a wealth of new crops for the modern Andes and even for the rest of the world.

FUTURE OF THE LOST CROPS

Today, almost all the native Andean foods are foreign to outsiders, and it is too early to predict the eventual extent of their worldwide acceptance. It is a long, hard, and very uncertain trail to make a little-known plant into an international crop.

However, at least some of these crops may soon become common household foods. It can be said with confidence that the basic qualities of these crops are sound—they can, for example, be cultivated to give acceptable yields and they offer good nutrition and interesting tastes. What is less certain is their adaptability to and profitability in new locations.

Whether the plants have been tried in new regions in the past is of little consequence. New technologies make it easier to develop and adapt new crops than at any time in history. Modern plant genetics is especially powerful for solanaceous species (plants of the nightshade

[8] Tapia et al., 1979.
[9] With support from the national government, IICA, IBPGR, and the IDRC.

family), of which the Andes has several, such as potatoes, peppers, pepino, tamarillo, goldenberry, and naranjilla.

All of the crops described in this book deserve investigation. For exploitation, most require very basic research, including the following:

• **Collection.** To preserve genetic diversity for the future, germ-plasm collections should be made, especially in isolated areas.

• **Selection.** The agronomic traits of the different germplasm should be characterized and important qualities noted.

• **Agronomy.** Analysis of cultural practices, plant establishment, and optimum plant density should be undertaken. Research into minimal fertilizer requirements is especially needed. For example, in Peru's Ayacucho region, even small amounts of fertilizer have boosted potato yields from 5 tons to 22 tons per hectare.

• **Genetics.** The plants' genetics should be investigated so that efficient plant-breeding strategies for their improvement can be devised. (With most of these crops, varietal improvement is in its infancy.)

• **Handling.** Improved harvesting, cleaning, and processing tech-niques are often needed, especially ones that lower labor requirements or enhance end-product value.

• **Nutrition.** Additional nutritional studies would be helpful in some species, especially to clarify the optimum dietary mix with other foods.

• **Pest and Disease Control.** Many of the plants now suffer from afflictions of viruses, bacteria, and nematodes, all of which are potentially controllable.

FUTURE BEYOND THE ANDES

It is in the Andes that the plants have their greatest potential, especially for developing food products for malnourished segments of the population. However, they also promise to become useful new crops for other developing regions of the world such as the tropical highlands of Asia, Central Africa, and Central America. In addition, they have notable promise for some industrialized regions such as the United States, Europe, Japan, and Australasia. In fact, one country outside the Andes already has had considerable experience and success with them—New Zealand.

The reason these plants could have this wide ecological adaptation is that although the Inca Empire stretched across the equator, a majority of its peoples actually dwelt more than three kilometers above sea level where bone-cracking cold descends at sunset, and the climate is more temperate than tropical. As a result, these crops in general have many characteristics that have adapted them for cultivation in regions well outside the heat of the tropics. However, additional uncertainties exist when a crop is to be transplanted from one part of the world to another—for example, daylength (photoperiod) dependence, which could be particularly troublesome.

Because the plants are native to latitudes near the equator (where the day and night lengths are equal year-round), some will not reach maturity during the long summer and fall days of the temperate zones. This difficulty has proved surmountable in potatoes, tomatoes, peppers, and lima beans, but it still could take growers some time to locate varieties or genes that can allow each of the crops described in this report to be grown as far from the equator as North America, Europe, Japan, and Australasia.

Difference in sensitivity to cold is another possible problem. Although the temperature variation in the Andean highlands often runs from a few degrees of frost at night to shirt-sleeve temperatures at midday, the frosts in the Andes are extremely dry, and they rarely form ice on the plants. Therefore, whether frost-tolerance data recorded in the Andes can be extrapolated to other areas is uncertain.

Nonetheless, the global promise of these plants is very high. In the last few centuries the tendency has been to focus on fewer and fewer species, but today many ancient fruits, vegetables, and grains are finding new life in world markets. This is heartening, because to keep agriculture healthy and dynamic, farmers everywhere need plenty of options, especially now when markets, climates, national policies, scientific understanding, and technologies are changing at a rapid pace.

The necessary next steps toward crop development and exploitation are often interdisciplinary, involving diverse interests such as genetics, processing, marketing, advertising, and technical development from the farm to the exporter.

Developing the lost crops of the Incas is the kind of research that scientists should undertake. In the process, they will lift the veil of obscurity and rediscover the promise of these crops the Spanish left behind. The Inca Empire's grains, tubers, legumes, fruits, vegetables, and nuts are an enduring treasure for the Andes and for the rest of the world. Millions of people should quickly be introduced to these neglected foods of a remarkable people.

A summary follows of the plants selected by the panel.

ROOT CROPS

Achira. Achira (*Canna edulis*, Cannaceae) looks somewhat like a large-leaved lily. Its fleshy roots (actually rhizomes), sometimes as long as an adult's forearm, contain a shining starch whose unusually large grains are actually big enough to see with the naked eye. This starch is easily digested and is promising for both food and industrial purposes. (Page 27)

Ahipa.   Ahipa (*Pachyrhizus ahipa*, Leguminosae) is a legume, but unlike its relatives the pea, bean, soybean, and peanut, it is grown for its swollen, fleshy roots. Inside, these tuberous roots are succulent, white, sweet, pleasantly flavored, and crisp like an apple. They are an attractive addition to green salads and fruit salads. They can also be steamed or boiled and have the unusual property of retaining their crunchy texture even after cooking. (Page 39)

Arracacha.   Above ground, this plant (*Arracacia xanthorrhiza*, Umbelliferae) resembles celery, to which it is related. Below ground, however, it produces smooth-skinned roots that look somewhat like white carrots. These roots have a crisp texture and a delicate flavor that combines the tastes of celery, cabbage, and roasted chestnut. They are served boiled or fried as a table vegetable or added to stews. They sell well throughout Colombia and have become popular in the big cities of southern Brazil. (Page 47)

Maca.   Maca (*Lepidium meyenii*, Cruciferae) is a plant that resembles a radish and is related to cress, the European salad vegetable. However, although its edible leaves are eaten in salads and are used to fatten guinea pigs, it is most valued for its swollen roots. Looking like brown radishes, these are rich in sugars and starches and have a sweet, tangy flavor. They are considered a delicacy in the high plateaus of Peru and Bolivia. Dried, they can be stored for years. (Page 57)

Mashua.   The well-known garden nasturtium was a favorite Inca ornamental, and at high altitudes in the Andes, its close relative, mashua (*Tropaeolum tuberosum*, Tropaeolaceae), is a food staple. Farmers often prefer mashua (also called "añú") to other tubers because it is easier to grow. It requires less labor and care, and it can be stored in the ground and harvested when needed. (Page 67)

Mauka.   Mauka (*Mirabilis expansa*, Nyctaginaceae) has thick stems and yellow or salmon-colored fleshy roots that make it a sort of cassava of the highlands. The plant was unknown to science until "discovered" in Bolivia in the 1960s, and it now has also been found in remote mountain fields of Ecuador and Peru. If placed in the sun and then put in storage, the tubers turn very sweet, like sweet potatoes. (Page 75)

Oca. An exceptionally hardy plant that looks somewhat like clover, oca (*Oxalis tuberosa*, Oxalidaceae) produces an abundance of wrinkled tubers in an array of interesting shapes, and in shades from pink to yellow. In the Andean highlands, it is second only to the potato in the amount consumed, and is still a staple for Peruvian and Bolivian Indians living at high altitudes. The firm white flesh has a pleasant, sometimes slightly acid taste. (Page 83)

Potatoes. The common potato became one of the 20 or so staple crops that feed the whole planet, but in the Andes are at least 5 other cultivated potatoes (*Solanum* species, Solanaceae). Collectively, these are adapted to a wide array of climates and provide a genetic source of diversity, disease resistance, and new crops. Many have unusual and marketable properties. Some are golden yellow inside, a number have a decidedly nutty taste, and almost all are more concentrated in nutrients than is the common potato. (Page 93)

Ulluco. Some of the most striking-looking roots in Andean markets are the ullucos (*Ullucus tuberosus*, Basellaceae). They are so brightly colored—yellow, pink, red, even candy striped—that their waxy skins make them look almost like plastic fakes. Once a staple in the Inca diet, ulluco is one of the few indigenous crops that has increased its range over the last century. In some areas, it vies with potatoes as a carbohydrate staple. Many consider it a delicacy, and it is commonly purchased in modern packaging in city supermarkets. It is usually prepared like potatoes and is used chiefly in thick soups and stews. (Page 105)

Yacon. Yacon (*Polymnia sonchifolia*, Compositae) is a distant relative of the sunflower. Grown in temperate valleys from Colombia to northwestern Argentina, it produces tubers that on the inside are white, sweet, and juicy, but almost calorie free. Because of their succulence, they are eaten raw and make a pleasant refreshment. They are also eaten cooked. In addition, the main steam is used like celery, and the plant also shows promise as a folder crop. (Page 115)

GRAINS

Kaniwa. This broad-leaved plant (*Chenopodium pallidicaule*, Chenopodiaceae) produces one of the most nutritious of all grains, with a protein content of 16–19 percent and an unusually effective balance of essential amino acids. It flourishes in poor rocky soil at high elevations, usually surviving frosts that kill other grain crops, and outyielding them in droughts. Incredibly, it thrives where frosts occur nine months of the year. Snowfalls or strong winds that flatten fields of barley or even quinoa (see below) usually leave kaniwa unaffected. (Page 129)

Kiwicha. The seeds of this amaranth (*Amaranthus caudatus*, Amaranthaceae), an almost totally neglected grain crop, have high levels of protein and the essential amino acid, lysine, which is usually lacking in plant protein. Kiwicha protein is almost comparable to milk protein (casein) in nutritional quality, and it complements the nutritional quality of foods that normally would be made from flours of corn, rice, or wheat. This makes kiwicha particularly beneficial for infants, children, and pregnant and lactating women. (Page 139)

Quinoa. Although the seed of this tall herb (*Chenopodium quinoa*, Chenopodiaceae) is one of the best sources of protein in the vegetable kingdom, quinoa is hardly known in cultivation outside its upland Andean home. However, experience in the United States and England shows that the grain is readily accepted by people who have never tasted it before. Quinoa can be grown under particularly unfavorable conditions, at high elevation, on poorly drained lands, in cold regions, and under drought. Already, much has been learned about this plant, which is becoming a commercial success outside the Andes. (Page 149)

LEGUMES

Basul. Basul (*Erythrina edulis*, Leguminosae) is a common leguminous tree of the Andean highlands. It is unusual in that it produces large edible seeds and is one of the few trees that produces a basic food. Accordingly, it has promise as a perennial, high-protein crop for subtropical areas and tropical highlands. Beyond its use in food production, it is also a promising nitrogen-fixing tree for use in reforestation, beautification, erosion control, and forage production. (Page 165)

Nuñas. The nuña (*Phaseolus vulgaris*, Leguminosae) is a variety of the common bean, but it is the bean counterpart of popcorn. Dropped into hot oil, nuñas burst out of their seed coats. The popping is much less dramatic than with popcorn—nuñas don't fly up into the air—but the product has a delightful flavor and a consistency somewhat like roasted peanuts. (Page 173)

Tarwi. This lupin (*Lupinus mutabilis*, Leguminosae) is one of the most beautiful crops, and its seeds are as rich, or richer, in protein than peas, beans, soybeans, and peanuts—the world's premier plant-protein sources. Also, they contain about as much vegetable oil as soybeans. Tarwi has been held back mainly because its seeds are bitter. The Indians soak them in running water for a day or two, to wash out the bitterness. Recently, engineers in Peru and Chile have developed machinery to do it more quickly and more easily. Also, geneticists in several countries have developed bitter-free varieties that need little or no washing. (Page 181)

VEGETABLES

Peppers. Chilies and sweet peppers (*Capsicum* species, Solanaceae) have become the most widely used spices in the world, but hidden in the Andes—the original home of all peppers—are several more domesticated peppers as well as some wild species. All of these are employed by local people, and they promise to add new pungency, new tastes, and new variety to many of the world's cuisines. (Page 195)

Squashes and Their Relatives. Several of the fruits that are variously known as pumpkins, squashes, gourds, or vegetable marrows have their origins or greatest development in the Andes. These (*Cucurbita* species, Cucurbitaceae) and some lesser-known botanical relatives are robust, productive crops, especially suitable for subsistence use. Many are little known elsewhere, and offer promise of new and better foods for scores of countries. (Page 203)

FRUITS

Berries. Along the length of the Andes are found several dozen localized berry fruits. These include relatives of raspberry and blackberry *Rubus* species, Rosaceae), blueberry (*Vaccinium* species, Rosaceae), and some small berries (*Myrtus* species, Myrtaceae) that are rather like mini guavas. Collectively, they represent a source of new and interesting fruits. (Page 213)

Capuli Cherry. The black cherries that are found throughout the Americas reach their best development in the Andes, where the capuli (*Prunus capuli*, Rosaceae) is a popular city and backyard tree. The cherrylike fruits are found in the markets three or four months of the year. Some are large, sweet, fleshy, and said to be at least as good as the traditional cherry of the rest of the world. (Page 223)

Cherimoya. Of all the Inca fruits, only the cherimoya (*Annona cherimola*, Annonaceae) is cultivated substantially outside the Andes. It is being grown commercially in Spain, Southern California, and a few other places. Such interest is understandable. Inside the thin greenish skin of the cherimoya is a delicious, sweet, and juicy flesh with a creamy, custardlike texture. Its unique flavor tastes like a subtle blend of papaya, pineapple, and banana. (Page 229)

Goldenberry. A relative of the North American husk tomato, the goldenberry (*Physalis peruviana*, Solanaceae) is fresh tasting and makes one of the world's finest jams. Under harsh conditions it provides a wealth of yellow, marble-sized fruits that are beginning to attract international acclaim for their flavor and appearance. (Page 241)

Highland Papayas. Although the papaya is one of the premier fruits of the world, its botanical cousins (*Carica* species, Caricaceae) of the Andes are all but unknown. They, too, have much promise, and they might allow the extension of the cultivation of papayalike fruits into cooler areas than is now possible. (Page 253)

Lucuma. This fruit (*Pouteria lucuma*, Sapotaceae) can be considered a "staple fruit." Unlike oranges or apples, its fruits are dry, rich in starch, and suitable for use as a basic, everyday carbohydrate. It has been said that a single tree can feed a family year-round. The fruits are often eaten fresh and are very popular in milkshakes, ice cream, and other treats. Dried, they store for years. (Page 263)

Naranjilla. Related to, but wholly unlike, tomatoes, this fruit (*Solanum quitoense*, Solanaceae) is highly esteemed in Peru, Colombia, Ecuador, and Guatemala, but virtually unknown elsewhere. Its delicious, refreshing juice is one of the delights of the northern Andes, and it could become popular in the African and Asian tropics, where the plant could conceivably flourish. (Page 267)

Pacay. Among the most unusual of all fruit trees, pacay (*Inga* species, Leguminosae) produces long pods filled with soft white pulp. This pulp is so sweet that the pods have been called ice-cream beans. Not only are the fruits attractive and popular, this nitrogen-fixing tree is extremely promising for reforestation, agroforestry, and for production of wood products. (Page 277)

Passionfruits. This exotic fruit (*Passiflora* species, Passifloraceae) is becoming popular in Europe, North America, and other places. With its concentrated perfume and flavoring ability, passionfruit "develops" the taste of bland drink bases such as apple juice or white grape juice. So far, all commercial developments have been based on a single Brazilian species. In the Andes are scores of other species, some of which are reputed to be superior to the Brazilian one. (Page 287)

Pepino. A large, conical, yellow fruit (*Solanum muricatum*, Solanaceae) with jagged purple streaks, pepino's mellow flesh tastes like a sweet melon. It is beginning to enter international commerce. Already gaining popularity in New Zealand and Japan, the delicate pepino seems destined to become a benchmark for premium fruit production. (Page 297)

Tamarillo (Tree Tomato). Inca gardens high on the mountainsides contained small trees that bore large crops of egg-shaped "tomatoes." Today these tree tomatoes (*Cyphomandra betacea*, Solanaceae) remain one of the most popular local fruits. They have bright, shiny, red or golden skins and can be eaten raw or cooked or added to cakes, fruit salads, sauces, or ice cream. Their succulent flesh looks somewhat like that of the tomato, but it is tart and tangy and has a piquancy quite its own. (Page 307)

NUTS

Quito Palm. The streets and parks of the city of Quito are lined with an elegant palm (*Parajubaea cocoides*, Palmae) that seems out of place because Quito is one of the highest cities in the world and has a cool climate. The palm produces many fruits that look and taste like tiny coconuts. They are so popular that only early risers can find any left on the streets. (Page 319)

Walnuts. While most walnut species are natives of the Northern Hemisphere, a few occur in the Andes. They are common backyard and wayside trees, and at least one of these (*Juglans neotropica*, Juglandeaceae) is a promising timber and nut tree. In New Zealand, this species has grown unusually fast for a walnut, and its nuts are of fine flavor. (Page 323)

PART I

Roots and Tubers

As food for humans, root crops are second in importance only to cereals. These plants—whose underground portions may be roots, tubers, rhizomes, or corms—feed hundreds of millions of people. For instance, the annual world production of potatoes has reached nearly 300 million tons, sweet potatoes and yams over 130 million tons each, and cassava at least 100 million tons.

Although pre-Columbian Indians of the Andes domesticated more starchy root crops than any other peoples, only one has become a world crop—the potato, which is now grown in some 130 nations and is the fourth largest food crop of the planet. The others have seldom been tried outside South America, yet they are still found in the Andes and represent some of the most interesting of all root crops. No other region displays such diversity.

The following chapters describe the "forgotten" Andean root crops: achira, ahipa, arracacha, maca, mashua mauka, oca, ulluco, yacon, and seven little-known species of potatoes. By and large, these are attractive and tasty. They come in myriad colors, shapes, and sizes. They belong to botanical families as different as those of mustard, legumes, and sunflower. They tend to be richer in vitamins and proteins[1] than today's conventional roots. And collectively, they show enormous adaptability to difficult conditions.

That this fascinating wealth of edible roots has been overlooked is a loss to the world and a particular loss to the Andes. From Venezuela to Chile, even minor efforts could earn big rewards. For example, some of the roots now being planted contain 10,000 years' accumulations of viruses. Although the viruses do not kill the plants, they greatly reduce vigor and yield. Removing them could bring enormous benefits, and it doesn't require fancy instruments or lots of money.

---

[1] Although protein levels are quoted throughout this section, the figures should be taken with skepticism. Most were derived by the traditional process of multiplying nitrogen values by 6.25. This has recently been shown to sometimes give readings that are too high because roots can contain nitrogen in a nonprotein form.

Each Andean nation should have at least one national center for cleansing propagation materials.[2] Moreover, with appropriate control procedures, reinfection of the plants can be minimized. Some plants, for example, take up to 10 generations to develop significant levels of reinfection.

Also, small agronomic improvements could have big effects. For example, in the Andes oca now has an average yield of 4.5 tons per hectare, but experimental plots, in which a little manure or fertilizer was used, have produced almost 10 times that amount. Similar improvements are likely with the other Andean roots.

All in all, it is important to give more emphasis to root crops. They are often the crops of the poor. They provide more calories per hectare than the major grains. Perhaps more than most types of food plants, they are vital in remote areas, removed from the mainstream of commerce and agricultural extension. In addition, root crops are in increasing demand throughout the world.

Together, the Andean root crops represent a new wealth of germ-plasm. It is the most promising source of new crops of this type. Considering that the Incas grew all these species along with potatoes, it seems irrational that only the potato has global promise. The 16 others described in this section at least deserve a chance. Given attention by plant breeders and other specialists, some, at least, could become important sources of food, not only for the Andes, but for dozens of countries where they are at present unknown.

[2] Peru has recently established such a center at San Marcos University in Lima.

HOW THE POTATO REACHED EUROPE

To provide perspective on the possible future adoption of the root crops described in the following section, it is instructive to consider the irrational reception the Europeans first accorded the potato.

When Columbus set foot in the New World, Europeans had no inkling of the existence of the potato. They lived on cabbage soup and mushes and gruels made of wheat, rye, barley, or dried peas. And, despite recurrent crop failures and repetitive famines, they seemed satisfied.

It was only in 1535, near Lake Titicaca in southern Peru, that Europeans—the Spanish conquistadores—first reported seeing this tuber that had been domesticated by Andean Indians thousands of years before. In his *Chronicle of Peru*, Pedro de Cieza de León wrote perhaps the first description. "...[T]he roots...are the size of an egg, more or less, some round some elongated; they are white and purple and yellow, floury roots of good flavor, a delicacy to the Indians and a dainty dish even for the Spaniards."

When it first appeared, the potato was classified as a form of truffle or underground fungus. This confusion led to several European names for the potato. *Kartoffel*, the German name for potato, as well as its Russian counterpart *kartochki*, both derive from the Italian word *tartufulo*, which means truffle. Even the word "tuber" comes from the same mistaken source. This drawing is reproduced from one of the earliest books on mushrooms: Franciscus Van Sterbeeck's *Theatrum Fungorum*, published in 1675. It was originally made for Clusius in 1601, but was not used because it was lost in the printer's office. It shows that even as recently as 300 years ago, Europeans still thought that potatoes were some form of fungus.

The potato reached Spain sometime before 1570. It was planted as an oddity in a monastery garden in Seville. In 1576, defying Spanish export restrictions, Charles de Lecluse (Clusius) smuggled two tubers and a seedling plant out of Spain. Later, this famous Austrian plant collector gave the potato special mention in his health food manual, *Rariorum Plantarum Historia*. But publicity was not enough. Most of Europe treated the potato with apathy and then with hostility. Peasants lived with starvation for two centuries before embracing the plant.

For example, in 1756 bad weather threatened to destroy Prussia's wheat, rye, and oat crops. Starvation, the recurrent age-old killer, once more stalked the land, and to offset disaster Frederick the Great decreed that all his subjects plant potatoes. That was nothing new: his father, grandfather, and great grandfather had made similar decrees over the previous 100 years or so—some even threatening to cut the nose and ears off anyone who refused. But, as before, it was all in vain. Convinced that potatoes caused leprosy, the Prussian peasants obeyed the king's troops but crept back at night and secretly pulled up the plants. Then, in self-righteous dignity they returned to their hovels to continue to suffer the agonies of empty bellies.

This was not an isolated incident. Similar scenes occurred in France, Sweden, Russia, Greece, and other nations. To Europeans of that era, the potato was dark, dirty, and highly sinister. Since it is not mentioned in the Bible, it was considered unfit for human consumption. Because it was not grown from seed, it was said to be evil. French experts insisted it would destroy the soil in which it was planted. Physicians all over Europe reported that it caused leprosy, syphilis, and scrofula. Botanists— even the great Linnaeus—cast suspicion on it because it is related to "devil's herb," the deadly nightshade. Several countries considered it a dangerous aphrodisiac that would send their people uncontrollably mad with lust.

It was not until the late 1700s that Europeans finally took up the potato with gusto. Thereafter, it revolutionized their eating habits. It came to feed millions to the exclusion of most other vegetables. Today, nine-tenths of the world's crop is produced in Europe. The greatest per capita consumers of potatoes in the world are Poland, Ireland, and East Germany. The largest overall producer is the Soviet Union. The Netherlands is typical of many: of the 5 million tons of vegetables it produces annually, 4 million tons are potatoes. From this former Inca crop, Scandinavia, France, Germany, and Russia eventually developed "national" dishes such as potato dumplings and potato pancakes, not to mention their renowned liquors aquavit and vodka.

The turnaround in popularity of the potato was due mainly to certain "crop champions"—individuals of vision who dedicated their talents, emotions, and egos to the crop's cause. The most famous was Antoine Parmentier. After convincing Louis XVI of the potato's qualities, he tricked French peasants into thinking potatoes were fit only for royalty. As a result, the people pilfered the king's potato fields, and the plant quickly ended up in gardens all over France.

In Germany, potatoes were also given royal cachet by a series of royal advocates, including Frederick the Great; and in Greece, by King Otto I. In Sweden, the potato's protagonist was Jonas Alströmer, who "stole" two sacks of them from England and (despite being chased by the British navy) got them safely to Stockholm.

Such crop champions are what the plants in the following chapters require today.

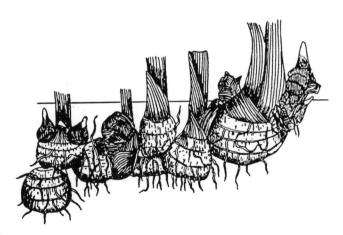

Achira

Achira (*Canna edulis*) looks somewhat like a large-leaved lily and is closely related to the ornamental cannas widely grown in both temperate and tropical zones.[1] It was probably one of the first plants to have been domesticated in the Andean region. Easy to plant and easy to grow, it develops huge, edible underground rhizomes[2] sometimes as long as a person's forearm.

Although little studied by modern scientists, these starch-filled rhizomes are produced throughout a vast region that extends from Mexico and the West Indies to Venezuela, through the Andes and the Amazon basin to Argentina, and along the Pacific coast to northern Chile. In much of this area achira is a market vegetable, but only in Peru and southern Ecuador is it a substantial crop.

Some achira is simply cooked and eaten. Most of the plants, however, are used to produce starch. In this process, the rhizomes are shredded, the grated material dumped into water, and the fibrous pulp separated from the heavy starch by decanting. The starch is then sold for use in foods as well as in other products, such as sizing and laundry starch.

This plant could have a much brighter future as both a food and a cash crop. Its starch has the largest granules ever measured. They can actually be seen with the naked eye and are three times the size of potato-starch granules, the current standard for starch-granule size. Because of its extraordinary proportions,[3] the starch settles out of solution in a few minutes, freeing it from impurities in little time and with minimum expense. The starch is clear and, when cooked, is glossy and transparent, rather than opaque like that of potato, cornstarch, or common arrowroot.[4] The cooked starch seems to be easily

---

[1] Indeed, some ornamental cannas are sold under the same botanic name as achira. However, the blossoms of true achira are much smaller than those of the ornamental cannas.
[2] Strictly speaking, these are corms; their growing tips are at the stem end of the swollen underground parts.
[3] Achira granules are about 125 micrometers long and 60 micrometers wide (see page 28).
[4] *Maranta arundinacea.*

digestible, an important feature for infants, invalids, the elderly, and people with digestive problems.

These attributes could make achira one of the most interesting of all carbohydrate resources. Its unusual starch is a possible complement to other starches now used in foods and industry, and it has the potential to be produced in quantity. Australians have mechanized the planting, cultivation, harvesting, and milling of the crop, thereby demonstrating that achira need not be restricted to areas where labor costs are low.

This adaptable plant is known and grown in a number of places outside Latin America. On the island of St. Kitts in the Caribbean, it has long been used and even exported. In Indonesia, Taiwan, the Philippines, and Australia, there has been small-scale commercial cultivation. In Madagascar, achira is common on the banks of rice paddies. It is also grown in Sri Lanka and Burma, and both Brazil and Hawaii[5] have produced it as fodder for cattle and pigs. However, in none of these widely scattered locations is it being taken seriously as an economic crop.

With research, achira may broaden the base of agriculture. In many areas, it possibly could be incorporated into patches of marginal (especially damp) ground now little used for crop production.

PROSPECTS

The Andes.   Given current knowledge, achira seems unlikely ever to become a major food of the region, but even so, there are several niches where it will remain a contributor to local diets and economies. It is a particularly good "safety net" for use when other crops fail, and it should be more widely planted for this purpose.

Future research may transform the region's use of this plant. It seems probable that current yields may be far surpassed. The application of fertilizer, alone, should boost production dramatically. In addition, the triploid forms (see later) may produce abundantly in now little-exploited locations.

If the unique starch proves to have widespread commercial utility, an export trade could result, to the benefit of the Andean region.

Other Developing Areas.   Achira is unlikely to replace foods based on other starchy roots (such as cassava or common arrowroot)

---

[5] In Hawaii it was formerly used in making "haupia," a traditional dessert (made of coconut milk, starch, sugar, and gelatin) served particularly at luaus.

Loja, Ecuador. Achira being prepared for starch. (J. Horton)

where these are staples and grow well.[6] However, in future more and more Third World people will come to depend on marginal lands to grow their food. Here, achira could provide the difference between hunger and health. Thus, in appropriate climatic zones—especially where prime land is already producing to its fullest or where conventional crops produce less than abundantly—the plant should be given immediate trials.

Industrialized Regions. It seems unlikely that achira will become a major crop of economically advanced countries. However, research may uncover agricultural niches for this robust species, as well as markets for its unusual starch. Indeed, achira starch has a good chance of finding markets in industry and perhaps also in specialty food products—such as baby food and livestock feed—where its easy digestibility and huge granules would be economic assets.

[6] In the lowland tropics, achira yields less than cassava, but at higher altitudes it yields more than cassava.

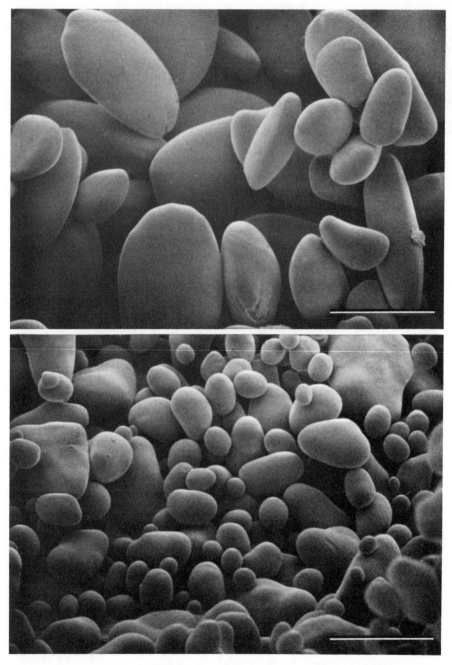

Achira starch granules (top) are by far the largest ever measured—twice the size of potato starch granules (bottom), the previous record holder. Along with its exceptional granule size, achira starch is unusually translucent. Other qualities of this promising carbohydrate probably await discovery. (R. Johnston)

USES

Although achira is one the few root vegetables that can be eaten raw, it is usually eaten cooked like potatoes, arrowroot, cassava, or taro. More often it is baked, whereupon it becomes translucent, mucilaginous, and sweet. A traditional Andean feast is baked achira, roast guinea pig, yacon, and quinoa beer.[7]

For use as starch, achira tubers are peeled, dried, and milled. In Colombia (notably in the departments of Huila and Tolima), the flour is used to make salted crackers in homes and in a factory for commercial distribution. It is also mixed with cheese (*colaciones*). In Vietnam, the flour is used to prepare a pastalike food.[8]

In addition to the tubers, the young shoots can be eaten as a green vegetable. In southern Ecuador, achira leaves are used to wrap foods for easy transport and for cooking.

Livestock eat both the crushed rhizomes and the foliage. Pigs relish the entire plant, readily munching the tops and rooting up the rhizomes. Near São Paulo, Brazil, farmers grow the crop extensively as a pig feed. They prefer it for its hardiness, high yields, and capacity to remain in the field without decaying long after reaching maturity.

In Ecuador, the achira plant, which grows 2 m tall or more, is commonly used as a living fence and as a windbreak to shelter other crops.

NUTRITION

Fresh tubers contain about 75 percent moisture. The dry matter contains 75–80 percent starch,[9] 6–14 percent sugar (mostly glucose and sucrose), and 1-3 percent protein. The potassium content is high; calcium and phosphorus, low.

The leaves and shoots are quite nutritious, containing at least 10 percent protein.[10]

AGRONOMY

One of the most robust of all root crops, edible canna grows well in a wide variety of climates, thrives in many soils unsuited to other tubers, and has few problems with diseases or pests.

[7] Yacon and quinoa are described later; guinea pig is dealt with in the companion volume *Microlivestock: L. !e-Known Small Animals with a Promising Economic Future.*
[8] Tu and Tscheuschner, 1981.
[9] The amylose content of the root starch is about 40 percent (Tu and Tscheuschner, 1981).
[10] They contain 70 percent carbohydrate, 10–14 percent protein, 2.5–5.0 percent fat, 20–25 percent fiber, and 12–17 percent ash. Information from J. Duke.

Los Baños, Ecuador. In many parts of the Andes, farmers commonly use the striking
achira plant as a "living fence" that serves as a windbreak, a boundary marker, and a
food supply. In cities such as Lima, achira and other cannas are common street
ornamentals. (N. Vietmeyer)

It is easy to propagate. Normally, rhizome tips (fairly large, immature
segments bearing at least two healthy, unbruised buds) are merely
stuck in the ground and covered over. (As a precaution against rotting,
the segments are sometimes first dipped in a dilute copper sulfate
solution.) Complete tubers also can be planted.

The crop is usually planted in furrows that help retain moisture.
The swelling rhizomes tend to emerge above the soil surface, so farmers
often earth them up, as they do with potatoes. An initial weeding is
usually necessary, but the spreading foliage quickly suppresses subsequent undergrowth.

HARVESTING AND HANDLING

Achira grows rapidly. It can be harvested starting about 6 months after planting, a point at which the tubers are most succulent and tender. Most of the crop, however, is harvested after 8–10 months, when about a third of the stems are flowering and the rhizomes have swollen to their maximum. At this time the milling qualities of the rhizomes are highest. Afterwards, there is little or no change in the tuber apart from an increase in fiber. Since achira shows no definite end-of-maturity, the plant can be left in the ground and harvested whenever needed.

Yields, of course, vary with rainfall, soil types, and other conditions. However, in some places achira outyields the common starchy roots, such as cassava and common arrowroot. The measured yields have ranged from 23 tons per hectare at 4 months after planting to 85 tons per hectare after a year. Average yields are probably 22–50 tons[11] per hectare. Starch yields are generally 2–5 tons per hectare and may be as high as 10 tons per hectare.

Achira is usually harvested by hand. However, in the 1940s, G.H. Burke of Queensland, Australia, developed a mechanical harvester. Its depth-controlled blade passed under the crowns, lifting rhizomes and soil via a conveyor to a cleaning drum. As the drum rotated, the "stool" of the plant was cleaned and the individual rhizomes broken off. This homemade implement could harvest about 5 tons per hour.[12]

LIMITATIONS

Achira is affected by both heat and drought, but it is largely unaffected by excessive moisture, light frosts, or snow.

For industrial use, the flour is traditionally obtained by grating the rhizomes under water. Thus, if achira is to be milled this way for starch, a plentiful supply of fresh water is needed.

Grasshoppers and beetles may feed on the foliage, and cutworms have been known to attack the rhizome. In South America, achira is sometimes attacked by a leaf roller, which also occurs in other parts of the world. In Peru, fungal diseases (especially *Puccinia cannae*, *Fusarium*, and *Rhizoctonia*) affect the crop.

---

[11] Throughout this report, measurements given in tons refer to metric tons.
[12] Information from P. Lloyd.

RESEARCH NEEDS

Among the several research needs, the most important are a general assessment of the crop and its status, agronomic research, starch characterization, and genetic improvement.

Assessment   One of the first needs is to analyze the crop and its uses worldwide. Despite its widespread use, there has never been a global review of achira. Little agronomic development has been reported in the recent international literature, and only a score of papers have been published on it in the last two decades.[13] Researchers should gather the experiences of various parts of the Andes, St. Kitts, Taiwan, Indonesia, Australia, Madagascar, Burma, and other areas. They should also assess the experiences of growing achira as a garden vegetable and an ornamental. Their published results could then be a basis for future judgment of the plant, its utility, and its prospects.

Agronomic Research   This plant has been so neglected by scientists that "everything" remains to be done, from basic physiology to modern genetics. For example:

• Collections of the plant's variability and all potentially useful germplasm should be made.
• Productive clones should be isolated and tested.
• Tissue culture should be attempted to explore its potential for mass-propagating the plant and for removing any viruses that may now infect it.
• The plant's pathology must be studied.

Starch Characterization   A fundamental research need is to further evaluate achira starch and its practical industrial and dietary uses. The starch needs to be tested in various products for palatability and market acceptability.

Genetic Improvement   Although the genetics and breeding of achira are poorly understood, the ornamental cannas (notably *Canna generalis* and *Canna indica*) have been intensively bred for horticultural traits. The insights of horticulturists could now be applied from these close relatives to improve this food crop.

Although the most productive and "domesticated" types fail to produce seed, seed-producing achiras are known and represent a genetic reservoir for breeding purposes. Investigations into them would seem warranted.

---

[13] An excellent series of reports was produced in Hawaii in the 1920s. Although they provide a baseline of scientific information, they were compiled prior to the development of many current methods of evaluating crops and nutrition.

Triploid forms of achira have been identified.[14] Because their chromosomes cannot pair up, the plants are sterile, which is an advantage in a root crop because the plant wastes no energy producing seeds. Various kinds of polyploids should be developed by interspecific crossing between achira and closely related species, and their qualities assessed.[15]

SPECIES INFORMATION

Botanical Name *Canna edulis* Ker-Gawler
Family Cannaceae
Synonyms *Canna achiras* Gillies
Common Names[16]
Quechua: achira
Spanish: achira; achera (Argentina and Bolivia); capacho (Venezuela); sugú, chisqua, adura (Colombia); luano (Ecuador); gruya (Puerto Rico); tolumán (Dominican Republic); tikas, punyapong, kaska, piriquitoya (Costa Rica); maraca, imocona, platanillo, cañacoros (West Indies)
Portuguese:[17] merú, birú manso, bery, imbiry, araruta bastarda, bandua de Uribe
English: achira, edible canna, purple arrowroot, Queensland arrowroot
French: tous-les-mois, toloman (West Indies), Conflor (Reunion)
Bahasa (Indonesia): ganyong, lembong njeedra, seneetra
Burmese: adalut
Malay: ganyong, kenyong, ubi gereda
Tagalog (Philippines): zembu
Thai: sakhu chin
Vietnamese: dong rieng

Origin. When and where achira was domesticated is unknown, but "wild" specimens are seen throughout the midelevations of the Andes. (Most occur at the edges of moist thickets—often in ditches.)

---

[14] The triploids are potentially valuable because their starch content is almost three times higher than normal. There is, however, no information on their yield. Information from T.N. Khoshoo.
[15] Achira itself might be a mixture of diploid, triploid, and tetraploid. Information from T. Koyama.
[16] The lists of common names throughout this book are included as a general guide and are not meant to be comprehensive or definitive. Also, because of the centuries-long associations among Quechua, Aymara, and Spanish, much mixing and borrowing of names has occurred. We have not attempted to sort this out.
[17] Throughout this book, the Portuguese names are in most cases Brazilian.

Cooked tubers appear in dry coastal tombs dated at 2500 B.C., indicating both an ancient origin and the fact that the roots were esteemed highly enough to be carried all the way from the highlands.

Description.   Achira is a perennial monocotyledon with clumps of purple fleshy stems and multibranched subterranean rhizomes. The large (30 cm x 12 cm) leaves are entire with a thick midrib. They are dark green, tinged with reddish brown or with reddish brown veins on the upper surface and purple on the underside. Achira grows to about 2.5 m in height. Dwarf forms are also known.

The beautiful, bright-red to orange bisexual flowers occur in long terminal clusters standing above the leaves. Some plants produce round, black seeds, but farmers propagate the plant exclusively by vegetative means.

The rhizomes may reach more than 60 cm in length; a single stool has weighed 27 kg. A single rhizome may consist of 12 segments representing five generations of growth. If the plant is defoliated, the rhizome will put up new shoots and leaves.

Horticultural Varieties.   In the Andes, two types are recognized: *verdes*, which has off-white rhizomes and bright green foliage, and *morados*, which has rhizomes covered with violet-colored scales. There are many variations in foliage color, stem height, rhizome size, earliness of flowering, and amount of seed production. No cultivars have been selected outside the Andes.

A collection of about 30 Peruvian clones is maintained at the University of Ayacucho.

Environmental Requirements

Daylength.   The plant is apparently daylength neutral, and it appears to grow under a broad range of light environments.[18]

Rainfall.   The plant withstands rainfalls from 250 to 4,000 mm, a huge range. At the lower levels, however, plants may be stunted and low-yielding. Achira does best with moderate, evenly distributed rainfall, although its rhizomes can survive periods of either flooding or drought. In Hawaii, annual rainfall of 1,120 m is said to be adequate.

Altitude.   Sea level to 2,900 m (at the equator).

Low Temperature.   Normal growth occurs at temperatures above

[18] Imai and Ichihashi, 1986.

9°C, although the plant can tolerate brief periods of temperatures down to 0°C without apparent harm. Light frosts will shrivel the leaves and concentrate starch in the tubers.

High Temperature. In Peru, achira is cultivated in the warm Andean valleys where temperatures of 20–25°C are normal. In the Brazilian plateau country (planalto), some achira cultivars survive at 30–32°C during the dry season.

Soil Type. Achira grows in most types of soils, including those with acidities from pH 4.5 to 8.0. The plant tolerates heavy soils and, reportedly, weathered, acidic, tropical latosols as well. Like most root crops, however, achira does best in loose, well-watered, well-drained, and rich soils. The rhizomes form poorly in compacted clays.

Ahipa

The ahipa[1] (*Pachyrhizus ahipa*) is a leguminous plant, but—unlike its relatives the pea, bean, soybean, and peanut—it is grown for its underground parts. Ahipa's fleshy, tuberlike roots can weigh up to 1 kg. Their white interior is succulent, flavorful, crisp like an apple, and can be eaten raw.

The roots of a close relative, the jicama[2] (*P. erosus*), are a favorite food of Central America and Southeast Asia and are becoming popular in the United States as a salad ingredient. Increasing amounts of jicama (pronounced *hee*-ca-ma) are imported from Mexico. Indeed, it has become the top selling specialty vegetable in the United States. Recently, its wholesale price reached $2.50 a kilo, an amazing figure for a root crop.

Ahipa (pronounced a-*hee*-pa) has received almost no agronomic attention, yet it produces a root similar to the jicama's, and could meet with the same enthusiasm. Indeed, ahipa may have even greater potential than its better known cousin. Unlike the jicama plant, the ahipa plant is small, nonclimbing, fast maturing, and unaffected by daylength.[3] Its rapid growth and low, sometimes dwarflike habit make it well suited for large-scale commercial cultivation. Ahipa could therefore be the key to a vast new root crop, even for temperate regions.

Today, however, ahipa is grown only in a few pockets of the Andean mountains. It is cultivated in Bolivia and Peru in fertile valley floors between 1,500 and 3,000 m elevation, and in the ceja de selva ("eyebrow of the jungle") area. It was once found in Jujuy and Salta provinces in northern Argentina, but no longer.[4]

[1] Usually spelled "ajipa." However, for this report we have chosen the "ahipa" spelling, the pronunciation of which is more obvious to the English-speaking public.
[2] Also known as yam bean or sinkamas (Philippines). This plant is described in the companion report, *Tropical Legumes*. It is not the "jicama" of Ecuador and Peru, which is better called "yacon" (see page 115).
[3] Samples from Bolivia and Peru have, under glasshouse conditions in Denmark, proved to be insensitive to daylength. Information from M. Sørensen.
[4] Information from M. Sørenson.

In contrast to most other root crops, the plant has the legume family advantage: rhizobia bacteria in its root nodules make nitrogenous compounds that nourish the plant. This fertilizer undoubtedly helps it grow vigorously in impoverished sites and enriches the soil in which it is planted. Other salient features are the crop's high yield and considerable disease and pest resistance.

PROSPECTS

Andean Region.   As more is learned about ahipa and its potential across the region, its use could become more intensive. However, its temperature and other climatic constraints need to be understood before its true potential becomes apparent.

Other Developing Areas.   Although the jicama is a favorite food in Mexico, Central America, and parts of tropical Asia, ahipa is little known outside the Andes. Nonetheless, in Asia, Africa, Oceania, and highland parts of Central America it could have much appeal.

Industrialized Regions.   In the United States (thanks largely to a growing Latin and Oriental population), jicama now appears regularly in supermarkets coast to coast. All of it is imported. Ahipa could earn similar popularity, and might be grown in the United States itself, as well as in Europe, Japan, Australia, New Zealand, and other nations. As a result, this round, brown root with the crisp, bright taste could be a new addition to millions of dinner tables, as well as a low-calorie food for the diet-conscious.

USES

Ahipa tubers are mainly eaten raw. The white[5] flesh is sweet and refreshing and is especially popular in summer. It is often sliced thin and eaten raw in green salads and fruit salads. Since it is slow to discolor, soften, or lose its crunch, it is particularly suited to garnishes or hors d'oeuvres.

Ahipa can also be cooked. It is often lightly steamed or boiled, and retains its crunchy texture even after cooking. It fries up much like a potato. For stir-frying and braising (briefly) it can be a replacement

---

[5] In Bolivia is found a purple and magenta striped form, which may be a particularly decorative addition to salads.

Ahipa is one of the least known, but most interesting, of the plant kingdom's edible roots. Its tubers are usually eaten raw and make a crunchy, delectable snack. The plants shown here were grown in Denmark, an indication that ahipa probably can be produced as a food crop in many places outside the Andes. (F. Sarup)

for water chestnuts. It absorbs sauces quickly and without softening. Even paper-thin slices seem to keep their characteristic freshness.

NUTRITION

The nutritional content of ahipa is unknown, but is probably similar to that of jicama. If so, it is low in sodium and calories (containing approximately 50 calories per cup raw) and is a good source of potassium and vitamin C. The starch of jicama is easily digestible.[6]

The protein content on a dry matter basis is higher than that of other root crops, but fresh tubers have a low protein content because their moisture content is extremely high.

---

[6] Approximately 80 percent of the starch particles are below 5 microns in diameter, and after a period of 16 hours in the digestive tract, 75 percent of the starch has been metabolized by glucoamylase as against 40 percent of the starch from sweet potato. K. Tadera, T. Tanguchi, M. Teramoto, M. Arima, F. Yagi, A. Kobayashi, T. Nagahama, and K. Ishihata. 1984. Protein and starch in tubers of winged bean, *Psophocarpus tetragonolobus* (L.) DC., and yam bean, *Pachyrrhizus erosus* (L.) Urban. *Memoirs of the Faculty of Agriculture (Kagoshima University)* 20:73-81.

AGRONOMY

The plants are easily propagated by seed and, except for good manuring of the soil before planting, require little attention. They can also be propagated using small tubers, which greatly reduces the growing time. In some areas, to encourage large, sweet roots, the flowers are plucked. This is said to double the size of the tubers.

Ahipa has a comparatively short growth period. It begins flowering about 2.5 months after planting; harvest takes place after 5–6 months.

HARVESTING AND HANDLING

In general, the roots are handled, stored, and marketed like potatoes. However, they can be stored in the soil—by cutting off the plant tops—until needed.

LIMITATIONS

Only the root is safe to eat. Leaves, stems, roots, ripe pods, and seeds contain insecticide[7] and may be toxic to humans.

No yield figures are available at present because no studies have been carried out.

RESEARCH NEEDS

The conservation of varieties and landraces of ahipa is of the utmost importance.[8] These almost certainly include potentially valuable germplasm, and many are threatened with extinction. For example, the landraces of the Argentinian provinces of Jujuy and Salta, as well as of the Bolivian Yungas, have almost disappeared. Ahipa should be collected from the Indian fields throughout the Andes before it is too late. The materials collected should be made available to institutes that deal with crop development and mutation genetics.[9] Special

[7] By analogy with jicama, the insecticide is almost certainly rotenone.
[8] A biosystematic research project examining the potential of the genus is currently being carried out at The Botanical Laboratory, University of Copenhagen, and The Botanical Institute of Crop Husbandry and Plant Breeding, both at the Royal Agricultural University, Copenhagen. A considerable collection of seed material of both wild and cultivated species has been obtained through various institutions, and collections were made in the original distribution area of the genus in 1985. Information from M. Sørensen.
[9] Because in tropical environments the seed remains viable for only three or four years, duplicates of all seed collections should be housed under controlled temperature and humidity at modern seed banks.

attention should be given to differences in protein content; because the species is leguminous, it seems likely that types with exceptional protein levels in their roots will be found.

Considerable variation in the size and quality of the roots, growth habit, leaf morphology, and ecological preferences have been recorded. There is a need for thorough tests involving a wide range of materials.

Methods used for growing ahipa throughout the Andes should be reviewed and concerted research programs organized to apply modern agronomic knowledge to boost production. Trials of spacing, fertilization, pest control, irrigation, and other cultural requirements are needed, with particular attention to the effect of intensive management on the culinary quality.

High moisture content makes ahipa tubers shrivel and lose condition more quickly than other root crops. Improved methods of storage and transportation are needed, as well as, perhaps, cultivars with a thicker epidermis.

The nodulation requirements should be studied in detail, along with identification of the specific symbiotic organisms.

Ahipa has promise for reducing the daylength sensitivity of related species. Hybrids between ahipa and jicama (which is very sensitive to daylength variations) might produce a valuable new man-made crop that expands the range of both parents and whose root growth might be independent of latitude and season.

The pods deserve research attention as well. It is thought that some varieties contain almost no insecticide, at least when green. There is the possibility that these could constitute a source of protein-rich food. Ahipa then would simultaneously provide a nutritious green vegetable and a valuable tuber crop. Research might also identify when insecticide develops in the pods or seed. If harvested before that, all types might be used as green vegetables.

SPECIES INFORMATION

Botanical Name   *Pachyrhizus ahipa* (Weddell) Parodi (Also spelled *Pachyrrhizus ahipa.*)
Family   Leguminosae (Fabaceae)
Synonym   *Dolichos ahipa* Wedd.
Common Names
   Quechua: ajipa, asipa
   Aymara: villu, huitoto
   Spanish: ahipa, ajipa, achipa (South America); dabau (Ecuador); fríjol chuncho (Bolivia, Peru), judía batata, poroto batata (Argentina)
   Portuguese: ahipa
   German: andine Knollenbone

Origin.   Ahipa has never been recorded in the wild state. Although the present area where this species is found in cultivation is restricted to a limited number of Andean valleys, the existence of archeological evidence from geographical areas outside its present distribution indicates that this crop was cultivated widely in the Andes at least 2,000 years ago.

Description.   This species is a nonclimbing, erect or semi-erect herb usually no more than 30–60 cm in height. Its trifoliate, pubescent leaves have asymmetrical and entire leaflets; they are wider than they are long.

The inflorescence is on short stalks (0.1–1.5 cm) with a few pale lavender or white blossoms. Its round to kidney-shaped seeds (0.8– 1.0 cm) are normally dull black, but can be black-and-white or brown in color and grow in 8–11 cm long pods.

Each plant has a single swollen root, which tapers towards both ends. The roots may be 15 cm (or more) in length, and usually weigh 500–800 g. Normally elongated or irregular in shape, they can also be nearly spherical. The pale yellow or tan skin encloses a white pulp that is interwoven with a soft fiber.

Horticultural Varieties.   None recorded.

Environmental Requirements

Daylength.   Apparently neutral for both flowering and root formation.

Rainfall.   Although the plants grow well in locations ranging from subtropical to tropical and dry to wet, for good yields they require a warm climate with moderate rainfall.

Altitude.   Sea level to 3,000 m.

Low Temperature.   They are sensitive to frost.

High Temperature.   Unknown.

Soil Type.   As with other root crops, the soil should be light and well drained so as not to restrict tuber growth or encourage fungal rot.

Related Species.   Ahipa has several relatives that produce edible tubers. Its Mexican relative, the jicama (*Pachyrhizus erosus*), has already been mentioned.

Another relative, the "potato bean" (*P. tuberosus*), is native to tropical South America.[10] Its home is thought to be somewhere in the upper basin of the Amazon River. It is similar to ahipa, but the plant is a large herbaceous vine climbing taller than 10 m. Its large, tuberous root is used like ahipa's. At present, this species is restricted to isolated areas in the Amazonian regions of Ecuador, Peru, Bolivia, Paraguay, Brazil, and possibly Venezuela and Colombia. It is grown by local Indian tribes in shifting cultivation or occasionally is collected from the wild.

Ahipa has undomesticated relatives, such as *P. panamensis* and *P. ferrugineus*, that produce roots, but whether they are edible is uncertain.[11]

---

[10] Some common names are nupe (Venezuela), jacutupé or macucú (Paraguay and Brazil), and dabau (Ecuador).
[11] In greenhouse trials, the wild species have produced yields of similar quantity and tubers of the size and weight of the cultivated species. Information from M. Sorensen.

Arracacha

Arracacha (*Arracacia xanthorrhiza*) is botanically related to carrots and celery, and it incorporates qualities of both. Below ground it produces mostly smooth-skinned roots that resemble white carrots. Above ground it produces green, sometimes purple-streaked stems that are boiled or eaten raw like celery.

At present, arracacha is known only in South America and a few parts of Central America and the Caribbean. However, like its famous relatives, it could become a familiar crop throughout much of the world.

Arracacha (pronounced ar-a-*catch*-a) is rich in flavors and is one of the tastiest foods to be found anywhere.[1] Native to the Andean highlands from Venezuela to Bolivia, it is often grown instead of potato because—although it takes longer to mature than modern potato cultivars—it is produced at only half the cost. Nonetheless, it was so overlooked in colonial times that it wasn't given a scientific name until 300 years after the Spanish Conquest.

Today arracacha is almost as little known scientifically as it was at the time of Pizarro, but it is eaten in most Latin American countries as far north as Costa Rica. Usually it is grown only in small gardens for local use. However, the roots are sold in considerable quantities in the larger cities of Colombia and the rural markets of northern Peru. It is also found in Cuba, Haiti, Dominican Republic, and Puerto Rico.

Recently, arracacha has gained popularity in southern Brazil and has become an established vegetable in the city markets.[2] It is grown in big fields using modern techniques. While characteristic of relatively high elevations in the Andes, the plant is being produced in Brazil at

---

[1] The late David Fairchild, dean of United States plant explorers before World War II, considered it "much superior to carrots." The great Soviet plant explorer, S.M. Bukasov, said, "There's nothing more tasty in the world than arracacha." Thousands of inhabitants of the Andes, as well as many visitors, agree.

[2] The species was introduced into Brazil at the beginning of this century, and in recent years its cultivation has become a big industry. More than 10,000 hectares are grown in the states of São Paulo, Paraná, Minas Gerais, and Santa Catarina. It is usually known as mandioquinha salsa. Information from A.C.W. Zanin.

low elevations with climates like those of many warm-temperate regions of the world. Thus, it seems probable that Brazil's experience is demonstrating arracacha's future potential for regions such as North America and southern Europe. People outside Latin America could soon be enjoying these underexploited roots just as the Incas did 500 years ago.

PROSPECTS

Andean Region. This crop is a good candidate for expanded cultivation in its native region. For example, it has been tested in the eastern valleys of the Andes, where it was previously unknown, and it yielded well there. Given research attention, it is likely to become a major product at intermediate elevations throughout the 4,000-km-long Andean region.

Other Developing Areas. Arracacha could become a valuable root crop in all tropical highlands, particularly if improved cultivars and cultural techniques are developed. The potato has already become successful in Nepal and Burundi. Arracacha (and the other Andean tubers) should now be introduced—using recognized quarantine procedures—to the highlands of Asia and Africa for experimental trials. Cultivation should be tested in the highlands and hill country of East Africa, Central Africa, India, Southeast Asia, and similar regions. Remnants of old introductions may still exist in the highlands of Central America and the West Indies; local agronomists should investigate.

As noted, arracacha has received little research attention, but, with modern technology, is being successfully cultivated in Brazil. Many countries of Latin America and elsewhere seem likely to reap direct benefit from this experience.

Industrialized Regions. In North America, Europe, Japan, and other temperate regions, arracacha is likely to become commonplace. The roots should prove highly acceptable to millions of consumers. In the United States, they are already found in Boston's produce markets (shipped from Puerto Rico),[3] and locally grown arracacha is available in a few markets in the San Francisco Bay area.[4] The plant has also been introduced to Australia.[5]

[3] Information from R. E. Schultes.
[4] Information from C. Rick.
[5] It is being grown at Nimbin, New South Wales. Current cultivars produce little root, but are relished for the huge celerylike stem. Information from M. Fanton.

Typical arracacha roots from the produce market at Medellín, Colombia. The roots resemble parsnips in form and color, but they have a mild flavor, sometimes reminiscent of celery. (W.H. Hodge)

USES

Young, tender arracacha roots are eaten boiled, baked, or fried, or are added to stews. They have a crisp texture; white, yellow, or purple flesh; and a delicate flavor that combines the tastes of celery, cabbage, and roasted chestnut. During cooking they emit a fragrant aroma.

These roots are a common ingredient in the typical Andean stew (called "*sancocho*") that is particularly popular in Colombia and some highland areas of Peru. Indeed, most soups in Colombia contain arracacha. In addition, much of Brazil's arracacha crop is made into dried chips that impart a pleasant and distinctive flavor to dehydrated soups. A famous Switzerland-based company uses it to flavor one dried soup that is popular throughout Brazil. In rural Costa Rica the roots have become particularly popular in wedding feasts.

Parts other than the roots are also used. The young stems, which are sometimes blanched, are used in salads or as a cooked vegetable. Although edible, the central root has a coarse texture and strong flavor, and it is usually fed to livestock. The foliage is also used as fodder.

NUTRITION

The roots have a starch content ranging from 10 to 25 percent. The starch granules are quite small, similar in many respects to those of cassava. The starch is easily digested and can be used in foods for infants and invalids. During storage the roots increase in sweetness, presumably because some starch hydrolyzes to sugars.

All parts of the plant have particularly high calcium content. The roots, with their bright yellow flesh, are undoubtedly rich sources of vitamin A.

AGRONOMY

Arracacha is traditionally propagated with offsets or shoots that are produced on the crown of the main rootstock.[6] After removal, the base of each offset is slashed repeatedly to stimulate the shoots to form and to encourage a uniform arrangement of lateral roots. The offsets are left to "heal" for 2–3 days and are then planted, usually in holes along furrows.

Although arracacha may be planted throughout the year, in southern Brazil it is generally set out in the early spring, and at the beginning of the rainy season in the Andes, where it is often interplanted with potatoes.

HARVESTING AND HANDLING

The tubers are normally harvested 300–400 days from planting. Immature tubers may be dug after 120–240 days. At harvest time there may be as many as 10 lateral roots (each about the size of a carrot) aggregated around the central rootstock.

---

[6] Although it is possible to propagate with seed, germination is normally less than 50 percent, and in some cases no seed is produced. Nonetheless, seeds are extremely useful for plant breeding.

Arracacha plant. Although the roots are the main product, the stems are also edible and look and taste rather like celery. (W.H. Hodge)

The farmer retards flowering by breaking the leaf stems, thereby increasing the root size. Harvest time is determined by snapping a finger against the lateral roots and judging the maturity by the sound. (Some growers harvest as the leaves begin to yellow, just prior to flowering.) At harvest, the entire plant is uprooted.

This is a productive crop. Yields normally vary between 5 and 15 tons per hectare; test plots have yielded as much as 40 tons per hectare.[7] One plant may produce 2–3 kg of edible lateral roots.

LIMITATIONS

Arracacha seems to have several agronomic limitations. Although exact photoperiod restrictions are not known, specific daylength requirements may explain why it is not more widely grown. The roots have a longer growing period than potato. The plant is not frost tolerant. Harvesting cannot be delayed past the flowering stage; roots left in the ground become fibrous and tough and develop a strong, unpleasant flavor.

Arracacha is particularly prone to spider mites and is susceptible to nematodes in some regions. Viruses (and perhaps mycoplasms) have

[7] Information from R. Del Valle.

ARRACACHA IN PUERTO RICO*

In a small area of Puerto Rico, farmers have been growing arracacha (under the name "apio") since at least 1910. Today, for many, it is a staple food. Between Barranquitos and Orocobis, in the highest parts of the island, at least 300 hectares of arracacha are currently grown. Rotated with cabbage, taro (dasheen), and especially cocoyam (tannia), arracacha provides an important food as well as some profit from sales in roadside stands and town markets. It is likely that the germplasm of this yellow-fleshed variety came from the Dominican Republic, where arracacha is also grown at higher elevations.

The crop has always been localized, unstudied, and unappreciated by most Puerto Ricans. However, in 1980, agronomist Reinaldo Del Valle of the University of Puerto Rico introduced 12 cultivars from Colombia. In replicated experiments in Barranquitas, these Colombian arracachas were compared with the local Puerto Rican "criollo" cultivar. The highest yield—21 tons per hectare from the Colombian cultivar "A"—was almost twice what is considered a good yield for the criollo. On the other hand, the criollo appeared to be more tolerant of local diseases and pests. The introduced plants were especially affected by root rot (caused by *Rhizoctonia*, *Pythium*, and *Fusarium* species) and insects, and some were also damaged by rats and snails.

In a second trial, preventive applications of insecticide and fungicide were made at two- or three-week intervals throughout the growing season. Cultivar A again produced the highest commercial root yield, 38 tons per hectare, while another Colombian cultivar yielded 23 tons per hectare, and the criollo yielded 15 tons per hectare. Del Valle suspects that yields of all types could be greatly improved, particularly by correcting micronutrient imbalances in local soils (deep clay) and by giving more attention to the problem of root rots.

been isolated from some roots, but their importance has yet to be evaluated.

The roots have a relatively short storage life—similar to that of cassava. Unlike carrots, they can be eaten only after cooking.

RESEARCH NEEDS

Arracacha is likely to prove a valuable root crop in many areas of the world if attention is given to determining its horticultural requirements, improving cultivars, and selecting good types.

Studies of the pathogens infecting the crop should be made, especially

Although it is not a big crop in Puerto Rico, arracacha is well established in the upland areas. Many farmers there depend on it, and the demand—especially in the mainland United States—is increasing. Because the region where arracacha grows is as low as 600 m elevation, Puerto Rico's experience indicates that the crop could have a big future in many tropical and subtropical regions. (N. Colón)

---

* Information from R. Del Valle and N. Colón.

of viruses,[8] before arracacha is introduced to new areas. Liberal introductions to new environments seem justified, but in areas where carrots and celery are important crops, information on possible disease and pest transmission is required before final decisions on its safety can be made. Tissue culture propagation seems a likely method for eliminating viral diseases.

It is necessary to explore seed physiology, viable seed production, and the variability obtained from sexual propagation for use in potential breeding programs. Wild varieties, such as *Arracacia aequatorialis*, *A. elata*, *A. moschata*, and *A. andina*, are found in southern Ecuador

[8] Information from A.A. Brunt.

and northern Peru. They should be sought and preserved for their possible use in future breeding programs.

Analyses of the relative nutritional merits of existing varieties should be carried out.

SPECIES INFORMATION

Botanical Name   *Arracacia xanthorrhiza* Bancroft
Family   Apiaceae (Umbelliferae)
Synonym   *Arracacia esculenta* DC
Common Names
   Quechua: laqachu, rakkacha, huiasampilla
   Aymara: lakachu, lecachu
   Spanish: arracacha, racacha, apio criollo (Venezuela); arrecate (Latin America); racacha, virraca (Peru); zanahoria blanca (Ecuador)
   Portuguese: mandioquinha-salsa, mandoquinha, batata baroa, batata salsa, batata cenoura
   English: arracacha, racacha, white carrot, Peruvian carrot, Peruvian parsnip
   French: arracacha, panème, pomme de terre céléri

Origin.   Arracacha has probably been cultivated as long as any plant in South America. Its wild ancestor is unknown, although there are many semidomesticated types that may include arracacha's progenitor. The greatest germplasm variation is in Ecuador and adjacent areas of Colombia and Peru.

Description.   This perennial is a stout herb, somewhat resembling celery in form. It is one of the largest of the cultivated umbellifers, and the crushed stems and roots have the aroma characteristic of the family. Stems and leaves usually attain a height of about 1 m and are ensheathed in dark green or purple leaves. Flowers are purple or yellow, small, and formed in flat clusters on stalks radiating from a central stem. Although many flowers are fertile, arracacha is generally harvested before completing a seed cycle.

The cylindrical central root bears numerous lateral roots that are 5–25 cm long and swollen to 2–6 cm in diameter. Their flesh ranges in color from white to yellow or purple, with a creamy white exterior. In some types, a cross section of the main root shows attractive rings of various colors.

Horticultural Varieties. Selections have been based mainly on the color of the root. In the Andes three main types are distinguished: blanca (white), amarilla (yellow) and morada (purple). Certain strains also differ in flavor, texture, and length of time to maturity. Types with golden roots and orange roots have been obtained by sexual propagation in Brazil. A type resistant to the bacterial disease *Xanthomonas arracaceae* is also being tested.[9]

Andean germplasm collections are held in Merida, Venezuela, and Cajamarca, Peru.

Environmental Requirements

Daylength. It is believed that arracacha needs short days for good production of roots, but the range of variation among specimens is unknown.

Rainfall. An even distribution of rainfall seems to be important; ideally, it should amount to 1,000 mm annually and never be less than 600 mm annually.

Altitude. Arracacha is cultivated at elevations from 3,200 m down to 600 m, or perhaps lower. In Colombia, it is said to grow best at altitudes between 1,800 and 2,500 m; in southern Brazil, between 1,000 and 2,000 m.

Low Temperature. A temperature range of 14–21°C appears to be required for best growth; lower temperatures delay maturity so much that the crop cannot be harvested before winter. As noted, the plant tolerates no frost.

High Temperature. Arracacha seems unable to tolerate extended periods above 25°C.

Soil Type. Sandy soils with pH of 5 or 6 are thought to be most suitable; these should be deep and well-drained. Yields are said to be enhanced by fertilizer high in phosphorus and low in nitrogen.

[9] Information from V.W.D. Casali.

Maca

Maca (*Lepidium meyenii*) is a largely unknown crop found at higher altitudes than perhaps any other crop in the world—for example, at altitudes up to 4,300 m in the northern Peruvian Puna near Lake Junín. Even most of the Indians of the Andes barely know this plant, which is so restricted in its distribution. Yet maca's enlarged tuberous roots are delicacies with a tangy taste and an aroma similar to butterscotch.

The area where maca (pronounced *mah*-kah) is grown is an environment of intense sunlight, violent winds, and bone-chilling cold. This area is among the world's worst farmland, especially in its upper limits, with vast stretches of barren, rocky terrain. Daily temperature fluctuations are so great that at sunset temperatures often plummet from a balmy 18°C to 10°C below freezing. Fierce winds evaporate more moisture than does the fierce sunlight, and carry away more soil than does the rain.

In this stark, inhospitable region, maca makes agriculture possible. Cultivated maca survives in areas where even bitter potatoes cannot grow (see page 99), and its wild ancestor grows even higher—just below the perpetual ice, on cold, desolate wastes where grazing sheep and llamas is the only possible land use, and the only other forage consists of coarse, sparse grasses lacking in nutritional quality.

Maca, a matlike perennial, is so small, flat, and inconspicuous that even visiting agronomists sometimes fail to realize they are standing in a farmer's field. Its tuberous roots resemble those of its relative the radish, and are yellow, purple, or yellow with purple bands. They are rich in sugars, starches, protein, and essential minerals—particularly iron and iodine.

To Andean Indians, maca is a valuable commodity. Dried, the roots can be stored for years. They are often exchanged with communities at lower elevations for staples such as rice, and they reach markets as far away as Lima. The sweet, spicy, dried root is considered a delicacy. Maca boiled in water is sweeter than cocoa. In Huancayo, Peru, maca pudding and maca jam are popular.

Maca is further valued because it reputedly enhances fertility in both humans and livestock. Whether this reputation has any validity is uncertain. However, soon after the Conquest, the Spanish found that their livestock were reproducing poorly in the highlands, and the Indians recommended maca. The results were so remarkable that they were noted by surprised Spanish chroniclers. Colonial records of some 200 years ago indicate that tribute payments of roughly 9 tons of maca were demanded from the Junín area alone.

Even if such medicinal effects prove invalid, the fact remains that where little else will grow, maca provides nutritious food that stores well. But for all its qualities, maca is in trouble. Its cultivation is declining. Once it was probably grown from Ecuador to northern Argentina, and hundreds of hectares of terraces apparently were devoted to its cultivation; now it is restricted to a few tiny scattered fields, and it is fast dwindling toward agricultural oblivion. Indeed, in 1982 maca was declared to be in danger of extinction as a domesticated plant.[1]

One reason for the decline is that maca is complicated to grow (see later). However, the main reason is neglect. This "poor person's crop" has been given little research or administrative support (even its growth cycle, flower biology, and chromosome number are not known). As a result, it is being displaced by foods imported into the region. To barter for these foods (mostly rice, noodles, sugar, and some canned goods), the local populations increasingly rely on raising sheep and llamas, as well as on using lower altitude fields, where corn can be grown.

But perhaps this Inca crop will not be lost. Scientists and governments are finally turning their attention to its merits. Its seeds—representing centuries of cumulative selections by farmers—are for the first time being collected, grown out, tested, and saved.

This attention is important, for maca shows potential for benefiting areas at extreme elevations both inside and outside the Andean zone. Its ability to thrive at unusually high altitudes means that large areas previously considered inhospitable to agriculture could be turned to productive use.

PROSPECTS

Andean Region. More attention to maca will almost certainly generate bigger markets and provide more income for the high puna, perhaps the most economically deprived part of the region, a place where few other crops can survive. It could be a nutritional complement to a diet weak in vegetables (other than potatoes) and probably lacking in vitamins and minerals, especially iodine. Moreover, the roots store

[1] International Board for the Protection of Genetic Resources (IBPGR), 1982.

Maca root. (S. King)

and transport well, and markets could expand throughout the Andean countries.

Other Developing Areas. Maca's future outside the Andes is unclear because the plant is currently so little known. Only research and trials will tell if it could represent a new contribution to the diets of people living in mountainous areas worldwide. Other than preliminary trials, this crop warrants no vigorous research attention outside the Andes at this point, although there are many areas (for example, the high Himalayas) where edible tubers are few and where it could be tested.

Industrialized Regions. Unlike most of the crops in this report, maca does not seem to have a major future in North America, Europe, or Australasia. However, it is worth some basic research, which might change the outlook. There is already interest in the United States in its reputed effects on human fertility.[2]

[2] Preliminary analyses have shown that it contains glucosinolates. Information from T. Johns.

USES

The fresh roots, which are considered a treat, are baked or roasted in ashes. The dried roots are mainly boiled in milk or water to create a savory, fragrant porridge. It also make a popular sweet, fragrant, fermented drink (*maca chicha*) that is often mixed with hard liquor to make "*coctel de maca*."

It is the dried roots that are most used. After sun-drying, they become brown, soft, and sweet, with a musky flavor. It is reported that the flavor remains strong for two years, and often for much longer.

No part of the plant is wasted. Even the leaves reportedly are eaten. (The plant is a close relative of cress, the European green whose pungent leaves are eaten in salads.) Maca is also a choice Andean feed for fattening guinea pigs for the table.

NUTRITION

In some areas of the Puna, maca is important in the diet. It has one of the highest nutritional values of any food crop grown there. The dried roots are approximately 13–16 percent protein, and are rich in essential amino acids. The fresh roots contain unusually high amounts of iron and iodine, two nutrients that are often deficient in the highland diet.

In addition to its nutritious ingredients, some antinutritional factors—alkaloids, tannins, and small quantities of saponins—have been reported.

During storage, the nutritional value stays high. Seven-year-old roots still retain a high level of calories as well as 9–10 percent protein.[3]

AGRONOMY

Maca husbandry is difficult, and the cropping system used to grow it is complex. To obtain seed, the strongest plants are left in the ground at harvest time. About a month later, when hard freezes have killed the tops of the plants, they are transplanted (with all their secondary roots) to special plots in unused sheep corrals or manure piles. There they are covered with soil and heavily manured.

Within a few weeks, new shoots appear. In a month or two, numerous flowers rise, and 3–4 months later, seed is set. The seeds are allowed

---

[3] Information from S. King.

Maca field near Lake Junín, Peru. Above ground the plant is all but invisible. The compact, low-growing habit protects the plant from the harsh growing conditions at high elevations. (S. King)

to mature and fall to the ground. The mixture of seeds, plant debris, and loose soil is used to replant the crop. Seeds remain viable 3–4 years.

Maca is usually planted in small plots, often surrounded by stone fences or earth ridges that protect the plants from desiccating wind and ground-creeping frosts. The tiny seeds, still mixed with plant debris and fine earth, are scattered on carefully worked soil. Sheep are then released to press the mixture into the ground as they walk around. Birds are especially fond of maca seeds, and people often watch over the newly planted fields to scare them away.

The seedlings are not usually thinned, and little further care is given. Weeds are no problem because little else can grow in the Puna, but as protection from the frequent frosts and snows, maca plants are sometimes covered with straw.

The roots are harvested 6–7 months after planting. However, in the harshest parts of the Puna, they may require up to 9 months to mature. The plants are dug, the leaves removed, and the roots cleaned and left to dry in the sun. Except for the seed stock, all the roots are harvested, even the small ones. Indeed, the smallest are preferred, as they are less fibrous.

Maca is usually grown in very small fields, where its yields average less than 3 tons per hectare. With use of improved agronomic techniques, however, yields equivalent to more than 20 tons per hectare have been achieved, even in nonirrigated plots.[4]

LIMITATIONS

As noted, domesticated maca is in danger of extinction. Its cultivation and use is little known outside of the Lake Junín area, where only scattered plantings remain.

The current methods of horticulture are complex and labor intensive.

Maca is considered a crop that exhausts the soil, and after it is harvested, the plots are left fallow for about 10 years to replenish themselves. Because land is plentiful, this is not a problem in the Puna.

The Puna is unusual in that the sunlight is extremely intense, whereas the temperatures are extremely low—a strange combination caused by its high altitude and proximity to the equator. Maca may be peculiarly adapted to this strange climate. If so, it may be difficult to cultivate outside equatorial highlands.

The dried roots are shriveled and brown and are not visually attractive.

RESEARCH NEEDS

The extent of maca cultivation in the Lake Junín and Huancayo regions should be investigated. Here the crop faces grave genetic erosion. Therefore, germplasm should be collected, broadly disseminated to germplasm banks, and cultivated in protected locations.

The plant's nutritional requirements should be determined, especially emphasizing tests to assess its reputation for soil depletion. A more likely cause is that Puna soils are poor to begin with, and that little fertilizer is used. The problem might also be due to allelopathy (the release of growth-regulating chemicals) because other *Lepidium* species appear to be strongly allelopathic.[5]

Other research needs include the following:

- Characterizing maca's reproductive biology;
- Refining and simplifying maca's agronomy;

[4] Information from C. Mantari C.
[5] Information from E. Rice. See, also, Bieber and Hoveland, 1968. Phytotoxicity of plant materials on seed germination of crownvetch, *Coronilla varia* L. *Agronomy Journal* 60:185–188.

THE PUNA

In referring to maca, the 16th-century chronicler Padre Cobo said: "this plant is born in the roughest and coldest of the sierra where no other plant, cultivated as food, grows."

Indeed, the Puna region of southern Peru, which is maca's native habitat, has an intensely cold climate that makes it all but impossible to cultivate other leafy plants. The fact that such an area was made habitable and self-sustaining is a demonstration of the Incas' agricultural skills, as well as of the potential to be found in maca.

The Puna is one of the most inhospitable places on earth. A treeless ecological zone between 3,800 and 4,800 m elevation, it is characterized by steppes, uncultivated fields, tundra, and barren alpine and subalpine plains. Its average temperature fluctuates between 5° and 10°C. At any hour of the day, but especially in the afternoons, strong winds blow. The most feared is the "phuku," a wind that, according to local lore, can lift a horseman off his mount and throw him to the ground.

In general terms, the landscape is extremely wild and captivating. Because of the luminosity at this high altitude, the mountain peaks are said to seem to be right at one's fingertips. There are few flat areas, and those are very small. Most of the region is undulating terrain, with rough slopes and freezing rocky areas. There are large stretches, barren of all vegetation and soil, with rocks already exposed on the surface.

• Identifying the variation of forms and collecting superior varieties;
• Analyzing the nutritional content and antinutritional factors;
• Investigating the reputed fertility enhancement, probably the primary economic value of maca at this stage;[6] and
• Testing maca outside the Puna to determine how the plant will perform under more benign conditions.

SPECIES INFORMATION

Botanical Name   *Lepidium meyenii* Walpers
Family   Cruciferae (mustard family)[7]
Common Names
   Quechua and Spanish: maca, maka, maca-maca, maino, ayak chichira, ayak willku
   English: maca, Peruvian ginseng

Origins.   Maca is a true Puna plant. It was widely grown during the pre-Columbian period. In Junín there are hundreds of square kilometers of ancient terraces that probably were used to cultivate it. Although it does not appear to be represented in the ancient Peruvian pottery, primitive cultivars have been found in archaeological sites dating as far back as 1600 B.C.[8]

Description.   The plant has 12–20 entire and scalloped leaves that lie close to the ground. This rosette is roughly circular, and is formed from the flat and fleshy central axis. As the outer leaves die, there is a continuous formation of new leaves from the center of the rosette.

The off-white, self-fertile flowers arise from a central stalk and are typical of the mustard family. The ovoid seeds are about 2 mm long.

The edible part is derived from the tuberous hypocotyl, that portion of the plant where the root joins the stem. These enlarged "roots" resemble inverted pears both in size (up to 8 cm in diameter) and shape. They end in thick, strong roots with numerous lateral rootlets, as in a radish. The flesh is pearly white and has a marbled appearance.

---

[6] In native Andean medicine, the potency of both maca and mashua (see next chapter) is traditionally judged by a particular odor yielded by the chemical p-methoxybenzyl glucosinolate, which both plants share. Johns, 1981.
[7] Cruciferae includes some of the most widely grown vegetable crops: radish, turnip, cabbage, mustard, and rape, for instance. Maca is the only *Lepidium* species whose roots are used as a food, but the leaves of other species are used as greens, especially the common cress, *Lepidium sativum*, which occurs naturally from Europe to the Sudan and to the Himalayas, and has been cultivated since ancient times as a green vegetable.
[8] Information from D. Pearsall.

It consists of two fairly well-defined parts: an outer region and a central cylinder. The outer section is creamy and rich in sugars; the inner section is firmer and particularly rich in starches.

Horticultural Varieties. There are four traditionally recognized types, all based on the color of the root: cream-yellow, yellow banded with a purple waist, purple, and black. The yellow ones are generally the most popular. Small collections have been made.[9] In Peru, INIAA (Huancayo) and Universidad San Cristobal of Huamanga (Ayacucho) are each caring for a few accessions (four varieties are maintained at Ayacucho).

Environmental Requirements

Daylength. Unknown

Rainfall. In the area near Junín, rainfall is seasonal, averaging 720 mm annually.

Altitude. 3,500–4,500 m, with most cultivation between 3,900 and 4,100 m.

Low Temperature. Frosts are common throughout the growing season. Resistance to night frosts of –10°C have been reported, although the plant is normally mulched to protect it from extreme cold (night temperatures of –20°C are not uncommon just before or after the harvest).

High Temperature. Unknown, but in the Puna temperatures usually reach 18°C (occasionally 22°C or higher).

Soil Type. The limits are unknown. Puna soils are often clayey; Junín is a limestone area.

[9] By the International Board for the Protection of Genetic Resources (IBPGR).

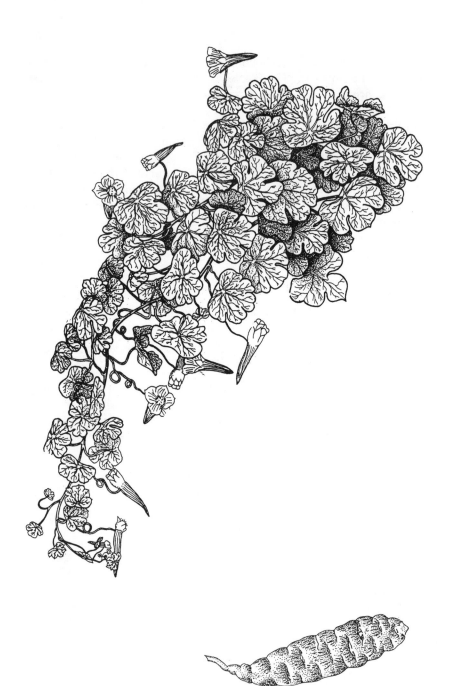

Mashua

Mashua[1] (*Tropaeolum tuberosum*) is probably the Andean region's fourth most important root crop—after potato, oca, and ulluco. It is a hardy plant, and in the poorest regions, where pesticides and fertilizers are too costly to use, mashua is sometimes the prevalent root crop. Its tubers can be found in almost any rural Andean market.

Mashua (pronounced *mah*-shoo-ah or *mah*-shwah) is closely related to the garden nasturtium, an Inca ornamental that is now well-known throughout most temperate zones. In fact, these two beautiful plants are often found together in Andean gardens, one grown for its edible tubers, the other for its pretty and edible flowers.

Among Andean tubers, mashua is one of the highest yielding, easiest to grow, and most resistant to cold. It also repels many insects, nematodes, and other pathogens, thus making it a valuable plant to intercrop with other species.[2] Yet, in spite of its productivity, pest resistance, and popularity, mashua is not widely commercialized—either in its native land or elsewhere.

The tubers—about the size of small potatoes—have shapes ranging from conical to carrotlike. Eaten raw, some have a peppery taste—reminiscent of hot radishes. But when boiled, they lose their sharpness and become mild—even sweet. Boiled mashua is used to add variety to other foods. It is popular in soups. In Bolivia and some parts of Peru, it is topped with molasses and frozen to make a special dessert.

Like oca, maca, ulluco, and bitter potatoes,[3] mashua can provide food at high elevations. This frost-tolerant crop is cultivated in small plots on hillsides—especially on ancient terraces—in cool and moist upland valleys of Argentina, Colombia, Ecuador, Peru, and Bolivia. Peru, it is reported, grows 4,000 hectares of mashua each year.

Within Andean communities, some families choose to plant mashua

---

[1] Also widely known as añu (pronounced *ahn*-yoo) and isaño (e-*sahn*-yo).
[2] In the Peruvian highlands, mashua, oca, ulluco, and native potatoes are often grown together in a multicrop system. Farmers believe that this is the best and cheapest way to control pests and diseases.
[3] These root crops are all described elsewhere in this section of the report.

because it is easier and less labor intensive to grow, and because its tubers are traditionally reserved for children and women. In the higher, colder altitudes, mashua functions like cassava in the tropical lowlands: a food that can be stored in the ground, harvested when needed, and is almost unaffected by poor management.

PROSPECTS

Andean Region.   In the Andes, mashua is associated with poverty. It is shunned by the upper classes because of its Indian origin and because it is eaten by poor country folk. It is disappearing rapidly and in a few years most people will not remember it.

Yet mashua is a vital, although still underrated, part of the Andean agricultural cycle. So little is known about it that its potential is almost certainly unrealized at present. It is a productive and robust plant, and its tubers are visually appealing. It could be selected for greater nutritional quality and palatability. It could also play an important part in pest control in intercropping situations because it suffers from almost no pest, is resistant to the Andean weevil that attacks potatoes and other root crops, and climbs over weeds and smothers them.

Although high yielding and particularly high in vitamin C, mashua is not as palatable as other tubers, and where people have access to rice, noodles, and sugar, it tends to be abandoned more readily than other traditional crops.

Recently, it has been found that mashua in the Andes carries virus infections that are probably extremely debilitating to the plants. Methods have now been developed to produce virus-free stocks. These healthy plants grow much more vigorously, and they represent a way for rapidly improving mashua throughout the Andean region.[4]

Other Developing Areas.   Mashua will probably never be widely grown outside the Andes, but it is worth trying in other tropical highland regions (for example, the Himalayas). Its pest and pathogen resistance alone may make it valuable. It is likely to be productive in areas with moderate temperatures and long growing seasons. Only virus-free germplasm should be introduced.

Industrialized Regions.   Mashua, along with more common nasturtiums, is grown as a flowering ornamental in Britain and, though

---

[4] Information from A.A. Brunt. A research team led by R. Estrada at the University of San Marcos Laboratories in Lima, Peru, is producing disease-free plants for growing by Andean farmers.

rarely, in the United States. The fact that it grows well so far from its Andean home suggests that, like the other Inca tuber crops, it, too, deserves much wider testing and recognition as a food plant. Coastal areas at high latitudes could well be ideal for production of mashua. It has already produced good yields in the Pacific Northwest of North America and in New Zealand. Nonetheless, daylength sensitivity may limit its widespread adoption until insensitive strains can be located.

USES

The sharp flavor of most mashua tubers makes them unsuitable for eating raw,[5] so they are usually boiled with meat to form a stew. (An ideal Andean stew contains meat, mashua, oca, potatoes, greens, quinoa, corn, rice, eggs, and herbs.) They are also eaten as a baked or fried vegetable and may be fried with eggs and onions. Near La Paz, they are soaked in molasses and eaten as sweets.

In New Zealand, where the plant has been newly introduced, it sets tubers well in open fields during a normal spring-to-autumn growing season. One grower cooks the tubers in soups and stews, to which they add a delicate, slightly fragrant flavor.[6] After boiling for five minutes, the tubers appear whitish with purplish spots at the nodes. Young tubers need no peeling, but older tubers are always peeled.

In addition to the tubers, the tender young leaves are eaten as a boiled green vegetable. The flowers are also eaten. (The blossoms of the garden nasturtium are used in restaurants all across the United States, for instance.)

Because mashua is high yielding and its tubers are rich in carbohydrates as well as other nutrients, it has been suggested that it could be grown as a feed for pigs and calves. It could become an especially valuable and cheap stock feed because of its high yield and the high protein content of its foliage.

NUTRITION

Mashua is quite nutritious for a root crop. Solids comprise about 20 percent and protein as much as 16 percent of the dry matter.[7] However,

[5] The hot taste is due to isothiocyanates (mustard oils), the compounds also responsible for the hot taste of radishes, mustard, and many crucifers, to which it is unrelated. Many mashua types are bland, however.
[6] Information from A. Endt.
[7] In one analysis, dry samples of the roots (per 100 g) contained: 371 calories, 11.4 g protein, 4.3 g fat, 78.6 g total carbohydrate, 5.7 g fiber, 5.7 g ash, 50 mg calcium, 300 mg phosphorus, 8.6 mg iron, 214 micrograms beta-carotene equivalent, 0.43 mg thiamin, 0.57 mg riboflavin, 4.3 mg niacin, and 476 mg ascorbic acid. Information from J. Duke.

the protein content is highly variable. One variety was found with tubers containing 14–16 percent protein (dry weight).[8]

Mashua traditionally has many folk-medicine uses. It is considered an anti-aphrodisiac and, hence, many Andean men recommend it for women while refusing to eat it themselves.[9] Male rats fed a tuber diet showed no decline in fertility, but did show a 45 percent drop in total levels of testosterone and dihydrotestosterone.[10]

AGRONOMY

Mashua is one of the common terrace crops of the Andes. Under traditional practices fields are small and often on precipitous slopes. It is planted much like oca and potatoes, using small tubers.

The plant often sprawls over the ground, but it has tiny, threadlike outgrowths that wind around anything they touch, and this allows it to twine up cornstalks or other supports. To improve yields, earth is mounded around the base of the stem as the plant grows.

As noted, mashua is extremely resistant to diseases and insects. It contains nematocidal, bactericidal, and insecticidal compounds (glucosinolates).[11]

HARVESTING AND HANDLING

The tubers are ready for harvest in 6–8 months. They form near the surface and are harvested like potatoes.

Because they have a high moisture content and no waxy surface, the tubers have a shorter storage life than other tubers. Nonetheless, they can be successfully stored for up to 6 months if cool (for instance, 2°C), well ventilated, and protected from strong light.

Mashua is high yielding. Even under conditions of almost no management, harvests are reported to be between 20 and 30 tons per hectare. Yields approaching 50 tons per hectare are reported from experimental plots near Cuzco.[12] A single plant may yield more than 4 kg of tubers.

[8] Originally from Bolivia, this variety was lost when terrorists blew up the research station in Ayacucho, Peru. Attempts to relocate the source are being planned. Information from J. Valladolid R.

[9] The Spanish chronicler Cobo stated that Inca emperors fed their armies on the march with such tubers, "that they should forget their wives."

[10] Johns, 1981.

[11] Tubers are antibiotic against *Candida albicans*, *Escherichia coli*, and *Staphylococcus albus*, the activity paralleled by benzyl isothiocyanate at 100 micrograms. The compound is also nematocidal (Johns et al., 1982).

[12] Information from H. Cortes.

Mashua tubers. In the high, cold altitudes of the Andes, mashua functions like cassava in the tropical lowlands. It requires little care and can be stored in the ground and harvested when the need arises. For this reason, therefore, mashua appeals to poor people and has been unjustly stigmatized as being an undesirable crop. To the thousands of highland people who know it best, however, it is a delicious food. (S. King)

LIMITATIONS

Like arracacha, oca, and ulluco, mashua is apparently heavily infected with plant viruses, most of which are undescribed.[13] One recent test identified it as a carrier of potato leaf roll virus.[14]

There is a possibility that consuming large quantities of mashua, combined with low intakes of iodine, could cause goiter.[15] This is unlikely in most diets; nonetheless, goiter is a problem in parts of the Andes—for instance, Bolivia.

RESEARCH NEEDS

Further collections of mashua germplasm are needed throughout the Andes. Relict populations from Argentina and Chile may provide germplasm with more adaptability to long daylengths.

Mashua sets seed freely and hybridizes well. Thus, there seems to be considerable potential for breeding new and improved types.

Much basic information on the plant's gross horticultural requirements is needed. Investigation of its ecology in field situations and evaluation of its intercropping potential could lead to higher yields and, consequently, to broader utilization.

[13] Information from A. B. Brunt.
[14] Information from J. Martineau.
[15] Glucosinolates are metabolized into isothiocyanates, thiocynates, and thioureas, a class of chemicals that is goitrogenic. Information from T. Johns.

The insecticidal, nematocidal, bactericidal, pharmacological, and other medicinal effects should be investigated. These might prove useful in practice.

Because it yields so prodigiously, mashua may be suited for industrial production of starch. Needed are analyses of the kinds of starch it contains and the amounts that might be produced under field, regional, and national conditions.

Selection for varieties with low levels of glucosinolates should be attempted.

SPECIES INFORMATION

Botanical Name   *Tropaeolum tuberosum* Ruiz & Pavón
Family   Tropaeolaceae (nasturtium family)
Common Names

Quechua: mashua, añu, apiñu, apiña-mama, yanaoca (black oca)
Aymara: isau, issanu, kkayacha
Paez (southern Colombia): puel
Spanish: mashua (or majua, mafua, mauja, maxua), mashuar, añú, anyú (Peru); cubios, navios, navo (Colombia); isaño, isañu, apilla (Bolivia); ysaño (South America)
English: mashua, anu

Origins.   Mashua has been cultivated since ancient times and its tubers show up in many archeological sites. The ancestral plant is uncertain. Weedy types are common in moist, wooded, brushy areas around 3,000 m elevation in Peru and Ecuador and may be representative of the ancestral type. Mashua may have originated in the same regions as the potato, but even today it is virtually unknown outside Bolivia, Peru, and Ecuador.

Description.   Mashua is a perennial, herbaceous, semiprostrate climber occasionally reaching above 2 m in height. Both erect and prostrate forms are known. It has circular, peltate, 3- to 5-lobed leaves, and glabrous, twining stems that attach themselves to other plants by tactile petioles.

The long-stalked, solitary, axial, bisexual, occasionally double flowers (favored by birds and insects) are orange to scarlet in color. Smaller than those of the garden nasturtium (*Tropaeolum majus*), they are borne profusely.

The fruit (schizocarp) has 3–4 lobes that contain joined seeds lacking endosperm. The abundant, viable seeds separate at maturity.

The tubers vary in color from white to yellow. Occasionally, the skin is purplish or red. Often, they are mottled or striped with red or purple, especially below the eyes. The flesh is yellow.

Studies have shown there is a high correlation between the yield of tubers on the one hand and plant height, tuber size, and number of tubers on the other.[16]

Horticultural Varieties. More than 100 varieties have been recognized; there are probably more. One reported in Colombia is var. *pilifera*, slender, long, deeply furrowed, and white, sometimes with pink-purple ends. Another, var. *lineovaculata*, in Colombia, Peru, and Bolivia, is white, streaked and spotted with red. Others may be yellow, orange, reddish violet, or dark purple, often stippled with bright red or purple dots and lines.

Color variants are recognized by a number of native descriptive names, among them in Peru are yana-añu (black), puca-añu (red), yurac-añu (white), sapullu-añu (yellow), and muru-añu (spotted).

Collections are maintained at Quito (INIAP, Santa Catalina), Ayacucho, Junín, and Huancayo.

Environmental Requirements

Daylength. The plant seems to require 12-hour days (or perhaps less) for tuber formation, although it has successfully developed tubers outdoors in Vancouver, Canada (in October when daylight was less than 12 hours),[17] and under glass in southeast England.[18]

Rainfall. The crop requires heavy rainfall; in its native range it receives between 700 and 1,600 mm. It seems to thrive in misty and cloudy weather.

Altitude. Mashua grows best between 2,400 and 4,300 m above sea level along the Andean cordillera. However, altitude may not be an important factor, considering its productivity in Canada, England, and New Zealand.

Low Temperature. It will tolerate light frost and is unaffected by temperatures as low as 4°C. In many parts of its range, it is regularly exposed to mild frosts.

High Temperature. Unknown. Probably above 20°C.

Soil Type. Mashua grows in soils ranging from pH 5.3 to 7.5. While it is tolerant of alkaline conditions, it performs best in fertile, organic soils. Good drainage helps inhibit soil fungi infestations.

[16] Delgado, 1977.
[17] Information from T. Johns.
[18] Information from A.A. Brunt.

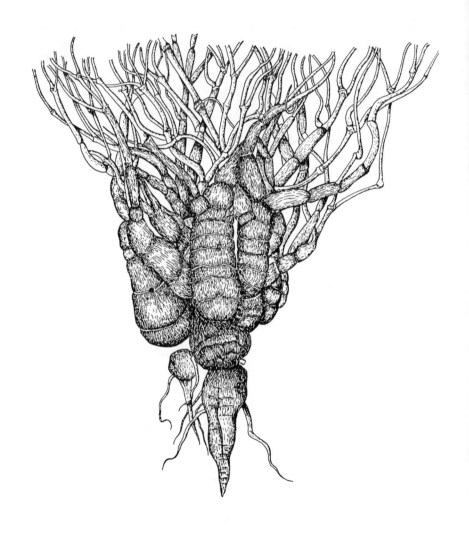

Mauka

It is rare that a new, domesticated food plant is discovered. Yet mauka (*Mirabilis expansa*) is only just now coming to light. In the early 1960s, Bolivian scientist Julio Rea first announced to the outside world that it was an important food of the Maukallajta Indians in the high valleys north of La Paz, Bolivia. In the 1970s, he found it being cultivated also in a few areas of the cold, dry uplands of Ecuador, where it is called "*miso*." Then, in late 1987, mauka was found growing in several locations near Cajamarca, Peru.[1] Outside these three remote areas—all above 2,700 m but separated by hundreds of kilometers—mauka has gone untasted for centuries.

The neglect of this plant is unfortunate, for mauka (pronounced *mah*-oo-kah) provides an abundance of succulent edible stems and tubers with an unusually high protein content. It is productive, cold tolerant, grows well at high elevations, and is relished by the local people who know it and grow it.

Mauka survives where constant winds and near-constant chill place heavy physical strain and moisture stress on plants. Other crops, including most varieties of potato, cannot withstand these harsh conditions.

Mauka appears to have the right qualities for a widely grown food. Its "tubers" can grow to be the length and diameter of a person's forearm. They are flavorful and have good keeping qualities. However, much study in the field and in the laboratory is needed before its potential can be understood.

PROSPECTS

Andean Region. On its merits, mauka would seem a candidate for introduction all along the Andes, where climates are similar to its

[1] This was discovered by J. Seminario C. In 1988 a detailed assessment of mauka cultivation in the Cajamarca area was made by S. Franco P. and J. Rodríguez C. Many of the facts they uncovered are incorporated in this chapter.

native habitat. It is possible that it could blend into the agricultural and culinary habits of many highland groups, providing them with a new food that is nutritious, tasty, and productive.

Other Developing Areas. Mauka seems worth testing in mountainous areas throughout the tropics. However, so little is known about this crop that as of now its future cannot be predicted. Exploratory experimental plots seem justified at this time, but anything beyond token trials should await further research in the Andes.

Industrialized Regions. Mauka is native to an upland region with a cool climate. Thus, in principle, it could grow in North America, Europe, Japan, and other temperate zones. Trials are worth attempting, but its commercial viability is far from assured. Years of research may be needed before its requirements for daylength, maximum production, and daily use are understood.

USES

The edible parts are the upper part of the root and the lower part of the stem. These swollen thickened clumps, much like those of cassava, are usually boiled or fried and served as vegetables.

When freshly harvested, the mauka roots grown in Bolivia contain an astringent chemical that can burn the lips and tongue. Exposing them to the sun, however, replaces the bitterness with a pleasant, sugary flavor. Traditionally, the sun-sweetened tubers are chopped, boiled, and mixed with honey or brown sugar and toasted grain. The combination makes a hearty meal, and the cooking water makes an especially flavorful drink.

It is said that the mauka grown in Ecuador is not astringent. It is prepared in two ways: salty or sweet (*de sal* or *de dulce*). For salty mauka, the tubers are cleaned, cooked, and peeled, and then eaten immediately. To make sweet mauka, the stems and roots are layered with barley or mauka stems in a hollow in the ground for about a month, by which time the starches have largely hydrolyzed to sugars. Both salty and sweet forms are commonly mixed with syrup or molasses and eaten with tomatoes and fish (particularly sardines or tuna).

Like other *Mirabilis* species in the Andes, mauka is used as a feed, mainly for guinea pigs. Animals consume it fresh or dried. (In feeding pigs and guinea pigs, the raw tubers, leaves, and stems are often mixed with corn and weedy vegetation.) People also consume the leaves in salads.

Mauka, one of the least known food crops in the world, is a traditional staple of the Maukallajta Indians. The starch-filled tubers are produced in sites where most potatoes and other root crops cannot survive. (S.D. Franco P.)

NUTRITION

Both the swollen stems and the roots are high in carbohydrates (87 percent on a dry-weight basis) with 7 percent protein (an appreciable amount for a root crop) and little fiber. Based on an evaluation of three separate ecotypes, mauka is richer than the other Andean tubers in calcium, phosphorus, and potassium.

The leaves contain about 17 percent protein. The level of digestibility is said to be higher than that of the other forages that can be grown in the upland Andes.

AGRONOMY

Mauka is generally cultivated as an annual, although it has the enlarged stems and roots of a perennial. It is propagated by portions of stem or root, as well as by offsets (which develop during the second growing season). Seed is also sometimes used, and could be useful in breeding programs, in freeing the plant from any viruses that may be present, and in facilitating the introduction of mauka to new areas. The seed remains viable for several years.

As the plant matures, the below-ground portion of the stem and the upper roots thicken into crowded clusters at and just beneath the soil surface. At harvest, the clumps are pulled up and the edible portions

MAUKA AND THE INCAS

The map of mauka's present production may tell much about the Inca system of subduing conquered peoples. Particularly when any of the empire's subjects became restless and portended trouble, the Inca rulers forcibly moved part of the group to a distant location and replaced them with loyal citizens. They were careful to move the tribe's plants as well, and to chose locations in which they would thrive. Mauka's present-day occurrence in three scattered locations perhaps results from the forced migration of people and plants from the Cajamarca region (known for its unrest in Inca times) to locations in today's Ecuador and Bolivia.

W.H. Prescott beautifully described the ingenuity of the Incas' peace-keeping policy in his classic 1844 account, *The Conquest of Peru.*

> *When any portion of the recent conquests showed a pertinacious spirit of disaffection, it was not uncommon to cause a part of the population, amounting, it might be, to ten thousand inhabitants or more, to remove to a distant quarter of the kingdom, occupied by ancient vassals of undoubted fidelity to the crown. A like number of these last was transplanted to the territory left vacant by the emigrants. By this exchange the population was composed of two distinct races, who regarded each other with an eye of jealousy, that served as an effectual check on any mutinous proceeding. In time, the influence of the well-affected prevailed, supported as they were by royal authority and by the silent working of the national institutions,*

to which the strange races became gradually accustomed. A spirit of loyalty sprang up by degrees in their bosoms, and before a generation had passed away the different tribes mingled in harmony together as members of the same community.

In following out this singular arrangement, the Incas showed as much regard for the comfort and convenience of the colonist as was compatible with the execution of their design. They were careful that the mitimaes, *as these emigrants were styled, should be removed to climates most congenial with their own. The inhabitants of the cold countries were not transplanted to the warm, nor the inhabitants of the warm countries to the cold. Even their habitual occupations were consulted, and the fisherman was settled in the neighborhood of the ocean or the great lakes, while such lands were assigned to the husbandman as were best adapted to the culture with which he was most familiar. And, as migration by many, perhaps by most, would be regarded as a calamity, the government was careful to show particular marks of favor to the* mitimaes, *and, by various privileges and immunities, to ameliorate their condition, and thus to reconcile them, if possible, to their lot.*

---

broken off. In traditional cultivation in the cold, harsh uplands, the plant may take up to a year to reach harvest size. The longer it is left in the ground, the greater the yield; after 2 years, it may produce more than 50 tons per hectare. However, a normal yield seems to be about 20 tons per hectare.

Commonly, mauka is interplanted with crops such as corn. Although sometimes planted in normal furrows, it is more often planted in shallow pits because the seedlings are somewhat delicate.[2]

LIMITATIONS

So little is known about mauka that it has only recently been collected, grown in observation plots, and studied by agronomists. Practically nothing is understood of its potentials or vulnerabilities. For example, it may have restrictive environmental requirements or be susceptible to particular pests and diseases. In the Cajamarca area, however, the mauka plots are remarkably free of pests and diseases. (The major pest seems to be the larva of a fly or butterfly that penetrates the subterranean parts, causing the foliage to wither.)

[2] Growing crops in pits is common in traditional Andean agriculture—pits provide protection from winds and from chill creeping close to the ground, and they help collect and retain water.

It is uncertain that the taste of the stems or tubers will prove widely appealing. The astringent effects of some types may hinder acceptance, but the selection of sweet types should eliminate this possibility.

RESEARCH NEEDS

Nearly everything remains to be done before this rustic crop's full potential can be characterized. Botanical investigations of the most basic kinds are needed. Because its cultivation is small scale, restricted to only a few areas, and declining, much genetic diversity has probably already been lost. Collections should be made to preserve the plant's variability. Seed should be conserved in germplasm banks. Variation among ecotypes should be sought and assessed; there is a particular need to find, multiply, and distribute elite genotypes.

Among agronomic features to be studied are water requirements, daylength sensitivity, and cold tolerance. It is uncertain at present if tuberization is dependent on daylength, temperature, moisture, some other factor, or a combination of these.

Complete studies of the nutritional qualities of all parts of the plant should be made. Amino acid analyses are particularly lacking. Also, the astringency should be studied. What causes it? Where is it found geographically? In what parts of the stems and tubers does it occur? How can it be eliminated? Also needed is more evaluation of the forage qualities of the foliage.

SPECIES INFORMATION

Botanical Name   *Mirabilis expansa* Ruiz & Pavón
Family   Nyctaginaceae (four o'clock family)
Common Names
  Aymara: mauka
  Pichincha (Ecuador): miso
  Cotopaxi (Ecuador): tazo
  Spanish: mauka, chagos, arricón, yuca inca, shallca yuca, yuca de la Jalca, pega pega, cushpe, arracacha de toro, camotillo
  English: mauka

Origin.   Although it didn't appear in the ethnobotanical literature until 1965, mauka is probably an ancient crop. Its wild ancestors are found in Peru, Bolivia, Ecuador, and Colombia.[3]

---

[3] Information from J. Rea, who recently found possibly wild relatives growing in Tunja, Colombia. Mauka presents an excellent opportunity for study by anthropologists, ethnobotanists, and others interested in the origins of cultivated plants.

Description. Mauka is a low, compact plant, not exceeding 1 m in height. The aerial part is a mass of foliage formed from the basal shoots. The stems are cylindrical, with opposite, ovoid leaves with reddish edges. Bolivian types seem to have uniformly purple flowers, but in Ecuador they range towards white. The inflorescences are terminal racemes covered with viscid hairs, to which small insects frequently become stuck.

The thickened stems below ground are white, salmon colored, or yellow. They are commonly smooth and fleshy, about 5 cm in diameter and 50 cm in length. The growth takes place on the outer surfaces, and the structure of the stem becomes more regular toward the cream-colored center, which is high in moisture, full of starch grains, and contains little fiber. The form preferred in Cajamarca has yellow skin with cream-colored flesh. The color may depend on age, young tubers being yellow and older ones being white.

Horticultural Varieties. There are no defined varieties, but there are different genotypes. In Bolivia, as noted, mauka has purple flowers and astringent tubers. In Ecuador, a full gradient of flower color from purple to white may exist, and not all tubers are bitter.

Environmental Requirements

Daylength. Unknown.

Rainfall. Mauka seems to survive in wet, cold areas as well as in seasonably arid regions. The limits of its moisture tolerance are unknown. However, in Cajamarca it thrives at 600-1,000 mm per year.

Altitude. Reported from 2,200 to perhaps 3,500 m within the central Andes. However, it has not yet been tested at other elevations.

Low Temperature. Unknown, although the plant may not be frost tolerant.

High Temperature. Unknown. The plant is probably sensitive to heat.

Soil. Not unexpectedly, mauka seems to yield best in loose, alluvial soils. The limits of its soil tolerances are unknown.

Oca

In the Andean highlands, only the potato is a more important root crop than oca (*Oxalis tuberosa*). But whereas the potato has spread to become the world's fourth largest crop, oca (pronounced *oh*-kah) is little known outside its ancestral home. This is unfortunate because oca tubers have great consumer appeal: brilliant colors and a pleasant flavor that many people find a welcome change from the potato.

Oca tubers look like stubby, wrinkled carrots. They have firm, white flesh and shiny skins in colors from white to red. Most varieties have a slightly acid taste—they have been called "potatoes that don't need sour cream." Others, however, give no perception of acidity. Indeed, some are so sweet that they are sometimes sold as fruits.

An attractive, bushy plant with cloverlike leaves, oca is easy to propagate, grows luxuriantly, requires little care, and is exceptionally tolerant of harsh climates—under which its yield can be twice that of the potato. Moreover, it prospers in poor soils and at altitudes too high for most food plants. From Venezuela to Argentina, oca is still a staple for Indians living at altitudes between about 3,000 and 4,000 m. For them, oca tubers are principally sources of carbohydrate, calcium, and iron.

Although scarcely known outside the Andes, oca has found a home in Mexico, where it has probably been grown for more than 200 years.[1] And in the last 20 years it has become popular in New Zealand, where the tubers—sold under the misleading name "New Zealand yam"— are now commercially cultivated. This provides an important glimpse of oca's potential future because the climate, latitude, altitude, and daylength regimes of New Zealand are similar to those of some farming regions of North America, Asia, and Europe. Thus, like the potato before it, oca could become a vegetable for temperate zones.

Wherever it is cultivated, this crop is likely to be readily accepted. It lends itself to many culinary traditions because it can be prepared

---

[1] Indeed, Mexico's big, bright-red tubers look better than most specimens found in oca's South American home. In the central highlands, where it is mainly grown, Mexicans call it "*papa roja*," red potato.

in numerous ways: boiled, baked, fried, mixed fresh with salads, or pickled in vinegar. New Zealanders now serve it with their national dish, roast lamb.

Although at present barely known beyond the Andes, Mexico, and New Zealand, it seems likely, during the coming decades, that oca will become a vegetable familiar to millions of new consumers. First, though, the crop needs improvement. The plants in the Andes are infected with viruses that depress yields and could infect other crops such as potatoes. Fortunately, simple ways to remove viruses are available, and now, before the plant begins to spread, is the time to apply them.

Daylength requirements may slow up the crop's acceptance in new areas. Most Andean oca varieties have specific photoperiod responses that limit their culture to equatorial latitudes. If grown elsewhere, they form no tubers. Before oca's potential can be achieved worldwide, varieties that are either daylength neutral or adapted to long days must be located.[2] The plants of New Zealand, the southern end of the Andes, and perhaps Mexico seem likely sources for these.

PROSPECTS

Andean Region. Despite the fact that oca is an important food and cash crop in upland Andean areas, it suffers unwarranted cultural scorn because it is considered a "poor-person's" plant. Education could rid oca of its "poverty food" stigma, and, given a change in attitude and better marketing, the plant is likely to become a major food, not just for highland Indians but for everyone in the region. For some countries, it also might eventually become a valuable export.

Oca already yields well, but research in Britain indicates that elite virus-free stocks give much greater yields. The use of these in the Andes could therefore bring rapid economic benefits to highland farmers, who are among the most destitute in the western hemisphere.

Other Developing Areas. Oca seems particularly promising for the highlands of Central America, Asia, and Africa.[3] It is also likely to become a valued crop in other cooler areas of the Third World,

[2] The potato probably had similar limitations initially and became a major crop in Europe and North America only after types that would tuberize during long days were selected (see next chapter).

[3] Currently, researchers are conducting tests in Nepal; British scientists are testing varieties for introduction to the Ethiopian highlands. Information from S. King and A.A. Brunt.

The oca plant. (S. King)

such as northern India, northern China, southern Africa, and the sprawling region from southern Brazil to Argentina.

The plant will probably perform poorly in tropical lowlands because it is susceptible to heat, and also because its soft, fleshy stems are easily infected by bacteria.

Industrialized Regions. Given New Zealand's experience, oca seems poised to become a commercial crop in warm-temperate areas of Australia, North America, Japan, and Europe. Products from oca "chips" to oca "fries" seem possible. It will never reach the potato's overall level of consumption, but it has the potential to become a well-known, widely enjoyed, and profitable crop.

USES

Oca can be used in many ways. In the Andes, a few types are eaten raw. Most, however, are added to stews and soups; steamed, boiled,

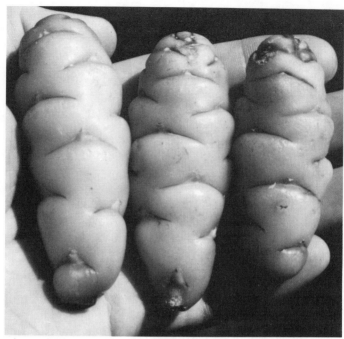

Oca tubers come in many colors and sizes, but they are all cylindrical and have distinctive wrinkles. (S. King)

or baked like potatoes; or served as a sweet, either plain or candied. In Mexico, oca is commonly sprinkled with salt, lemon, and hot pepper, and eaten raw. It is also made into bottled preserves (often in vinegar).

In the Andes, the tubers are often placed in the sun for a few days, during which they become sweet, and the amount of glucose can nearly double. Bitter varieties are almost always converted into dry products (*cavi* or *caya*), during which the bitterness disappears to leave bland-tasting products that can be stored without refrigeration. The freeze-drying *chuño* process—which involves soaking the tubers in water, exposing them repeatedly to freezing night temperatures, and squeezing the water out by stamping on them—also removes the bitterness.

Because of their high dry-matter content (normally about 20 percent, but sometimes as high as 30 percent[4]), the tubers may have potential for producing starch or alcohol. Theirs is a quality starch with promise for the food industry.

Oca plants can also be used as stock feed. Livestock—notably pigs—relish both tubers and foliage.

[4] Information from N.W. Galwey.

NUTRITION

Oca tubers show high variability in nutrition levels. However, by and large they have a nutritional value as good as or better than that of potatoes. On average, they contain 70–80 percent moisture, 11–22 percent carbohydrate, and about 1 percent each of fat, fiber, and ash. Protein levels vary greatly among different types; certain high-protein tubers contain more than 9 percent on a dry-weight basis. This is excellent for a root crop, and the protein is of high quality, with a good balance of essential amino acids (valine and tryptophan are the limiting ones). The carbohydrate is usually rich in sugar and is easy to digest.

The sour or "bitter" tubers contain amounts of oxalic acid varying up to 500 ppm. Some sweet types, on the other hand, have only an

Produce auction, Auckland. In New Zealand, oca has become popular, and its production and marketing are well advanced. (G. Samuels)

insignificant trace (79 ppm)—merely one-twentieth that found in standard potatoes.[5]

AGRONOMY

Oca is usually propagated by planting whole tubers; however, aerial stems are sometimes employed. Apparently, it is never propagated by seed.

Like potatoes, the edible tubers form on subterranean outgrowths of the stem, called stolons. Farmers mound dirt over the base of the plants to encourage stolon formation, which starts usually about 4 months after planting and peaks at about 6 months. As more stolons appear higher up the stem, more dirt is heaped over them.[6] The tubers normally take another 2–3 months to mature, after which the aboveground part of the plant usually dies back. Mexican types apparently mature more rapidly—6 months from time of planting to time of harvest.

Photoperiod and temperature both influence the rate of plant growth and tuber formation. Although many types collected from Peru and Ecuador are daylength sensitive, the ones in New Zealand (most likely originating from southern Chile in the 1860s[7]) are apparently unrestricted by daylength. They are grown commercially, for example, at Feilding (latitude 40.13°S) and Invercargill (latitude 46.24°S).[8]

Oca tubers are harvested like potatoes, but they tend to be more fragile, and they must be dug and handled carefully. Yields average about 5 tons per hectare under traditional Andean husbandry. Commercial yields average 7–10 tons per hectare in New Zealand and Peru. These figures probably do not indicate the plant's true potential, however. Reports from Cuzco indicate that, under experimental conditions in small plots, some clones yielded 40 tons per hectare.[9]

Oca seems less affected by pests and diseases than potatoes. However, this is probably because of the small scale of its current cultivation. Even so, problems do occur. In the Andes, the principal insect pest is a tuber-boring beetle related to the potato beetle.

---

[5] Information from S. King. Spinach, by comparison, can contain 5,000 ppm of oxalic acid.
[6] In the Andes, farmers usually do it at about 5 weeks after planting and then again at about 9–10 weeks as new stolons appear higher up the stem.
[7] Information from P. Halford.
[8] There is a possible relationship between photoperiod and temperature. For example, in some instances low temperatures may reduce a plant's sensitivity to short daylengths. Indeed, a few researchers think that low temperatures might sometimes be more important than daylength for stimulating tuberization. Information from A.J. Martínez.
[9] Information from H. Cortes Bravo.

Nematodes also affect the crop. As noted previously, viruses cause chronic yield reduction. Also, mycoplasma-like organisms have been identified in severely diseased Bolivian samples.

Although various fungi occur on oca plants, they seem unimportant in the field. After harvest, however, molds can cause major losses. Although more perishable than potatoes, oca tubers, if properly handled, can be stored at room temperature with little deterioration for several months. Dry tubers easily store over winter, and they will sprout precociously when temperatures rise in the spring.

LIMITATIONS

The major limitations have already been mentioned. They are viruses, daylength restrictions, and the presence of oxalates.

Viruses. Virus infections are a major constraint and must be removed before the plant can be used outside its current locations.

Daylength. For cultivation in many new locations, types that are daylength neutral or that come from an equivalent latitude must be used. In this regard, ocas from Mexico and New Zealand are important sources of germplasm for nontropical latitudes. Also, the plant is found in Chile at least as far south as the island of Chiloe, and this, too, is a promising source of daylength-insensitive clones.

Oxalic acid. Types for use in new areas must be carefully chosen. Whereas some have been shown to have much lower levels of oxalic acid even than potatoes, many traditional varieties in the Andes accumulate oxalate. All levels, however, are far below those of spinach and some other widely eaten green vegetables.

RESEARCH NEEDS

Oca offers superb research opportunities for root crop specialists, graduate students, the Centro Internacional de la Papa (CIP), and other agricultural research facilities throughout the Andean region and the world. Now is the time to seize the opportunity to develop a thorough understanding of the crop and to explore its promise.

To help oca achieve its potential, researchers should assess types gathered from throughout the Andes, New Zealand, Mexico, and

Europe.[10] Existing collections need to be maintained, and the diversity of oca from distinct geographical areas systematically evaluated. In this, as in many Andean crops, there is great potential for international cooperation.

To aid in the wider testing, procedures for "virus cleaning" should be publicized,[11] and selections of virus-free clones made available for direct use in the Andes, Mexico, and New Zealand, as well as for experimental trials elsewhere. Institutes, corporations, and nurserymen, both inside and outside the oca-growing nations, can foster this.

Attention should be paid to advances already made by Andean researchers. Individual plants should be closely examined for uniformity, growth patterns, and desirable tuber qualities such as size, shape, shallow eyes, and color.

There is much genetic diversity in this crop. However, flowering and seed set are uncommon, which limits the opportunities for genetic improvement. Techniques for seed production are needed. Without them, breeding will be slow and difficult.

Trial shipments of New Zealand oca have been turned away by U.S. agricultural inspectors on the grounds that the tubers "look like potatoes." If export trade is to be developed, the possibility of introducing diseases to the potato industry will have to be resolved. Oxalidaceae and Solanaceae are not closely related families, and oca viruses probably will not affect potatoes. Research is needed to settle the question, one way or the other.

SPECIES INFORMATION

Botanical Name   *Oxalis tuberosa* Molina
Family   Oxalidaceae (oxalis, or wood sorrel family)
Synonym   The name *Oxalis crenata* is used in some older literature, but is now assigned to another species.[12]
Common Names

Quechua: O'qa, okka
Aymara: apiña, apilla, kawi
Spanish: oca, ibia (Colombia); quiba, ciuba, ciuva (Venezuela); huisisai, ibias (South America); papa roja (Mexico)
English: oca, sorrel; kao, yam (New Zealand)
French: truffette acide
German: Knollen-Sauerklee

---

[10] For over 100 years, oca has been grown in Britain and continental Europe as a home-garden ornamental. It was also once grown in the south of France as pig feed. Until a decade or so ago, it sometimes appeared in Paris produce markets. Although available through nurserymen, few people realize that the tubers are edible.

[11] Researchers in Britain have successfully propagated virus-free plants using meristem (tissue) culture. Initial observations indicate that these outgrow and outyield normal (infected) stocks. Information from A. A. Brunt.

[12] *O. crenata* is a diploid with $2n = 14$; oca is an octoploid with $2n = 64$. Information from A.J. Martínez.

Origin.   Oca may be one of the oldest Andean crops. Tubers have been found in early tombs on the coast, hundreds of kilometers from its native highland habitat.[13] Although wild relatives exist throughout much of South America, the ancestral plant is unknown.

Description.   Oca is a compact, perennial, tuberous herb, usually 20–30 cm high, with cylindrical, succulent stems that vary in color from yellow and green to a purplish red. The stems normally rise from the base of the plant. Oca has an efficient plant "architecture" for photosynthesis because of its extremely high leaf area (due to its growth form and leaf angle, shape, and thickness).

Under long days, the stolons grow as above-ground stems; under short days, they penetrate the soil and form tubers. As days shorten, the stolons swell into rhizomic tubers that generally range in length from 3–20 cm and are produced in abundance. As in the potato, tiny scale leaves border the deep-set eyes.

Horticultural Varieties.   The Andean Indians recognize about a dozen cultivars and more than 50 distinguishable types. The Colección de Ocas—over 400 accessions—is housed at Cuzco, Peru. There are also major collections at Puno and Huancayo, Peru, and Quito, Ecuador.

Environmental Requirements

Daylength.   The common Andean types generally require days shorter than 12 hours to initiate tuber formation; in most cases longer days promote only foliage development.

Rainfall.   In the Andes, the crop is grown where annual rainfall is 570–2,150 mm, distributed evenly throughout the growing season.

Altitude.   Oca grows near sea level in New Zealand, but in the Andes of Peru, Bolivia, and Ecuador, it is found at 2,800–4,000 m elevation.

Low Temperature.   Although oca is resistant to low temperatures and thrives in moderately cool climates, freezing kills back its foliage. However, the plant's tubers have exceptional regenerative capacity.

High Temperature.   Temperatures above about 28°C cause the plant to wilt and its leaves to die; resprouting can occur, but tuber production is consequently delayed.

Soil Type.   Oca seems indifferent as to soil and is reported to tolerate acidities between about pH 5.3 and 7.8.[14] Not unexpectedly, a light, rich soil is best for tuber production.

---

[13] There are even some tantalizing suggestions that oca may have reached Polynesia before the arrival of European ships.
[14] Information from J. Duke.

Potatoes

During the approximately 8,000 years that potatoes have been cultivated in the Andes, farmers have selected types to meet their particular local needs and preferences, as well as to thrive in the myriad microenvironments scattered throughout South America's 4,000-km-long mountainous backbone. This vast and long-standing selection process has resulted in thousands of distinct types, and Andean Indians sometimes grow up to 200 different kinds of potatoes in a single field.

Most of these Andean potatoes (various *Solanum* species) are quite unlike what people elsewhere take to be "normal" for a potato. They can have skin and flesh that is often brilliantly colored (sometimes bright yellow or deep purple). Some have eye-catching shapes, often being long, thin, and wrinkled. And most have a rich potato flavor and a high nutritional quality.

These "odd" potatoes deserve much more recognition. Many have appealing culinary qualities and could fill specialty niches in the huge worldwide potato industry. For example, they can be less watery than common potatoes or have nutlike tastes and crisp textures. Moreover, most of these little-known potatoes are adapted to marginal growing environments and possess considerable resistance to various troublesome diseases, insects, and nematodes, as well as frost.

There has never been a better time to investigate these lesser-known crops. New markets for small or unusual potatoes are springing up. In North America, for instance, the food industry is avidly exploiting miniature vegetables of all kinds, and demand is increasing for small and colorful potatoes in particular.[1,2]

---

[1] In 1986, for example, the state of Maine sold almost 400,000 kg of potatoes, ranging in size from golf balls to billiard balls. The wholesale price was about one-third higher than for normal-sized potatoes. Sold as gourmet delights, these "Baby Maines" are packaged in designer boxes. Sales have climbed each year since the program began in 1983.

[2] An entrepreneur in California has had remarkable success selling golden and purple-colored potatoes as premium specialty vegetables from coast to coast. Her marketing is based solely on their color.

In addition, it is important that these potatoes be assessed and used because most of these species are becoming rare in the Andes—phased out in favor of modern varieties, which have undergone vastly more agronomic development. Indeed, some are so close to extinction that it is vital to focus scientific attention on them before they are lost.

However, it should not be assumed that all one needs to do is to gather these Andean potatoes and distribute them to the world. On the contrary, they have grave limitations. Many are grown only on high mountain slopes and may be restricted to such environments. Most seem to be less vigorous and to yield fewer and smaller tubers than modern commercial potatoes, especially when grown under commercial conditions. Most have deep eyes and irregular shapes that make them harder to process and handle in bulk than regular potatoes. Also, many, if not most, have strict daylength requirements and currently yield poorly in temperate zones because they need short days to induce tuberization.

Despite some apparent geographic and daylength limitations, these potatoes have potential for commercial success; the technical constraints to their wider adoption seem likely to be overcome through diligence, conventional breeding and tissue-culture techniques, and the improved disease-indexing techniques now available. They could perhaps usher in a new era in potato cultivation.

Even the low yield may not be inherent. Under the marginal conditions where they now grow, many of these native potatoes are not reaching their potential because of soil infertility, inadequate moisture, poor management, soil nematodes, viruses, and the poor quality of "seed" available.[3]

SPECIES

Many of the little-known potatoes of the Andes belong to different species from the common potato elsewhere, but one is its ancestral form. This one and seven others are described below.[4]

Pitiquiña.   Widely considered the most primitive of the domesticated potatoes, this species (*Solanum stenotomum*[5]) produces tubers that are long; cylindrical; knobbly; red, black, or white; and small

---

[3] Information from J.S. Niederhauser, who reports that in the highlands of Bolivia, andigena varieties (see later) have yielded the equivalent of 15–20 tons per hectare when there was a source of good seed, plenty of fertilizer, and control of nematodes, insects, and weeds.

[4] The scientific names used in this chapter are for identification purposes only. The taxonomy of the potato is complicated in the extreme. No endorsement of one set of claims over another is intended or implied.

Selling limeña potatoes on the streets of Lima. In the homeland of the potato, where there are literally hundreds of types to choose from, limeña potatoes are among the most popular because of their flavor. The larger, more watery types that are best known elsewhere are among the least popular. (Z. Huamán)

with deep eyes. Some are spiral in shape. They have a good, nutty flavor and unusually high amounts of protein and vitamin C.

This diploid (see page 102) is grown intermixed with common potatoes in traditional fields. Bolivian farmers, for instance, often plant a few rows of "*collyu papa*," a pitiquiña (pronounced pee-tee-*keen*-ya) variety, for their own consumption, and andigena varieties (see below) for the market.

The plant is becoming rare and is not now grown outside the Andes. Some strains are fairly frost resistant. It produces fertile seed. The tubers require a dormant period before they will sprout. Typically, they are stored 4-5 months between crops.

Limeña.   Known in the Andes as limeña (pronounced lie-*main*-ya) or papa amarilla ("yellow potato"), this species (*Solanum goniocalyx*) produces a potato with deep-yellow flesh of exceptional flavor. It is fried and sold as a culinary specialty in the streets of Lima, Peru, for

[5] Thought to be the original progenitor, from which all other cultivated potatoes sprang. It is extremely close to such wild species as *Solanum leptophyes* and *S. canasense*, which are Andean weeds commonly found in vacant fields and along roadsides. It may have arisen from them by selection.

The potatoes of the Andes come in a variety of shapes, colors, sizes, and flavors. Shown on these pages are the little-known potatoes highlighted in this chapter. All are different species from the potato used in the rest of the world, except for the andigena potato, which is a different subspecies. (Z. Huamán)

instance. Most varieties have white flowers and yellow tubers (both flesh and skin) that are the basis of a tasty yellow soup that is a traditional Peruvian food and an important part of the noon meal in many Andean countries.

The plant is a diploid and is closely related to pitiquiña, of which it may be just a variant or subspecies. It is still widely grown in temperate-climate areas of the Andes because people are willing to pay a premium for its quality and taste. It is unknown outside the Andes. It produces fertile seed.

Phureja.   The phureja (pronounced foo-*ray*-ha) potatoes (*Solanum phureja*)[6] are also small, irregular, and tasty. They are grown mainly at lower altitudes (2,200–2,600 m) on the warm, moist eastern slopes of the Andes from Venezuela to northern Argentina. Some yellow-fleshed types are fried and sold in city streets (for example, in Bogotá where they are called "*papa criolla*") and in the markets of La Paz, Bolivia. Although rarely seen outside the Andes, the phureja potato

---

[6] For simplicity, we refer to it as a separate species, but some scientists consider it a subspecies, a variety, or a cultivated group of *Solanum tuberosum*.

Ajanhuiri

Rucki (S. x juzepczukii)

Rucki (S. x curtilobum)

Andigena

has become popular in the Netherlands, because of its resistance to disease.

The plants exhibit good heat tolerance, and their genes have been incorporated into the two most heat-tolerant varieties of the common potato. But phureja tubers lack dormancy: most of them are already sprouting when they are harvested.[7] This is a useful trait for growers who expect two or three crops a year, but it causes problems in storage and handling of the commercial harvest.

This diploid plant probably arose from *Solanum stenotomum* when ancient peoples selected it for its short dormancy. At least 500 named varieties are known. Most are deep-eyed and highly pigmented (often purple), with spindly, twisted shapes. In the Andes, they are mostly boiled, but they can be baked or fried. They are high in protein and vitamin C and have a stronger flavor and a firmer texture than the common potato.

Andigena. This is the potato[8] immediately ancestral to the potato of commerce. To most botanists the two are the same species, but to nonspecialists they look vastly different. In Latin America, this is not

[7] Normal potatoes must remain dormant for 2–3 months at room temperature or 4–5 months refrigerated before they will sprout.
[8] Some designate this as a species in its own right (*Solanum andigenum*); others classify it as a subspecies, or a horticultural variant of *S. tuberosum*.

a "lost" crop: from northern Argentina to Venezuela, as well as on mountainsides in Central America and the Mexican cordillera, it is perhaps the best-known potato. However, there is little or no commercial cultivation of it anywhere else.[9]

Of all the traditional potatoes of the Andes, andigena (usually pronounced an-*di*-je-na in English) potatoes produce the largest tubers. They are rounder, shallower eyed, and more uniform in shape than those of the other neglected species. They come in a range of pigments from yellow to black. They are firm and nutritious: protein levels up to 12 percent on a dry-weight basis have been recorded, which is higher than that of modern commercial varieties (about 8–10 percent).[10] Like all potatoes, they are high in vitamin C.

This overall superiority in culinary properties and nutritional values, however, is offset by susceptibility to late blight. Although yields are often low, there are varieties yielding up to 30 tons per hectare.[11]

Of all Andean potatoes, this species shows the greatest diversity, with 2,500 distinct native varieties. It is a tetraploid, believed to have sprung from *Solanum stenotomum* through chromosome doubling or by hybridization with another wild species, *Solanum sparsipilum*. It produces fertile seed.

Chaucha. A hybrid between the two cultivated species pitiquiña and andigena (*Solanum stenotomum* and *Solanum andigenum*), the chaucha[12] potato (*Solanum* x *chaucha*) is widely distributed from Colombia to northwestern Argentina. It is an early-sprouting potato that needs no rest period. As it is a sterile triploid, the plant produces no seed; propagation is exclusively vegetative. New genotypes are produced only occasionally as a result of natural mutation in the field.

The *huayro*, one of the major commercial potatoes of the Indian populace, is a chaucha potato. At least in Peru, this species has spread from the central highlands to almost the entire country and has much potential. Its tubers tend to be larger than those of many native potatoes, possibly because of hybrid vigor.

Ajanhuiri. This highly frost-resistant potato (*Solanum ajanhuiri*) is extensively cultivated at altitudes of 3,800–4,100 m in the Andean Altiplano of the Lake Titicaca basin. This area is an inhospitable windy

---

[9] Spaniards introduced the andigena potato to Europe at least as early as 1570. Although no longer grown in Europe, it is believed to be the ancestor of the modern potato. The actual sequence of events is a matter of debate and conjecture. Some researchers believe that modern potatoes are direct descendants of *Solanum tuberosum* gathered in Chile in the 1500s. This is now disputed by most authorities.

[10] Protein percentage is a function of tuber size, which has to be taken into account.

[11] Information from J.S. Niederhauser.

[12] Pronounced *chow*-cha. The word means "early."

plateau located in southern Peru and northwestern Bolivia. Frost severely limits agriculture there, restricting the choice of crops to this potato and a handful of other plant species. Some evidence shows that ajanhuiri (pronounced a-han-*hwee*-ri) was crucial to the survival of the Aymara Indians who live in the area. Subsistence farmers still grow it in small plots as an "insurance" crop, in case the andigena potato crop should fail owing to unpredictable heavy frost.

The tubers have high contents of dry matter and vitamin C, and they store well. Only one clone, called "sisu" (pronounced *see*-soo), can be eaten without preparation. It is sweet, floury, and tasty. Its two main varieties are *azul* (blue tubers) and *jancko* (white tubers). The other clones are bitter and are made into *chuño* (see below).

This species matures early and withstands temperatures as low as − 5°C, as well as hail and drought. It is also resistant to viral diseases and round-cyst nematode, and is immune to *Synchytrium* black wart as well. Daylength requirements may reduce its usefulness for high latitudes, but if this restriction can be overcome, ajanhuiri is a potato with great international promise. Although known to taxonomists and plant breeders for some 50 years, it has not yet been widely utilized in potato breeding.

This species is a diploid. It rarely produces fertile seed, and even then only in small amounts.

Rucki. The rucki (pronounced *rue*-kee) is perhaps the most frost resistant of all potatoes. Actually, the name covers two species (*Solanum* x *juzepczukii*[13] and *Solanum* x *curtilobum*), which are grown in central to southern Peru and in northern Bolivia at altitudes up to 4,200 m. At this rarefied height they are often subjected to heavy frosts, even during the growing season.

Both plants are hybrids between a cultivated and a wild species. (Such crosses occur along the margins of farmers' fields, which probably explains how they arose.) Both contain genes from *Solanum acaule*, a tiny wild species found at altitudes so high that in some cases it grows along the edges of permanent snowbanks.

The cultivation of these plants long predates the Inca period, and although it continues to the present, farmers generally obtain low yields because neither plant has been subject to agronomic improvement. The farmers grow them as "security" crops because in the altiplano frosts can occur during 300 days of the year.

Tubers from both plants are usually bitter, and can be eaten only after processing. They are used mainly to produce *chuño*, a freeze-

---

[13] Pronounced jo-sep-*soo*-kee (English) or yu-sep-*chu*-kee (Spanish).

dried food that is bland, not bitter. This dry, white product can be stored almost indefinitely and is widely used in soups and stews.

Solanum x *juzepczukii* is a triploid and gives no seed; *Solanum* x *curtilobum* is a pentaploid and produces fertile seed.

Solanum hygrothermicum. This potato[14] is cultivated by Indians of the warm and humid lowlands of Peru's Amazon basin region.[15] It is the only potato traditionally grown under the climatic regime of a warm rainforest (2,000–3,000 mm of rain a year). It may be of value both as a hot-climate potato or for breeding purposes to impart heat tolerance. It has shown resistance, for instance, to bacterial wilt or "potato black-leg." Unfortunately, it is so rare that it is close to extinction, and in fact may already be extinct. At present, there are no living collections, but the plant probably can still be found in the jungles of the Amazon basin. Explorations should be undertaken.

PROSPECTS

Andean Region. In the Andes, there exists enormous potential for the improvement and economic exploitation of these various, flavorful species. They usually sell for much higher prices than the common potato and continue to be widely (even if sparsely) grown. They are important as a buffer against the variability and unpredictability of environmental conditions. For some of the world's poorest populations, they increase farming options and reduce the risk of disastrous crop failure, especially that caused by frost.

Given increased research attention, it seems probable that these lesser-known potatoes will be greatly improved and will find farmers eager to grow them. Yields can probably be increased merely by use of virus-free seed, for instance. Tissue culture propagation and the use of true (botanical) seed are also promising new technologies for developing inexpensive, virus-free plants.

These species are important also for genetically changing the major cultivated potatoes. Their exceptional variability provides a rich source of genetic traits for incorporation into commercial potato cultivars. The cold tolerance of some species is of extreme importance. Also, there are potential nutritional advantages.

Increased screening of native varieties should be seen as comple-

---

[14] Information on this species from C. Ochoa.
[15] For example, the Campas and Aguarunas in the central Amazon basin and the Machigangas in southeastern Peru.

THE PROMISE OF UNDOMESTICATED POTATOES

In this report we have not described wild potatoes—of which there are several hundred species in the Andes. But this is not to suggest that the wild species lack utility. Indeed, some of them have unusual genetic qualities. A few, for instance, are virtually immune to the most formidable pests in potato farmers' fields.

British agronomist R.W. Gibson has found two species of Bolivian wild potato whose leaves are veritable minefields to insects. Even a tiny aphid—one of the potato's major enemies—crawling over the surface breaks open minute, four-lobed hairs that cover the leaves. This releases a sticky material that clings so firmly that the aphid's legs become glued to the leaf and it dies. The glue will also catch the Colorado potato beetle, potato leafhoppers, and both tarsonemid and tetranychid mites. These particular wild potatoes are unsuitable as food crops, but already researchers are beginning to breed them with the common potato to give it glandular hairs with which to ensnare its insect enemies.

The photomicrograph reveals an aphid that has become stuck to the leaf of *Solanum berthaultii*. (Rothamsted Experimental Station)

mentary to the ongoing effort to breed new varieties. For a subsistence farmer, reduction of risk overrides all other factors. If a crop is dependable, he will probably prefer it to something that has a high yield only in "good" years.[16]

Other Developing Areas. In recent decades, the common potato has become a glowing success in the highlands of Central Africa and the Himalayan region. The Andean experience suggests that, with research, certain native varieties can compete successfully with the common potato. Thus, in time, these little-known potatoes could also be good contributors to the welfare of Africans and Asians.

There could well be localized environments where even now the lesser-known species may prove superior to common potatoes. The heat-tolerant *Solanum hygrothermicum*, for instance, is native to a climate that is almost inimical to most varieties of the common potato. Also, some of the extremely cold-tolerant species may find valuable niches high on African or Asian mountainsides where few other crops can survive. Expanded research to evaluate such possibilities could open new vistas.

As noted, some of the species are valuable sources of germplasm for enhancing the common potato's culinary quality, productivity, and resistance to pests, disease, and harsh environments.

Industrialized Regions. In the United States, specialty vegetables are becoming a driving force in the multibillion dollar produce industry, and commercial interest in unusual potatoes is rising. Golden and purple potatoes are already selling at premium prices, and demand for more striking variants is probably endless. All the lesser-known potatoes should be intensively investigated and the adaptable cultivars promoted. A few cultivars of pitiquiña and phureja potatoes are currently being grown in Western Europe and North America. In North America, andigena potatoes would appear to be ideally suited to the specialty market.[17]

In the Andes, most of these species are cultivated at fairly high altitudes, so the cool autumn weather of high latitudes (in Europe, North America, or Australasia, for instance) may pose little problem. However, the long daylength of summer poses a giant problem. The Andean potatoes come largely from equatorial latitudes and tuberize

---

[16] It has been said that a new high-yielding, frost-resistant potato cultivar could do more for the political and economic stability of Bolivia than any other single factor.

[17] One interesting marketing strategy would be to maintain the diversity inherent in these species and raise them as multicolored and multishaped populations. Baskets of rainbow-colored potatoes would be a trendy produce-marketer's delight.

HOW THE MODERN POTATO DEVELOPED

The potato that reached Europe in the late 1500s was the andigena (page 97). But how it became the modern potato is a matter of debate. When grown in Europe today, andigena's stolons are very slow to swell to form tubers, and it produces little or no yield. Differences in daylength between the short days of the central Andes and the long days of a northern European summer are the cause.

It is probably accidental that the andigena was transformed into a useful crop for Europe. In the 1600s and 1700s, some people propagated potatoes by planting botanical seeds. The resulting seedlings were highly variable; virtually every plant differed from all the others. This allowed a vast number of genes to be combined and expressed, and among the types that arose were some that could tuberize during long days.

This is the explanation believed by most potato geneticists. There is, however, a possible alternative: that an unrecorded ship introduced "long-day potatoes" from southern Chile. Chilean potatoes are almost certainly also derived from andigena, but for centuries they have been adapted to long-day production.

Whichever method transformed andigena, it was one of the most valuable genetic developments of all time; it gave the world what is now its fourth largest food crop: the modern potato.

---

only under short day conditions. Such restrictive daylength require-ments have been overcome in other crops by selection and undoubtedly could be done again. Appropriate varieties usually appear when large numbers of plants are grown under long-day conditions—only the few adaptable ones produce a crop. Also, long-day types are most likely to be found in the southern limits of the range of each species in the Andes.[18] The use of true seed to bring out daylength variability seems highly promising.

The bitter potatoes are unlikely to create much interest outside the Andes, where *chuño* would be hard to make under natural conditions.

As with other new produce (kiwifruit, for example), catchy names could be the key to consumer acceptance. For lesser-known potatoes, market-oriented names might, for instance, play up the brilliant colors, firm texture, bizarre shapes, or nutritive quality.[19]

---

[18] The Chilean regions of Temuco and Chiloe seem likely to produce daylength-neutral types.
[19] Reviewers of this chapter came up with the following provocative suggestions: rainbow, gemstone, jewel, hotdog, spiral, golden-delicious, corkscrew, early bird, or brillante potatoes.

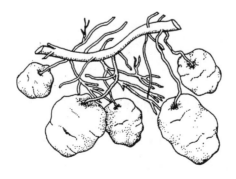

Ulluco

In many highland areas of the Andes, ulluco (*Ullucus tuberosus*) is a staple, and in a few it is the predominant root crop. One of the most striking foods in the markets, its tubers are so brightly colored—yellow, pink, red, purple, even candy striped—and their waxy skins are so shiny that they seem like botanical jewels or plastic fakes. Many are shaped like small potatoes but others are curiously long and curved like crooked sausages.[1] Their skin is thin and soft and needs no peeling before eating. The white to lemon-yellow flesh has a smooth, silky texture with a nutty taste. Some types are gummy when raw, but in cooking, this characteristic is reduced or lost. Indeed, a major appeal of ulluco is its crisp texture, which remains even when cooked.

The future of ulluco (pronounced oo-*yoo*-koh) seems particularly bright. In the Andes, demand is on the increase, and its attractive tubers are likely to prove popular elsewhere. The plant is easy to grow, resists frost, is moderately drought tolerant, and produces reasonable yields in marginal soils.

Although it has attracted little modern agronomic attention, ulluco is one of the few Indian crops to have been enthusiastically accepted by those of Hispanic descent. This ancient tuber is now sold (usually under the name "papa lisa") in modern packaging in supermarkets in Lima, Quito, Cali, and other big cities. Throughout the Andean region, it is considered a delicacy. It is one of the few native crops that is more widespread in the Andes now than it was 100 years ago; production is estimated to have doubled in just the past 20 years. In Peru in 1983, for example, more than 15,000 hectares were under cultivation; estimated production there is now more than 60,000 tons per year, and continues to increase. Around Cuzco, as well as in southern Colombia, ulluco is outranked only by oca, potatoes, and maize; near some urban areas, it is grown in almost every plot of suitable ground.

---

[1] One of the bent types, splashed with maroon streaks, is known as "Christ's knee." A small, pink, curled variety is called "shrimp of the earth." A slightly curved yellow variety in Paucartambo is called "cradled baby." The types with pink and magenta spots are sometimes called "*pica de pulga*" (flea bites) in Peru.

105

Ulluco is popular with Andean farmers because it has few pest and disease problems. However, even though the plant seems virtually disease free and its yields are considered high, recent studies in England indicate that it is probably invariably infected with viruses. Techniques for eliminating viruses (using meristem culture) have been developed, and disease-free clones may soon provide impressive increases in productivity. Preliminary observations indicate that virus-free plants show yield increases of 30–50 percent. With these vigorous, virus-free plants, ulluco yields comparable to those of potatoes can be expected.[2] Thus, it seems that removing viruses could transform this crop.[3]

PROSPECTS

The Andes. Ulluco has greater potential than most people realize. The availability of elite virus-free stocks could push it into the mainstream of commercial agriculture throughout the Andes. Because many diseases and pests are now appearing on the potato crop, potato production is becoming increasingly expensive. Ulluco is an excellent alternative for the small farmer.

Ulluco seems particularly promising as a cash crop. Although most is grown for home use, some growers produce it mainly for the markets, and Ecuadorian farmers already consider it a prime cash crop Thus, dramatically improving yields could benefit both the diet and economic situation of the highland farmers.

Other Developing Areas. This crop is now virtually unknown outside the Andes, but, like the potato, it seems to hold promise for temperate zones and tropical highlands. Resistant to frost as well as heat, ulluco grows vigorously and particularly thrives in moist conditions. It is high yielding in terms of tubers per plant, and is adapted to high altitudes. It could be grown in many upland regions of the tropics, and it has already fared well in Sri Lanka.[4] It also seems

[2] An additional observation was that, under glass, the infected plants became dormant during winter, but the virus-free plants continued growing and, moreover, formed tubers more rapidly.
[3] Information from O.M. Stone and A.A. Brunt. Field trials remain to provide conclusive evidence that these results will carry over to farm conditions. The most important virus, ullucus mosaic virus, is spread by aphids. Two others, ullucus mild mottle and papaya mosaic viruses, are probably spread mechanically. The fourth, ullucus virus C, is tranmitted by beetles. Regular distribution of clean planting stock is one solution to reinfection.
[4] Information from J. Duke.

Chinchero, Peru. Although the potato has become the fourth largest crop of the world, ulluco and some of the other "unknown" tubers are equally popular in its homeland in the Andes. Here, potatoes (in bowl) and ulluco (in basket) are being prepared together for a stew. (N. Vietmeyer)

promising for the uplands of Africa and China. All in all, the Andes has much to give the high-altitude regions of the tropics, and ulluco is a good example.

Industrialized Regions.   Ulluco will probably be popular wherever it becomes available. In Europe, North America and Japan, for example, it would be beautiful in supermarket displays, and could prove to be a profitable specialty crop. The tuber's small size could be a marketing asset—rather than a drawback, as with other roots. Because of the variety and brilliance of their colors, they could be sold as a mixed blend rather than as a uniform product.

Despite this potential, ulluco may not be an easy crop to transplant to high latitudes. Most of the easily accessible types in Peru and Ecuador probably are limited by virus infections and daylength restrictions. Nonetheless, given concentrated research, ulluco could almost certainly be cultivated in high-latitude regions; researchers have already cultivated it in greenhouses at the latitudes of Vancouver, Canada (50° N), and Helsinki, Finland (60° N), as well as in open field conditions in England.[5]

---

[5] Information from T. Johns and A. Rousi, respectively.

Insular climates (such as in New Zealand or Hawaii) may lend themselves particularly well to ulluco. When days shorten as winter approaches, such places still have weeks of growing season left, during which time the tubers can form and swell. Continental summers, on the other hand, tend to end abruptly and may not leave enough short, cool days for the tuberization process to occur. In Hawaii, ulluco might thrive in the higher, cooler, moister elevations.

USES

Ulluco tubers have a wide range of culinary uses. Because of their high water content, they are most often boiled—sliced, shredded, grated, mashed, or whole—rather than baked. There is, however, some loss of color on boiling. The skin is about as thick as that on new potatoes, and is easily removed. Inside, the flesh is either white or yellow, and crisp like young potatoes, but slightly gummy until cooked.

Ulluco tubers are also pickled or mixed with hot sauces. Generally, however, they are used to thicken soups and stews. For this they are preferred to potatoes, because they yield a smooth, silky soup rather than a grainy one. In urban households and restaurants in Ecuador they are frequently boiled and served cold as salad.

The tubers have a good shelf life. They are stored for up to a year at ambient temperatures in the cooler areas of the Andes.[6] They must, however, be kept in the dark for their skins turn green in sunlight.

Peru now exports canned ulluco to the United States, where it is often found in Hispanic markets. Unlike many vegetable products, the tubers retain their original taste and texture when canned; only a little color is lost.

In the Andes, the tubers are sometimes freeze-dried (in the way potatoes are made into *chuño*) into a long-lasting product called "*llingli*." This is usually ground into flour and added to cooked foods. The dried tubers have a much stronger taste than the fresh ones.

The green leaves of ulluco are also nutritious. The plant is related to Malabar spinach (*Basella rubra*), which is widely eaten in the tropics as a potherb. In Colombia and Peru, the mucilaginous ulluco leaves are occasionally eaten in salads and as a vegetable. They are also sometimes boiled to make soup, or used like spinach, which they resemble in taste.

---

[6] They have sprouted successfully after being stored two years in a refrigerator. Information from C. Sperling.

NUTRITION

Ulluco is a good source of carbohydrate. Fresh tubers are about 85 percent moisture, 14 percent starches and sugars, and 1–2 percent protein. They are unusually high in vitamin C, containing 23 mg per 100 g fresh weight. They contain a gum, but no fat and almost no obvious fiber.[7] There is considerable nutritional variation, especially in protein content, which has been reported as high as 15 percent dry weight.[8]

The leaves contain 12 percent protein dry weight.

AGRONOMY

The crop is normally propagated by planting small tubers. However, the plant is also easily propagated by stem cuttings or pieces of tuber (indeed, as long as they include a node, chopped-up pieces will root with weedlike robustness) without the help of hormones or other special treatment. The tubers sprout and grow readily when temperatures rise above about 18°C.

Ulluco is grown much like oca, mashua, and potatoes. In fact, all four species are often planted together, with the tubers separated after harvest.

As daylength shortens, stolons begin to grow out of the stem, and then tubers begin to develop on the ends of the stolons. This process can occur at any level on the stem, and ulluco is usually "earthed-up" to increase the number of stolons formed. Cultivars vary greatly in the time they require to reach maturity. The growing cycle may be as short as 5 months, but 6–8 months is more common; at high elevations (above 3,750 m) 9 months or more is the norm.

Ulluco tubers are dug by hand. They resist bruising but, like new potatoes, they scar easily. Although fully mechanized harvesting has not been developed, it seems feasible. However, machines are likely to scuff the shiny coats, and in a crop whose appearance is of major importance, this could be a problem.

Yields average 5–9 tons per hectare under traditional conditions. The largest tubers, from Colombia, may be fist size, but elsewhere they tend to be smaller than the size of an egg. Actually, it is the smallest tubers that are most sought after in the market.

The tubers are stored almost year-round in the Andes. As noted,

---

[7] On a dry-weight basis, 100 g of tubers may contain 364–381 calories, 10–16 g protein, 72–75 g total carbohydrate, 4–6 g fiber, 3–5 g ash, and 0.6–1.4 g fat. However, there is likely to be great variation between plants and growing sites. Information from S. King.

[8] Information from S. King.

they are best stored in the dark. If exposed to the sun, the colors fade and eventually (because tubers are inherently stem material) they turn green.

Most diseases that are known to attack ulluco seem to be specific to the plant. Andean farmers usually intercrop it with potatoes and there appears to be little or no interchange of diseases.

LIMITATIONS

Ulluco, as now produced, usually has lower yields than potato, but it seems unlikely that this would be a long-term limitation. Already, production sometimes reaches 10 or 15 tons per hectare, and at higher altitudes it can equal or surpass the potato yield. Give virus-free stock and improved growing conditions, low yield should no longer be a limitation.

The tubers have a high water content and are not good for frying. They also shrink more than oca and potatoes during cooking.

Because all ulluco appears to be tainted with viruses, strict cleansing and quarantine procedures must precede any introductions to new areas, thus limiting the availability and variability of germplasm in the short term.

RESEARCH NEEDS

Despite its importance to millions of people, little agronomic information concerning ulluco is readily available. Following are some of the research areas to be explored.

Controlling Viruses   Production and use of virus-free stocks and control of virus infections will probably lead to immediate, dramatic, and sustainable increases in production. This merits rapid and substantial research and operational effort.

Removing Photoperiod Restrictions   The plants are very sensitive to photoperiod. Cultivars need to be screened for their daylength requirements, and a survey made of their responses to different photoperiods. These variations will help indicate areas in which ulluco might be grown. Clones from the southern limits of ulluco's range should be evaluated for tuberization under longer daylength conditions and possible adaptation to temperate regions.[9]

Inducing Seed Production   In the past, it was thought that the plant never produced viable seed. However, researchers in Finland have

[9] The plant occurs to 27°S latitude in northern Chile and northern Argentina.

obtained fertile seed under controlled circumstances.[10] This should greatly expand the potential for breeding and hybridizing; ulluco's genetic improvement should be speeded up as a result. In particular, true seed can be used to remove viruses and locate daylength-neutral types.

Because viable seed would help with breeding and genetics, clones should be checked throughout the Andes and efforts made to develop fertile plants. In addition, the seed-producing capacity of wild relatives should be further assessed. Overcoming the sterility barrier would increase the variety of ulluco's colors and other variability to genetic manipulation.[11]

Shortening the Cropping Cycle   There is a need to find types for use in temperate latitudes that mature in a growing season of five months or less, yet are insensitive to daylength.

Increasing Adaptability   Clones should be screened for their relative characteristics by rotating them among different growing sites.

Mechanization and Postharvest Handling   Methods of mechanized cultivation and harvesting could prove useful. Selection for growth forms adaptable to mechanization is also desirable (sprawling types might not be acceptable, for instance).

Methods of minimizing the scarring of tubers, loss of color, and for minimizing sprouting are also important. Ways to decrease harvesting, storage, and transportation problems could greatly increase the range of markets.

Tuber Quality   Consumer acceptability could be enhanced through selection of tubers that, when fresh, have reduced gumminess.

Fertilizer Experiments   The plant's fertilizer requirements are little studied, and it seems likely that substantial yield increases can be obtained in the Andes merely through the modest application of manure or fertilizer.

SPECIES INFORMATION

Botanical Name   *Ullucus tuberosus* Caldas
Family   Basellaceae

---

[10] Rousi et al., 1988.
[11] Because there is much variation in spite of no known seed formation, there seems a good likelihood of frequent mutation in the vegetative tissues of ulluco. Researchers should keep a watchful eye for unusual tubers.

Somaclonal variation selection is being carried out via tissue culture by R. Estrada in Peru.

Synonyms *Ullucus tuberosus* Loz., *Ullucus kunthii* Moq., *Basella tuberosa* HBK; *Melloca tuberosa* Lindl. and *Melloca peruviana* Lindl.[12]

Common Names

Quechua: ullucu

Aymara: ulluma, ullucu

Spanish: melloco (Ecuador), olluco, ulluco, rubas (Colombia, Ecuador); rubia, ruba, tiquiño, timbós, mucuchi, michuri, michiruí migurí (Venezuela); camarones de tierra, ruhuas, hubas, chuguas, chigua (Colombia); papa lisas, lisas, olluco, ulluco (Peru, Bolivia); olloco, ulluca, ulluma (Argentina); papa lisa (Peru, Spain)

English: ulluco, melloco

Origin.   Ulluco is a completely domesticated crop. It is often represented in pre-Columbian art, and tubers have been found in 4,250-year-old ruins in coastal Peru—far from the area in which it currently grows.

Wild forms (for instance, *Ullucus tuberosus* subsp. *aborigineus* Brücher) occur in Peru, Bolivia, and northern Argentina. They are mostly vinelike, with long internodes and reddish stems. Their spherical white, pink, or magenta tubers are about the size of small peas or marbles, and are more bitter than those of domesticated varieties.

Description.   Ulluco is a low-growing herb. All parts are succulent and mucilaginous. On long petioles from the angular stem are borne alternate, heart-shaped leaves, the color of which depends upon the cultivar. Wild forms are prostrate. Cultivated forms come in a gradient of types from prostrate or semiclimbing vines to dense, compact, bushlike mounds up to 50 cm tall.

The small, green-yellow to reddish flowers are borne in clusters arising from the forks of the branches. Seed set has never been shown in either wild or cultivated forms in the Andes.[13]

The plant forms tubers on long stolons both below and above the ground. Most arise below ground from the mass of fibrous roots, the ends of which thicken and swell.

The tuber skin is thin and soft, with inconspicuous buds. Cultivated tubers can be elongated (2–15 cm) or curved. Some in southern Colombia are as big as normal potatoes. The most common are spherical and lemon yellow. However, coloration may be white, pink, orange, red, or magenta—a common, popular form has magenta spots speckled on a yellow background. Inside, the tubers are yellow or

---

[12] Lindley recognized two species of ulluco, based on flower color and petal shape.

[13] Viruses may contribute to their failure to set seed. As noted, researchers in Finland, using virus-free stock, have succeeded in getting ulluco seed. Information from A. Rousi.

white, with a clear distinction between skin and interior. They lack noticeable fibrous material.

Horticultural Varieties. There is enormous variation in this crop: a single market may display six distinct types, and a market nearby may have six entirely different ones. Based on tuber appearance, some 50–70 distinct clones exist.[14] Some have been transported throughout the length of the Andes.

Environmental Requirements

Daylength. Daylengths of 10–13.5 hours are needed for tuber production in the varieties most commonly grown in the central Andes. However, some ulluco is grown in northern Argentina at 27°S latitude, and it seems likely that daylength-neutral types can be found in such places.[15]

Rainfall. Moisture requirements are unknown, but probably are in the range of 800–1,400 mm during the growing season in the Andes.

Altitude. Ulluco is an important mid- to high-altitude crop from Venezuela to Chile.[16] However, it has also been grown at sea level in Canada, England, and Finland.

Low Temperature. The plant grows well in cool, moist conditions and is frost resistant.

High Temperature. Although they thrive under high light intensities, ulluco plants produce tubers poorly in hot climates.

Soil Type. This crop tolerates a wide range of soil conditions. Not unexpectedly, however, it does best in a fertile, well-drained loam with a pH between 5.5 and 6.5.

---

[14] Information from C. Sperling.
[15] This considerable variation may be an indication of ulluco's inherent diversity.
[16] The uppermost elevation varies with latitude and location. Ecuadorian production is concentrated at elevations between 3,000 and 3,500 m; the highest report is of cropping at 3,700 m. In the Sierra Central of Peru, there are mixed crops of ulluco, oca, and bitter and nonbitter potatoes at about 4,000 m.

Yacon

Yacon (*Polymnia sonchifolia*) is a distant relative of the sunflower, but this Andean crop is grown not for seed but for its edible tubers.[1] These enlarged storage organs have a clean, crunchy crispness, set off by a refreshing sweetness. They have been described as being like a fresh-picked apple with mild, sweet flavor reminiscent of watermelon.

Yacon (pronounced ya-*kon*)[2] should prove agreeable to a wide range of palates, and it also has a future as an industrial crop. Most other roots and tubers store carbohydrate in the form of starch—a polymer of glucose; yacon, on the other hand, stores carbohydrate in the form of inulin—a polymer composed mainly of fructose.[3] Yacon, therefore, may possibly be a fructose-sugar counterpart of sugar beets.

Yacon tubers also may have potential as a diet food. The human body has no enzyme to hydrolyze inulin, so it passes through the digestive tract unmetabolized, which means that yacon provides few calories. This could be an attractive marketing feature to dieters and diabetics.

In addition, the main stem of the young plant is used as a cooked vegetable. The species also shows promise as a fodder crop because the leaves contain 11–17 percent protein on a dry-weight basis.

From Colombia and Venezuela to northwestern Argentina, yacon is found at elevations below about 3,300 m. Children, in particular, consider its roots a special treat. In some areas, almost everyone has a few plants in the family garden plot. Much is grown in northern Argentina, for instance, and in Latacunga, Ecuador, yacon is sold in large quantities, especially on the traditional Day of the Dead.[4] On the other hand, in other areas, it is seldom abundant in markets; in some places it is almost unknown.

---

[1] Strictly speaking, these are not tubers, but an integrated mass of root and stem.
[2] In Ecuador it is frequently called "jicama." Internationally, that name is used for another plant (see page 39).
[3] Yacon shares this form of carbohydrate storage with most members of the Compositae, or sunflower family. Rarely, however, does inulin appear so abundantly or in so pure a form.
[4] Information from R. Castillo.

115

The plant grows fast and easily, and survives even in poor soil. It is not restricted to upland areas, and has shown excellent growth at sea level.

Outside the Andes, yacon is almost unknown. However, in New Zealand a few nurserymen now offer it for home gardeners and commercial planting, and the tubers are being packaged like carrots for sale in stores. It has been successfully introduced into southern Europe, but is not widely known. It has only recently been introduced to the United States, and amateur gardeners have found that it thrives in many parts of the country—in California, Oregon, New Mexico, Florida, Alabama, and northern Virginia, for instance. There seems a good likelihood that it could be viable in most parts of the temperate and subtropical zones.

PROSPECTS

The Andes.   Yacon is little exploited even in its native habitat. There is probably an untapped demand in many urban areas, both among immigrant highlanders and urbanites themselves. Thus, given promotion and consumer education, this crop has a future throughout the Andes.

The region is the logical center for the selection and development of cultivars, and it seems likely that researchers will discover varieties with unexpected qualities. (Those now available have not been improved and are considered landraces at best.)

Yacon could prove to be a profitable source of high-fructose sweeteners as well as a fresh snack vegetable. It might also be a useful, perennial fodder crop.

Other Developing Areas.   Yacon seems to have promise worldwide. It is already popular in some South American regions outside the Andes, as well as in parts of Southeast Asia. Although fresh yacon is not nutritious, it is an easy-to-grow sweet treat that could become popular in many areas of the tropics and subtropics.

Industrialized Regions.   Yacon is easy to grow, widely adaptable, and seems to be unrestricted by differences in daylength. It is refreshing, low in calories, and can yield an industrial sweetener. Any plant with these features seems destined to become commercially valuable. Like jicama and jerusalem artichoke before it, yacon might find its way into upscale markets in the United States, Europe, Japan, and other industrialized areas as a food for dieters.

Yacon is a root crop that can be eaten raw. It has been called "apple of the earth" because of its sweet taste and crunchy texture. (N. Vietmeyer)

In addition, yacon could be a source of inulin for use in sucrose-free foods for diabetics. As noted, the high-fructose sugars produced from inulin might also eventually compete with other, less-efficient sources.

USES

Yacon is usually eaten raw. The sweet, crunchy tuber is often chopped and added to salads, imparting flavor and texture. The tubers are also consumed boiled[5] and baked. In cooking, they stay sweet and

---

[5] If boiled "in the jacket," the skin separates from the flesh and can be peeled off like a boiled egg.

THE RISING DEMAND FOR FRUCTOSE

In the past, crystalline sugar was the main commercial sweetener, and fructose (which is hard to get in crystalline form) was of little interest. But now, syrups dominate the industrial use of sugar, and this has brought fructose to the forefront. With today's greater interest in high-fructose sweeteners, yacon might be economically more suitable than ever before. Indeed, "super-high-fructose" syrups from inulin could have value on their own account in addition to diluents to raise the fructose level in normal syrups.

In recent years, fructose has gained much attention as a sweetening agent. It has twice the sweetening power of normal sugar (sucrose). So far, it has been made by using enzymes that transform the glucose in corn syrup. Most of the resulting "high-fructose corn syrups" contain less than 60 percent fructose. With yacon, however, no transformation of one sugar to another would be needed, and the resulting syrup would likely contain more than 90 percent fructose.

The solids in a yacon tuber may be 60–70 percent inulin, which is readily hydrolyzed (by acid, or by the enzyme inulase) to fructose. In spite of previous interest in using inulin to produce fructose, there has been little effort to do it in practice. Yacon might be the key to making the process economically attractive.

A few other crops, mostly of the family Compositae, contain inulin. The best known are jerusalem artichokes (*Helianthus tuberosus*) and chicory (*Cichorium* spp.). Yacon tubers, however, are like big fat fingers with smooth skin, unlike the knobby irregularity of jerusalem artichokes, and are thus more easily processed. Yacon yields are also higher than those of jerusalem artichokes, and the fresh tubers contain almost 19 percent inulin.* The strong flavor of chicory roots limits their usefulness as a sweetener source (they are used as a coffee substitute). Compared to the competition, yacon has the advantages of providing both a specialty food and animal fodder, giving high yields, and probably being easy to harvest and process.

* Calvino, 1940.

slightly crisp. They usually weigh 180–500 g, but some are said to weigh as much as 2 kg each. In the Andes, they are often grated and squeezed through a cloth to yield a sweet, refreshing drink. Sometimes this is concentrated to form dark-brown blocks of sugar called *chancaca*.

The skin can have a resinous taste, so the tubers are usually peeled before eating. Undamaged tubers keep well, and in Spanish Colonial times yacon was used as a food for sailors.

As previously mentioned, the main stem of the plant is used as a cooked vegetable.

Yacon might have potential as a forage crop. The foliage is luxuriant, and the leaves have a protein content of 11–17 percent (on a dry-weight basis). When cut, the foliage sprouts again from the underground stems. The tubers may also be good cattle feed, for inulin is rapidly metabolized by ruminants. Additionally, the plant may be useful in agroforestry, because it grows well beneath a canopy of trees.

NUTRITION

The food value of the tubers is low and consists chiefly of carbohydrates. Fresh tubers have been analyzed as containing 69–83 percent moisture, 0.4–2.2 percent protein, and 20 percent sugars. The sugars consist mainly of inulin. Dried tubers vary from 4–7 percent ash, 6–7 percent protein, 0.4–1.3 percent fat, 4–6 percent fiber, and approximately 65 percent sugars. The tubers are said to be high in potassium.

The inulin molecule in yacon is uncharacterized, but in related species (other Compositae) it has a molecular weight of 3,000–5,000. It is a polymer of fructose but the terminal unit is a glucose sugar. Thus, inulin contains a small amount of glucose.

The dried herbage contains 11–17 percent protein, 2–7 percent fat, and 38–41 percent nitrogen-free extract.

AGRONOMY

Yacon is propagated with offsets (small "plantlets" taken from the base of the above-ground part of the main stem) with a few cylindrical roots attached. Single-node stem cuttings root readily. Moreover, the storage tubers can be easily divided. In addition, selected clones and disease-free materials can be derived from tissue culture propagation.[6]

Yacon is planted throughout the year, providing there is adequate soil moisture and warmth. Early growth is rapid, and it requires little attention apart from weeding.

The plant reaches maturity in 6–7 months. Having flowered, the tops wither and die back, at which time the tubers are harvested.

---

[6] Unlike many Andean root crops, a tested sample of yacon was found free of several common tuber viruses, including the potato leaf-roll virus, and potato viruses X, Y, S, M, and A. Information from J. Martineau.

HARVESTING AND HANDLING

The plant must be dug carefully to prevent breaking the brittle tubers. These tubers are separated from the central stem, which is often fed to livestock. Yields of 38 tons of tubers per hectare have been reported.[7]

Once soil is removed, the roots can be stored in a dark, dry place for months. (Thus, unlike sugar-beet growers, yacon producers could spread out the time they harvest and process the crop.)

LIMITATIONS

Yacon provides little in the way of human nutrition. It is consumed for flavor and variety rather than for sustenance.

Outside its native region, yacon is little known either in agriculture or as food.

Whitefly and looper caterpillars have been reported to be pests.

RESEARCH NEEDS

As a first step, an international effort is needed to scout out the available types, evaluate them, and store them in gene banks.[8]

Yacon could be an attractive crop for producing alternative sweeteners, and research to determine this should be undertaken at a university or industrial research laboratory.

There is a need to develop strains that produce tubers of uniform flavor. At present, one plant may be as sweet as candy, whereas its neighbor is scarcely sweet at all. Also needed is research to determine how flavor is related to growing conditions.

Research on improving the storage of the tubers is vital. (One question, for example, is whether yacon can be stored without breakdown or alteration of its inulin.)

The plant's potential as fodder has been little explored in the last 40 years. As in other *Polymnia* species, the leaves may contain sesquiterpene lactones that make them of little use as foodstuffs.[9]

The prevalence of diseases—especially viruses—needs to be determined. Virus-free material could potentially expand yield and is *essential* for the movement of germplasm. In particular, quick inexpensive procedures (such as the ELISA test used in potatoes) need to

---

[7] Kay, 1973.
[8] There is a collection of about 25 Peruvian clones at the University of Ayacucho.
[9] Information from T. Johns.

YACON IN ITALY

Before World War II a far-sighted Italian agronomist, Mario Calvino, came across yacon while working in the Dominican Republic. He took some tubers to northern Italy, hoping the plant would make a palatable high-protein forage, as well as a possible source of sugar for producing alcohol for fuel.

From Calvino's fields, yacon was introduced to other parts of southern Europe; however, war brought this work to an abrupt halt. After the war, Calvino and his plant were forgotten, but the fact that yacon grew vigorously in this temperate lowland region, so far from its Andean homeland, demonstrates to us, 50 years later, that, like the potato before it, this is an Inca crop with worldwide potential.

The photograph, reproduced from one of Calvino's papers, illustrates the magnificent growth of yacon at Sanremo, Italy, on December 20, 1939.

be available for monitoring the presence of viruses. As of now, however, yacon seems to be virus free.

Capabilities to produce elite clones inexpensively need to be greatly expanded. Apparently, tubers are especially amenable to meristem tissue culture, as they are composed of stem material with numerous buds.

SPECIES INFORMATION

Botanical Name   *Polymnia sonchifolia* Poeppig & Endlicher
Family   Compositae (sunflower family)
Synonym   *Polymnia edulis* Weddell, *Smallanthus sonchifolia*[10]
Common Names
  Quechua: yacón, llakuma
  Aymara: aricoma, aricona
  Spanish: yacón, jacón, llacón, llamón, arboloco, puhe, jícama (not the common jicama of commerce, see page 39), jíquima, jíkima, jiquimílla
  English: yacon, yacon strawberry, jiquima
  French: poir de terre Cochet
  German: Erdbirne
  Italian: polimnia

Origin.   Yacon grows wild in Colombia, Ecuador, and probably Peru, and it is commonly naturalized at medium altitudes in South America. It has been found in pre-Incan tombs in Peru, indicating a wide dispersal in early times.

Description.   Yacon is a handsome, compact, herbaceous plant with dark-green celerylike leaves. The aerial stems can reach 2 m in height, and are hairy with purple markings. Small, daisylike yellow or orange flowers are packed close together at the top of the plants and on additional stems arising from the lower leaf axils.

Yacon tubers are irregularly spindle-shaped to round (somewhat resembling those of the garden dahlia) and can vary considerably in shape, size, and sweetness. Fused to the swollen stem (4–5 or even 20 in a bunch), they splay out like fat spokes from a hub.[11] On the outside, they are tan to purplish brown, but inside they are white, yellow, purple, orange, or yellow, sometimes with magenta dots. A tuber usually weighs 200–500 g, but can reach 2 kg.

---

[10] The genus *Smallanthus* has been suggested for yacon and many of its relatives (H. Robinson. 1978. Studies in the Heliantheae (Asteraceae). XII. Re-establishment of the genus *Smallanthus*. *Phytologia* 39(1):47-53.)
[11] Yacon actually produces two types of edible underground portions—rhizomataceous stems (used by the plant for vegetative reproduction) and tuberous roots (used by the plant for food storage). The swollen roots are preferred for eating as they are sweeter, juicier, and not fibrous. The stems, although succulent when young, coarsen (lignify) as they mature.

Environmental Requirements

Daylength.   The plant is daylength neutral for stem- and root-tuber formation, at least for some clones.

Rainfall.   The annual foliage and perennial underground stems make yacon adaptable to seasonal cycles of drought or cold.

Altitude.   Generally between 900–2,750 m in the Andes, but it has been grown at sea level in New Zealand and the United States and reported at elevations up to 3,500 m in Ecuador.

Low Temperature.   Although foliage is damaged or killed by frost, apparently the underground tissues are not affected unless frozen.

High Temperature.   Tolerant of a wide range of temperatures.

Soil Type.   Although it grows in a wide range of soil conditions, yacon does best in well-cultivated, rich, well-drained soil.

PART II
Grains

To the Incas, corn (whose origins are in Mexico and Central America) was a sacred grain and a major food, but climate restricted its cultivation to lower elevations. At altitudes where corn could not thrive, it was replaced by kaniwa, kiwicha, and quinoa.

In pre-Hispanic times, these three native grains ranked as staple foods, but they were superseded by the wheat, barley, and oats brought by the Spaniards. Subsequently, the production of native grains fell drastically. In recent decades, it has dropped even further as imported grains—particularly U.S. wheat—have been available at subsidized prices. Also, the increasingly urbanized and media-dominated lifestyles in the Andean region have lured people away from their cultural roots. As a result, the traditional grains are now mostly cultivated in marginal lands and are consumed by people on the fringes of the predominant society.

But for all that, kaniwa, kiwicha, and quinoa have outstanding qualities and are beginning to return to general favor. The Peruvian government, for example, is fostering their increased production—including tariff supports for exports and promotion of their use in processed products—especially for combating childhood malnutrition in the highlands.

All three native grains are broad-leaved plants rather than grasses like wheat, rice, corn, and the other conventional cereals.[1] They are highly tolerant of marginal conditions. Kaniwa and quinoa are adapted to cold and drought; kiwicha to drought. They are cultivated at elevations up to the rarefied heights of the Andean snow line. There, where few crops can survive the dry air, short seasons, and cold soil, their cultivation goes back thousands of years.

---

[1] One South American grass domesticated as a food grain was *Bromus mango*. Not a true Andean plant, it comes from Chiloe Island and the adjacent Chilean coast. Cultivation was abandoned during the last century after the introduction of wheat. Until recently, domesticated varieties were believed extinct, but a dramatic rediscovery of the plant in Argentina was reported in 1987. Information from R. Reid.

These three grains are also outstandingly nutritious. Their food value relates particularly to their unusual proteins. All food proteins are composed of differing amounts of 20 amino acids. A few of these cannot be manufactured in the body and are generally referred to as "limiting," because once the body runs out of any one of them, it stops synthesizing proteins. In most plant proteins the three limiting amino acids are lysine, methionine, and tryptophan. But in the protein of kaniwa, kiwicha, and quinoa, the levels of lysine and tryptophan are excellent and the methionine level is adequate. This makes these three grains unusual among plant foods. Indeed, they approach animal foods such as milk or meat in their protein quality.

The current limitation of native grains is their low productivity. This can be overcome. Researchers have shown, for example, that substantial yield increases can be achieved through applying a little fertilizer and through the use of better selected seeds. Given such attention, the ancient grains of the Incas seem capable of yielding as much as the best products of modern science—especially under marginal conditions. In overall production, none seem likely to rise to the heady heights of wheat, rice, and corn (the world's top three cereal crops), but they all can play a much bigger role in feeding the world than they do today.

KIWICHA IN PERU

Kiwicha is described in a later chapter (page 139), but here we present the history of its recent resurrection in the Andes. Our purpose is to provide an example of what can be accomplished with the now little-known grains (as well as the other types of crops) that once fed the Incas.

After hundreds of years of lying dormant, kiwicha is being reborn. The "midwife" pulling it into the modern world and breathing new life into it is Luis Sumar Kalinowski, a scientist from the ancient Inca capital, Cuzco. Sumar and his colleagues proudly call themselves "kiwilocos"— because, they say, they're crazy about kiwicha.

In global terms, today's kiwicha plantings are still small, but they are astounding considering that just 10 years ago the plant had all but vanished. Peru's commercial kiwicha cultivation had risen to more than 700 hectares by 1988, not including several hundred hectares cultivated by farmers for their own use. These days, people looking down into the Sacred Valley of the Incas—the Vilcanota—can see it studded once more with brilliant red kiwicha fields. Peruvian farmers produced about 1,200 tons of kiwicha grain in 1988, most destined for a children's breakfast program in the Cuzco schools.

Now, in Peru's village markets and city supermarkets, the grain can be found in both raw and processed forms, ranging from breakfast foods to

Kiwicha "champion" Luis Sumar Kalinowski with a seedhead of one of his advanced lines of kiwicha. (N. Vietmeyer)

baby foods. Newspapers and government pronouncements have created a widespread awareness of the grain's nutritional value. To the public, it is a new food associated with health, vigor, and strength.

The efforts of the kiwilocos are a model for others interested in developing new food crops. One key feature is that they are developing and promoting the crop simultaneously. Thus, while selecting improved strains and breeding new varieties, they are also visiting farmers and designing suitable implements such as planting and harvesting machines. Moreover, they are investigating the grain's chemistry, processing methods, and culinary uses.

Public interest is now so high that kiwicha is one of the most lucrative crops in Peru. In 1988, farmers were paid 40 intis per kg for kiwicha grain compared with 20–26 intis for corn, 14 intis for barley, and 15–16 intis for wheat. In the Quillabamba area, at least one farmer earned twice as much per hectare from kiwicha as from coca leaves.

And excitement for the ancient Inca grain is spilling over beyond the borders of Peru. For instance, the United Nations Children's Fund (UNICEF) has provided funds for Sumar to stimulate kiwicha-growing in Bolivia and Ecuador.

In agronomic qualities, the kiwicha plant has made remarkable advances. When the researchers started their work in the early 1980s, top yields were around 1,800 kg per hectare and the average was much lower; now top yields are commonly 5,000–6,000 kg per hectare in experimental plots; average yields are about 3,000 kg per hectare.

A major advance in yield improvement came when types with large, upright seedheads were found. Before that, the plants had "dangling" heads that tended to spill their seed on the ground. (Sumar says, "Kiwicha was asleep five years ago.") Today's erect types can be cultivated mechanically because few of the seeds fall out when the plants are bumped or shaken.

None of this came easily. The researchers initially gathered 400 lines from different sites and altitudes throughout Peru. After four years of evaluation, 16 lines appeared to have good disease resistance and the potential for high yields in a broad range of environments. After three more years, however, only two lines proved promising for profitable production in farmers' fields.

While searching for types with good field performance, the researchers also located types likely to show good market performance. These had large white seeds with good cooking characteristics. The big breakthrough came when they perfected a blower that could separate large, heavy seeds from small, light seeds (for example, those that were incompletely filled or had loose seed coats). The bigger seeds showed much better germination and vigor, were easier to harvest and handle, and were much more acceptable in the marketplace.

By simultaneously tackling market requirements and farmers' requirements, the researchers have stimulated modern interest in a truly lost crop of the Incas and laid the foundation for its long and lasting future.

Kaniwa

Kaniwa (*Chenopodium pallidicaule*) is a remarkably nutritious grain of the high Andes that has been described as helping to "sustain untold generations of Indians in one of the world's most difficult agricultural regions."[1] Kaniwa (pronounced kan-*yi*-wa) reigns in the extreme highland environment where wheat, rye, and corn grow unreliably or not at all because of the often intense cold. Even barley and quinoa (see page 149) cannot yield dependably at the altitudes where kaniwa grows. In its native area, for example, year-round temperatures average less than 10°C, and frost occurs during at least nine months a year, including the height of the growing season.[2]

Kaniwa is so cold hardy that in the high Andes it serves subsistence farmers as a "safety-net." When all else fails, kaniwa still provides food. Indeed, it is perhaps more resistant than any other grain crop to a combination of frost, drought, salt, and pests—and few other food plants are as easy to grow or demand such little care. Moreover, although its grains are small, few cereals can match their protein content of around 16 percent.

Although kaniwa produces a cereal-like seed,[3] it is not a true cereal but a broad-leaved plant in the same botanical genus as quinoa. At the time of the Conquest, kaniwa grain was an important food in the high Andes. It is still widely grown, but only in the Peruvian and Bolivian altiplano—a lofty, semiarid plateau hemmed in by high ranges of the central Andes (see map page 133). Most kaniwa is consumed by the family that grows it, but some can be bought in Andean markets, especially near Puno.

The plant is not completely domesticated, and it often grows almost like a weed, reseeding itself year after year. (Farmers like it, however,

---

[1] Gade, 1970. During Inca times, however, it reportedly was restricted to the Inca emperor himself and to his court; the general population was forbidden to eat this "royal food."

[2] Kaniwa possibly may resist cold because a special anatomical structure protects its flowers from damage at temperatures as low as −3°C.

[3] As harvested, the "seed" is actually a hard-walled fruit (achene) containing the true seed.

and encourage this "weed" to grow in their plots of potatoes, quinoa, or barley.) Its seeds—unlike those of quinoa—contain little or no saponins and can be eaten without elaborate processing. However, harvesting and dehusking them is laborious.

Kaniwa requires much scientific attention before it reaches its true potential. At present, it exhibits many of the adverse characteristics of semidomesticated plants: for example, great variation in appearance and time to maturity, and failure of plants from the same seed to ripen at the same time. It also exhibits the favorable characteristics of a rustic crop: self-sufficiency and adaptation to widely varying habitats, for instance. With selection for plant type, nonshattering seedheads, uniformity, and higher yield, kaniwa would prove to be a valuable "life-support crop"[4] for extreme highlands throughout the world. Indeed, as an almost fail-proof backstop for conventional grains, it may open a more efficient agricultural use of the world's highest cultivated terrain.

PROSPECTS

The Andes. For people who live on subsistence agriculture in the altiplano, kaniwa is extremely important. Given promotion[5] and research support, its production could increase greatly. Because broad climatic fluctuations are the norm throughout the highlands, this extremely resilient plant should be tested as a food, feed, and cash crop over a much wider area.

Although it is unlikely ever to be a substantial food of the whole Andean region, kaniwa will continue as a vital agricultural support that sustains the lives of many highland peoples, especially during the most difficult times. By standing between total crop failure and starvation—especially in high-altitude, marginal areas—it will always be important to the well-being and stability of the region. For this reason alone, it deserves far greater research attention.

Other Developing Areas. Because of its adaptability to cold and aridity, kaniwa could expand the amount of cultivable land in some marginal tropical highlands.[6] However, this is a distant prospect

[4] This term was coined by Promila Kapoor, an expert on Himalayan chenopods that are related to kaniwa.

[5] At present, many Indians are reluctant to grow kaniwa because it is so associated with extreme cold that they fear its cultivation will encourage frost. Information from R.T. Wood.

[6] A related species, *Chenopodium album*, has been used for centuries in the Himalayas for its seed and leaves, and it is deeply entrenched in traditional horticulture and foods. See T. Partap, 1985. The Himalayan grain chenopods. I. Distribution and ethnobotany. *Agriculture, Ecosystems and Environment* 14:185–199.

Fields of kaniwa, Patacama, Bolivia. (M. Tapia)

because even in the Andes the crop is not well understood, nor have its various types been fully collected and compared. Moreover, its acceptability in diets outside the Andes is uncertain.

Industrialized Regions. Kaniwa seems to have little immediate potential as a cash crop for North America, Europe, or other industrialized areas. The lack of knowledge of its productivity and mechanized cultivation would make it a risky commercial undertaking. Nonetheless, kaniwa is one of the most nutritious grains and most resilient plants known. It could perhaps prove useful as a forage crop or as a specialty grain for nutritionally conscious consumers. For instance, kaniwa could become popular among vegetarians and "health-food" consumers, as is happening with quinoa.

USES

The seed is usually toasted and ground to form a brownish flour (*kañihuaco*) that is consumed with sugar or added to soups. It is also used with wheat flour in breads, cakes, and puddings. And it is made into a hot beverage, similar to hot chocolate, and sold on the streets of cities such as Cuzco and Puno.

The leaves are especially high in calcium, and the plant is valued for soil improvement. It also provides a forage that is especially important for animal survival during droughts, when other forage is scarce. Farmers sometimes grow it above 3,800 m in the altiplano, where its biomass is comparable to, or greater than, that of other forages. Fresh forage yields of 24 tons per hectare have been reported.

The grain is also a potential feed. In one test, a mix of 80 percent kaniwa grain, 9 percent fishmeal, and 6 percent cottonseed meal yielded results equal to those of a commercially produced poultry ration.[7]

NUTRITION

Kaniwa seed adds high-quality protein to meat-scarce diets. This is particularly important because in the Andes, as well as in other tropical highlands, millions of people survive primarily on starchy tubers. The protein content of the grain is extremely high. Moreover, it has an exceptional amino-acid balance, being notably rich in lysine, isoleucine, and tryptophan. This protein quality, in combination with a carbohydrate content of nearly 60 percent and a vegetable oil content of 8 percent,[8] makes kaniwa exceptionally nutritious.

Kaniwa foliage is also nutritious and can be used as a potherb. The leaves of young plants (at about a month and a half after planting) have protein contents as high as 30 percent (dry weight). The crop residue is highly digestible, mineral rich, and valuable for livestock feed.

AGRONOMY

To plant kaniwa, farmers usually broadcast unselected seed over the land. Often they choose soils loosened by previous tuber crops. The seed also can be successfully planted using mechanical equipment.

After sowing, weeding and thinning are beneficial, but the growing plants are normally given little or no attention until harvest (this is mainly because weeds are scarce where kaniwa is grown, since weather conditions limit the growth of most other plants). Kaniwa responds well to nitrogen and phosphorus, although in the rural Andes fertilizers are rarely used. Because it has a short, stout stem, the plant resists strong winds and heavy rains. By the time it is 5 cm high, it seems quite resistant to drought. In addition, it is more resistant than barley and quinoa to unusually low night temperatures.

[7] Tapia et al., 1979.
[8] This is also a remarkably high figure for a grain, but it could be a problem, for in milled grains the fat oxidizes and makes flour production and storage more difficult.

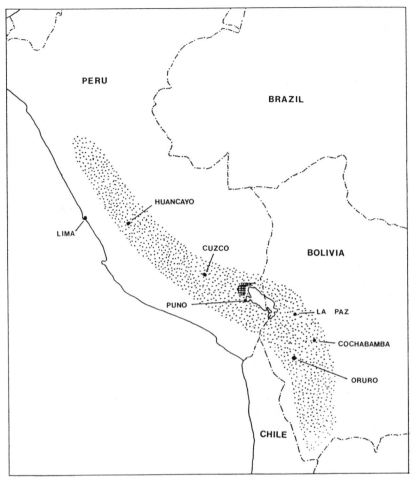

Kaniwa is grown in scattered plots on marginal land throughout the altiplano region (dotted), mainly at altitudes over 3,800 m. However, only in a small area (crosshatched) north of Lake Titicaca is it grown intensively on a large scale. (Map courtesy J. Risi C. and N.W. Galwey)

HARVESTING AND HANDLING

The most common varieties take about 150 days to reach maturity; however, at least one quick-maturing type can be harvested 95 days after sowing. Current varieties must be harvested before they mature fully; otherwise the seeds scatter on the ground. Several harvests are usually needed to get the whole crop, because the fields contain mixtures of types, and different plants ripen as much as several weeks apart. Normally, the entire plant is uprooted bodily.

The plants are threshed immediately after harvesting and again after air-drying, by which time more seeds have matured and loosened. Mechanical threshers have been tried with success. Under field conditions, seed yields of 2,400 kg per hectare have been reported; in experimental plots, twice that has been obtained.[9]

Kaniwa seed is tedious to prepare, for it is enclosed in a papery covering (a remnant of the flower called a "perigonium") that must be removed. The covering is loosened by soaking, following which it is rubbed off.

As now grown, kaniwa seems resistant to major plant diseases (this may be largely because in the cold, arid regions where it grows, the diseases are, relatively speaking, not serious). The flowers and leaves are sometimes infected with mildew, but this generally disappears and apparently has little lasting effect on seed yield. Some quinoa pests are also found on kaniwa. Minor damage is done by insects such as cutworms, beetles, and aphids.

LIMITATIONS

With such a little-studied crop, the uncertainties are manifold. For instance, kaniwa may have strict latitudinal limits; it may be restricted to high altitudes; it may succumb to diseases, pests, and weeds when grown outside its now almost competition-free environment; and it may require the intense sunlight (insolation) of its native home.

Known limitations include the fact that kaniwa must be harvested and threshed several times, and that preparing its seed is laborious. Also, its seed is not white. Currently, there are two colors: black and dark brown; most types are brown-seeded.

The plant's small, closed, normally hermaphroditic flowers make cross-breeding for crop improvement even more difficult than usual in a grain crop.

RESEARCH NEEDS

Kaniwa needs a great deal of experimental work, especially concerning its agronomy. So far, plant improvement has been limited to a small amount of germplasm evaluation, cytological studies, and the production of tetraploid specimens to increase the size of the seeds.

[9] In Puno, Peru, small-plot trials identified three ecotypes that yielded 5,000 kg per hectare even after being subjected to three frosts of $-8°C$. In the surrounding area, quinoa, barley, potatoes, and winter wheat had all suffered severe frost damage. Information from J. Risi.

As a first step, a major germplasm collection and evaluation should be made. Seed should be collected both from the wild and from farmers' fields. Variation in color, sensitivity to daylength, uniformity of maturation, yield, susceptibility to pests, and adaptation—particularly to cold and salinity—should be noted.

It is vital when breeding improved kaniwa to get plants with seeds of uniform maturity that stay in the seedhead as it dries out and that can be easily husked. Such nonshattering plants with easily husked seeds would usher in a vastly expanded future for kaniwa. In some circumstances, mechanized threshing would also help.

The plant's potential as a fodder deserves intense investigation. Because it grows readily at high altitudes where traditional forage crops fare poorly, it could help extend the usefulness of many now-marginal lands. It is highly digestible, nutritious, and mineral rich, and can be left standing in the field as a reserve for use when pastures dry up and forage is scarce.

Kaniwa's value as a source of genes for other chenopods is worth investigating. For example, although quinoa is not closely related, kaniwa genes might prove transferable using modern techniques, and they might contribute increased hardiness, dwarf stature, and saponin-free seed coats to the quinoa crop.

Trials should be performed outside kaniwa's native region to measure the plant's geographical adaptability. Together with other "life-support crops" from the Andes, the Himalayas, and elsewhere, kaniwa should be put into high-altitude trials and tested for its potential to sustain life in Asia and Africa as it has been doing for millennia in the Andean heights.

SPECIES INFORMATION

Botanical Name   *Chenopodium pallidicaule* Aellen
Family   Chenopodiaceae (the family of lambs-quarters)
Synonym   *Chenopodium canihua* Cook
Common Names
   Quechua: kañiwa, kañawa, kañahua, kañagua, quitacañigua, ayara, cuchi-quinoa
   Aymara: iswalla hupa, ahara hupa, aara, ajara, cañahua, kañawa
   Spanish: cañihua, cañigua, cañahua, cañagua, kañiwa
   English: kaniwa, canihua

Origins.   Kaniwa's origins are uncertain, but it is almost certainly an Andean native. It has a strong tendency to "volunteer" itself in highland fields, which may explain how it came to be adopted for cultivation. At the time of the Conquest it was cultivated over a much wider area than at present.

Description. Kaniwa is a highly variable, weedy annual that is normally between 20 and 60 cm high. It is erect or semiprostrate, highly branched at the base, with a vigorous but shallow taproot. Red, yellow, or green patches and streaks occur in the stalks and leaves, increasing in size and width towards the base of the plant.

The hermaphrodite flowers are inconspicuous, and are formed along the forks of the stem. Because at fertility the flower is closed, kaniwa is almost exclusively self-pollinating.

The numerous seeds (achenes) are approximately 1 mm in diameter (about the size of amaranth grains or half the size of quinoa grains), and have a clasping, papery covering. Most seed coats range in color from chestnut brown to black. Compared with conventional grains, the embryo is large in relation to the seed size.

Horticultural Varieties. Agronomic classifications have been devised based on plant shape and seed color. There are two "ecotypes": an erect plant (*saihua*) with 3–5 basal branches and determinate growth, and a semierect type (*lasta*) with more than 6 basal branches and indeterminate growth. Each of these types is further classified by the black or brown color of the seed.

The erect types usually grow faster for about 70 days, at which time dry-matter production ceases and the plants flower. The semierect types continue to grow throughout the season, and eventually produce more stems and dry matter than the erect types.

Some 380 accessions have been collected and are under evaluation in Puno, Peru.

Environmental Requirements

Daylength. All genotypes tested have been daylength neutral, and kaniwa has produced seed in England. In field trials in Finland, 35 ecotypes (collected from Puno, Peru) produced mature grains at latitude 60°49'N, and 5 ecotypes matured grains at 64°41'N.[10]

---

[10] Although seed production was apparently unaffected by daylength, plant growth was poor, probably because of low light intensity and weed competition. Information from J. Risi. In field trials carried out in Finland, 35 ecotypes from Puno, Peru, produced mature grains at 60°49'N, and 5 at 64°41'N. Carmen, 1984.

Rainfall. 300–1,100 mm around Lake Titicaca; requires moisture at the early growth stages, resists drought after establishment. The plants seem susceptible to excess humidity.

Altitude. Today, kaniwa is rarely cultivated below 3,800 m (below this, quinoa predominates). The upper limit is nearly 4,400 m in protected areas.

Low Temperature. Kaniwa is remarkably cold tolerant. It will germinate at 5°C, flower at 10°C, and mature seed at 15°C. Adult plants are unaffected by nightly frosts.

High Temperature. Midday temperatures in the altiplano are usually only 14–18°C, but kaniwa can withstand relatively warm conditions (up to 25°C) given sufficient soil moisture and air movement. It tolerates broad swings in temperature and high insolation.

Soil Type. Prefers an open, friable soil. Because of its short taproot, it seems particularly suited to shallow soils. It is successfully cultivated in soils ranging from pH 4.8–8.5, and shows some salt-tolerance.

Related Species. Kaniwa was long considered a weedy variety of quinoa, but chromosomal studies have confirmed that the two belong to separate species complexes (kaniwa has a chromosomal designation of $2n = 2x = 18$; quinoa has $2n = 4x = 36$).

The nearest morphological relatives are *Chenopodium carnosulum* and *C. scabricaule* from Patagonia.[11] The greatest diversity of related species in South America (assignable to the same subsection of *Chenopodium* as kaniwa) is centered around Argentina's pampas and western highlands. However, the most common and widespread species is *C. petiolare*, which extends far into the Andes, and often grows interspersed with kaniwa. Although it produces little grain, it is "semiperennial," and sometimes will produce seed over several seasons. *C. petiolare*, along with other kaniwa relatives, merits much further attention from botanists and agronomists.[12]

---

[11] Information from L. Giusti.
[12] Information from H. Wilson and L. Giusti.

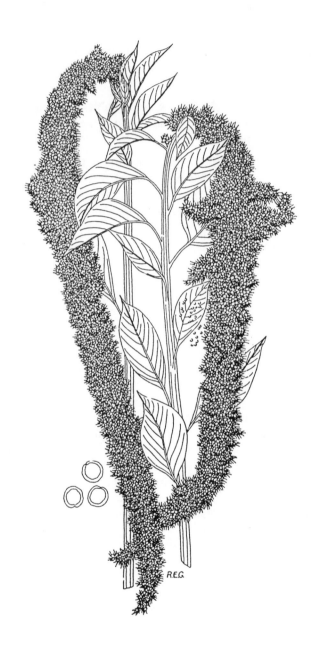

Kiwicha

A staple grain of the Incas, Aztecs, and other pre-Columbian peoples, amaranth was once almost as widely dispersed throughout the Americas as corn.[1] The most important Andean species is *Amaranthus caudatus*. In Quechua, the ancient Inca language that is still spoken in the Andes, it is called "kiwicha" (pronounced kee-*wee*-cha).[2]

Kiwicha is one of the prettiest crops on earth; the beautiful colors of its broad leaves, stems, and flowers—purple, red, gold—create fiery fields that blaze across the mountainsides. The plant grows vigorously, tolerates drought, heat, and pests, and adapts readily to new environments, including some that are inhospitable to conventional grain crops. Nonetheless, it is little known outside the highland regions of Ecuador, Peru, Bolivia, and northwestern Argentina.

Kiwicha's grains are scarcely bigger than poppy seeds. However, they occur in huge numbers—sometimes more than 100,000 to a plant. Like other amaranth grains, they are flavorful and, when heated, they pop to produce a crunchy white product that tastes like a nutty popcorn. Light and crisp, it is delicious as a snack, as a cold cereal with milk and honey, as a "breading" on chicken or fish, or in sweets with a whisper of honey.[3] The grain is also ground into flour, rolled into flakes, "puffed," or boiled for porridge. Because of its high nutritional value, it is considered especially good for children, invalids, and the elderly.

These seeds are one of the most nutritious foods grown. Not only are they richer in protein than the major cereals, but the amino acid balance of their protein comes closer to nutritional perfection for the human diet than that in normal cereal grains.

Five hundred years ago kiwicha helped feed the Incas. After the conquest it was nearly forgotten, like so many other ancient Andean

---

[1] For more information on amaranth in general, see our companion report: *Amaranth: Modern Prospects for an Ancient Crop*.
[2] In this chapter, we use "kiwicha" to refer specifically to *Amaranthus caudatus*.
[3] Prepared this way, amaranth is a favorite confection in Mexico, where it is called "*alegría*," which means "happiness." In northern India, a similar product called "*ladoos*" is popular.

Cuzco, Peru. In the Peruvian Andes kiwicha is returning to farmers' fields after being absent since the time of the Incas. In some places (for example, Limatambo), farmers have switched to kiwicha because its resistance to drought and pests makes it more profitable than the crops grown previously. Every market in the Cuzco area now carries the tasty and nutritious grain. (N. Vietmeyer)

crops. Now kiwicha (as well as other amaranths) is undergoing a renaissance, and in the last few years this ancient grain is returning to compete with modern crops (see page 125).

In other parts of the world, amaranths have caught the imagination of farmers as well. About 500 hectares of amaranth (Mexican species) were grown in the United States in 1987 and the harvest was sold to bakeries and food markets. Indeed, cookies and breakfast foods made of amaranth are already in health food stores and some supermarkets from New York to San Francisco. Given research, kiwicha might also find a place in world agriculture, although so far it has not performed as well in the northern hemisphere as the Mexican species.

PROSPECTS

Andean Region. As an indigenous crop, kiwicha is well adapted to the Andes. Given more attention, it can play an increasing role in Andean nutrition. Already the present small program in Peru has had notable effects; products made from kiwicha are appearing in open markets and supermarkets, and their nutritional punch has become known to millions.

As kiwicha and other grain amaranth species become more popular worldwide, consumption will probably also increase among all levels of South American society. This, in turn, will boost kiwicha's attractiveness as a cash crop, and should also encourage even more sustained long-term research. The crop could then be reestablished in many places throughout the region after an absence of almost 500 years. Eventually, it could become a vital nutritional complement to the diets and incomes of millions of traditional farmers, as well as to the rural and urban poor. Kiwicha requires less processing than many Andean crops—beans, quinoa, and tarwi, for example—which is particularly important where fuel is limited or expensive.

Other Developing Areas. Various amaranth species are used as grains, greens, fodders, or ornamentals around the world. This is especially true in the lower elevations of the Himalayas, where they are well established in the nonirrigated croplands. So far, kiwicha amaranth is barely known outside the Andes, but the work done in Peru and nearby countries could be extended to other regions.

During the past five years there has been a marked increase in the research and production of amaranth. Substantial plantings are reported in China, Nepal, India, Kenya, and Mexico, where amaranth often occupies the rainfed croplands. The work done in the Andean countries could benefit the expansion of kiwicha amaranth in these countries.

Amaranth can undoubtedly be used to raise the nutritional quality of foods that are normally made from other grains such as corn, rice, or sorghum. In such blends, its food value is particularly beneficial for infants, children, and pregnant and lactating women. Nutritionists have compared amaranth favorably with milk. By introducing amaranth into diets based on cereals and tubers, a much-improved nutritional balance is obtained. In sum, kiwicha could become an important source of protein, vitamins, and minerals in many areas, particularly in tropical highlands.

Industrialized Regions. Amaranth is becoming an established specialty crop in the United States. Only Mexican varieties have been used so far because kiwicha types have performed poorly. However, selection of genotypes that set seeds under long daylength conditions seems likely to uncover better adapted forms. A killing frost is required to dry down the plants sufficiently to permit direct combine harvest. Amaranth is of particular interest to farmers growing dryland crops in areas of the Great Plains. The falling aquifers and increasing water costs make it an attractive alternative to crops with a high water requirement.

Interest in amaranth (Mexican types) has spread to New Zealand and a few other such nations in what appears to be the beginning of a wider acceptance of kiwicha and its sister species.

USES

The meal or flour from kiwicha grain is especially suitable for unleavened breads, where it can be used as the sole or predominant ingredient. The flour of other amaranths is used in Latin America and in the Himalayas to produce a variety of flatbreads such as tortillas and chapatis.

For making yeast-raised breads or other leavened foods, kiwicha meal or flour must be blended with wheat meal or wheat flour because it lacks functional gluten. Blends of 80 percent wheat and 20 percent kiwicha give normal leavening to breads, and the high lysine content of the kiwicha greatly improves the nutritional quality over that of breads made with wheat flour alone.

In the form of whole grain, flour, toasted grain, popped grain, flakes or sprouted-grain flour—kiwicha can be used in many other foods, including soups; pancakes; breakfast cereals; porridges; breads, rolls, muffins, and similar baked foods; and salads.

Kiwicha may have great promise as a vegetable crop as well. In much of the world, young leaves and stems of several other amaranth species are boiled as greens. Although they are virtually unlisted in agricultural statistics, the various amaranths may actually be the most widely grown vegetable crop in the humid tropics.

From the red varieties of kiwicha is obtained a food coloring (called betalaina) that is nontoxic. It is slowly degraded by light, but nonetheless has promise because synthetic red dyes are suspected of being health hazards. Simple methods for extracting the brilliant red coloring have been developed in Peru.[4]

After the grain is threshed, the kiwicha residue (stover) can be used as a source of fodder for cattle. Research in Peru has demonstrated that it is much better in nutritional value than the residue of other Andean crops. Andean farmers traditionally maintain their livestock on crop residue during the dry season, when forage is limited.

Kiwicha also has potential as a forage crop. It can rapidly produce a large amount of biomass with a high protein content, especially in the tropics where many high-protein forages yield poorly.

[4] Information from L. Sumar.

NUTRITION

Kiwicha produces mild-tasting, cereal-like seeds that have protein contents of 13–18 percent, compared to about 10 percent in corn and other major cereal foods. Moreover, the seeds have high levels of lysine, a nutritionally essential amino acid that is usually deficient in plant protein. For instance, they have nearly twice the level of lysine found in wheat protein. Popping and flaking seem to have no major effect on protein digestibility or utilization; however, heat may damage the protein unless care is taken.[5]

Amaranth grain is also high in calcium, phosphorus, iron, potassium, zinc, vitamin E, and vitamin B-complex. Its fiber, especially compared with the fiber in wheat and other grains, is very soft and fine. It is not necessary to separate it from the flour; indeed it may be a benefit to human health.

The starch grains vary in diameter from 1 to 3.5 microns, comparable to those in quinoa starch, and much smaller than those in wheat starch or cornstarch. Their tiny size gives them possible uses in industry. They are, for example, small enough to be fired through the nozzles of aerosol cans, and may therefore be suitable substitutes for talc, which is under investigation as a possible health hazard.

AGRONOMY

Kiwicha and other amaranths adapt to many environments and tolerate adversity because they convert the raw materials of soil, sunlight, and water into plant tissues by using an especially efficient type of photosynthesis. Known technically as the C4 carbon-fixation pathway, this process is particularly efficient at high temperature, in bright sunlight, and under dry conditions. Plants using it tend to require less water than those that use the more common C3 carbon-fixation pathway.

The crop is easy to establish. The seed is either broadcast into the field or sown in rows.[6] Mechanical planters, such as those used to plant wheat, can be adapted to plant the tiny seed. The seeds may germinate in as little as three days, but the seedlings are slow starters and are easily overwhelmed by weeds.[7] Once established, however, they grow quickly and their maintenance is relatively easy.

[5] Pedersen et al., 1987a.
[6] A simple planter, devised in Peru, consists of a disposable plastic coffee cup fitted on the bottom of a short piece of plastic pipe. The pipe is filled with seed, and a nail hole in the bottom of the cup allows the seed to dribble out as the farmer walks along.
[7] Because they are C4 plants, atrazine-type herbicides show promise.

HARVESTING AND HANDLING

Most types mature in 4–6 months. However, in some highland regions they may take as long as 10 months. Yields of 1,000–3,000 kg of seed per hectare are not uncommon; up to 6,000 kg per hectare has been achieved in research plots. In 1987, a Peruvian farmer (using improved varieties and traditional farming methods) obtained a harvest of 5,000 kg per hectare in a 6-hectare farm field.[8]

Most traditional kiwicha varieties are harvested just before maturity. This is because the seeds are not held tightly in the seedhead, and they will scatter on the ground when the seedheads dry out. Researchers in the Andes have developed strains with upward-facing seedheads that cradle the seed until harvested (see picture, page 126). These are suitable for mechanical harvesting. Also, the researchers have had success with making simple modifications to threshers developed for conventional small grains such as wheat and rice.[9]

LIMITATIONS

A few amaranth species are serious weeds, which can cause concern when any amaranth is being introduced to a new area. However, the weedy types are distinctly different from kiwicha, which is not a persistent plant.

Kiwicha seems to have daylength sensitivity. As noted, the varieties so far tested in the United States have not grown well, apparently because of daylength incompatibility.

Diseases such as damping off and root rot can be very damaging. However, resistance has been found in certain types in Peru. The biggest problem has been weeds (kikuyu grass is particularly troublesome). Hand hoeing has so far been employed, but the use of dense plantings shows promise to suppress most weeds. Also, it has recently been found that rotating the crop with potatoes seems to solve the weed problem.

The seeds are similar in size to the chaff and to impurities such as sand. This makes it difficult to winnow the crop to obtain clean seed.[10]

The small seed size increases the difficulty of establishing a good plant stand. Small-seeded crops require a shallow planting depth; thus the seed zone of the soil is prone to drying out during germination and emergence.

[8] Information from L. Sumar.
[9] Losses during combine harvesting and threshing amount to only about 5–8 percent, a huge improvement over the 50-percent shattering losses of just a few years ago and approaching the levels acceptable for other commercial grains.
[10] A pneumatic winnow has been designed in Peru and is used to remove dust and chaff from the grain. Sand is still difficult to separate. Information from L. Sumar.

Compared with other grains, research on amaranth is quite limited. Additional work on cultural requirements, plant breeding, food technology, and nutrition is needed to determine the potential and to discern the appropriate niche of amaranth in relation to other grains.

RESEARCH NEEDS

Agronomists have already improved kiwicha by breeding plants of uniform height with sturdy, wind-resistant stalks and high-yielding seedheads that hold onto their seeds until they can be harvested. The interrelated responses to a complex variety of climates, soil conditions, pests, and diseases have also been worked out for several cultivars. Among research yet to be done is the following:

• The further collection and evaluation of germplasm to assess the range of genetic diversity. Daylength-neutral types should be sought, especially in the most southern regions of kiwicha's occurrence.

• The expansion of adaptability trials to study the effects of environmental variation on different landraces. One test, for example, is needed to determine if red in the plant's coloration is correlated with frost resistance, as it is in some other crops.

• The determination of optimal growing practices under various environmental and cultural conditions.

• The improvement of harvesting techniques for increasing grain uniformity and reducing contamination.

• The improvement of processing methods, such as those for cleaning and grinding the seeds.

• The development of ways to substitute amaranth in popular food products that have poor nutritional quality.

• The breaking of the "daylength barrier." Perhaps crossing kiwicha with other species, such as *Amaranthus hypochondriacus* or *A. cruentus*, could expand the range of latitudes where kiwicha can be grown, and maybe even increase the seed size.

SPECIES INFORMATION

Botanical Name   *Amaranthus caudatus* Linnaeus
Family   Amaranthaceae
Synonyms   *Amaranthus edulis*, *Amaranthus mantegazzianus*
Common Names
   Quechua: kiwicha, quihuicha, inca jataco; ataco, ataku, sankurachi, jaguarcha (Ecuador), millmi, coimi
   Aymara: qamasa

Spanish: kiwicha, amaranto, trigo inca, achis, achita, chaquilla, sangorache, borlas.[11]
Portuguese: amaranto de cauda
English: amaranth, love-lies-bleeding, red-hot cattail, bush green, Inca wheat (normally used for quinoa)
French: amarante caudée

Origin. White-seeded and apparently domesticated kiwicha has been found in Andean tombs more than 4,000 years old. It is not found in the wild, and although until recently it has been a "rustic" crop, it is believed that it has long been fully domesticated.

Although never as important or imbued with such special attributes as the pre-Columbian Mexican amaranths, kiwicha undoubtedly played an important nutritional role in pre-Conquest Andean society, particularly in the region of the Incan and Aymaran homelands in southern Peru and Bolivia.

Description. Kiwicha is an annual, broad-leaved dicotyledon. Its central stem can reach 2–2.5 m at maturity, although most varieties are shorter. Usually, leaves and side branches form on the central stalk (depending on the density of plants in an area). These may start as low as the base of the plant (depending on variety), which, in general, is shaped like an irregular cylinder.[12] The taproot is short and enlarged, with secondary roots penetrating downwards into the deeper soil.

The often spectacular flowers and seed are in panicles that arise from lateral buds and—especially—from the main stem. In some types the inflorescences can be 90 cm long, and often look like a long, red cat's tail. They can be erect, semi-erect, or lax. Each panicle has male and female flowers and is self-pollinating (the flowers can also be wind pollinated).

The fruits (pyxidia) each contain a single seed. The seeds are seldom larger than 1 mm in diameter but occur in massive numbers. Color ranges from black through red to the more common ivory or white. The seed covering is shiny, and the embryo is curved around the small endosperm (perisperm), much as in quinoa. Unlike quinoa, however, amaranth seeds contain no bitter saponins.

The chromosome number is usually $n = 32$ and occasionally $n = 34$.

---

[11] The names "bledo" and "bledos" ("wild amaranth" in Castilian) are also used. The plant has also been confusingly called "quinua," "quinua de Castilla" (Ecuador), and "quinua del valle" because of the superficial similarity to quinoa.
[12] These side branches are prone to breaking away from the plant (lodging), and harvesting heads of grain low on the plant is also more difficult. A major goal of selection has been a plant with its seed concentrated at the top of the central stem (see page 126).

Polyploidy has been induced, but results in only minor morphological changes.

Horticultural Varieties. Numerous landraces have been found in the Andes, generally distinguished by panicle form and the color of stem, leaves, fruit, and seed. True varieties have been selected in Peru. These include, in particular, "Noel Vietmeyer" and "Alan Garcia." The first is tall; "rustic"; resistant to mycoplasms, sclerotinia, and alternaria; and yields 3–3.5 tons per hectare. Its seed is translucent and makes good flour and flakes. The second is short and susceptible to diseases, but yields 3–5 tons per hectare under good conditions.

Over 1,200 accessions are being maintained in the Andean region, with many duplicates in other areas.

Environmental Requirements

Daylength. Most kiwicha varieties are short-day. Cultivars exist, however, that flower at daylengths ranging from 12 to 16 hours.

Rainfall. Grain amaranths have set seed in areas receiving as little as 200 mm. Some estimates place their moisture requirements at about equivalent to those of sorghum, or about half those of corn. Although kiwicha will tolerate dry periods after the plant has become established, reasonable moisture levels are critical for proper germination. Also, some moisture is needed during pollination.

Altitude. Kiwicha appears to be the only grain-amaranth species to thrive above 2,500 m. In the Andes, most is grown between 1,500 and 3,600 m. Commercial varieties are being successfully cultivated at sea level near Lima, Peru.

Low Temperature. Although more cold tolerant than most grains, it cannot tolerate frost. Some "cold-tolerant red lines" have been found that can stand temperatures of 4°C.[13]

High Temperature. 35–40°C. The plant grows best at 21–28°C.

Soil Type. Kiwicha grows well on soils containing widely varying levels of nutrients, although it does best in loose, sandy soils with high humus content. Genotypes that tolerate alkaline soils with pH as high as 8.5 have been discovered. In addition, some with an apparent ability to withstand mild salinity have been identified.[14] Other *Amaranthus* species are renowned for their tolerance to acid soils and aluminum toxicity; kiwicha probably is similar.

---

[13] Information from L. Sumar.
[14] In trials, kiwicha has grown well at salinity levels up to 8 mmhos per cm, a level that most conventional cereals cannot withstand. Information from L. Sumar.

Quinoa

To the Incas, quinoa (*Chenopodium quinoa*) was a food so vital that it was considered sacred. In their language, Quechua, it is referred to as *chisiya mama* or "mother grain." Each year, the Inca emperor broke the soil with a golden spade and planted the first seed.[1]

In the altiplano especially, quinoa (pronounced *keen*-wa or *kee*-noo-ah) is still a staple. For millions it is a major source of protein, and its protein is of such high quality that, nutritionally speaking, it often takes the place of meat in the diet. Outside the highlands of Argentina, Bolivia, Chile, Colombia, Ecuador, and Peru, however, the cultivation of quinoa[2] is virtually unknown.

Quinoa's large seedheads and broad leaves make it look something like a cross between sorghum and spinach. Its grain is rich in protein and contains a better amino acid balance than the protein in most of the true cereals. In earlier times this grain helped sustain the awesome Inca armies as they marched throughout the empire on new conquests. Today, it is made into flour for baked goods, breakfast cereals, beer, soups, desserts, and even livestock feed. When cooked in water, it swells and becomes almost transparent. It has a mild taste and a firm texture like that of wild rice, a popular gourmet grain of North America. Traditionally, quinoa is prepared like common rice or is used to thicken soups, but some varieties are also popped like popcorn.

The seeds of most varieties contain bitter-tasting constituents (chiefly water-soluble saponins) located in the outer layers of the seed coat. Because of this, they need to be washed—a tedious, time-consuming process—or milled to remove the seed coat. Practical, commercial methods for both processes have been developed in recent years.

Quinoa is beginning to attract scientific attention. In South America, governments and international agencies are extending research support. Moreover, in recent years, seeds have been distributed to more than

---

[1] Today, the cultivation of quinoa almost exactly coincides with the limits of the Inca Empire, although in Chile quinoa extends into the territories of the Araucanians, an extremely creative culture never conquered by the Incas.

[2] To conform with the most common English usage, we have chosen to use the spelling "quinoa" rather than "quinua," which is more usual in Spanish.

149

50 countries beyond the Andes. As a result, the cloud of uncertainty that has enveloped this grain for more than four centuries is beginning to disappear.

Indeed, quinoa seems like a grain of the future. Already demand is rising in the United States. Boxes of grain, flour, or pasta can now be bought in health food stores and supermarkets from Los Angeles to Boston. More than 750 tons of quinoa grain were sold in 1988, most of it imported from South America. Quinoa has been widely featured in newspaper and magazine accounts of promising new foods for the American dinner table. It is generating similar enthusiasm among consumers in Switzerland, and seems likely to do so in many more countries.

PROSPECTS

Andean Region. With recent advances in commercial methods for removing the bitter ingredients, a major impediment to expanding quinoa utilization is being overcome. The plant is well adapted to many parts of the Andes where the need for more food and a better nutritional balance is great. It seems likely that quinoa will become ever more important in diets of both the highland villagers and urban settlers. Because it is now primarily a food of campesinos and poorer classes, increasing its production is a good way to improve the diets of the most needy sector of society.

Quinoa also shows export potential. The most desirable varieties are currently best adapted to cultivation in the Andean region itself, and the long-term prospects for quinoa exports, although uncertain, seem promising.

Increased foreign demand for quinoa has not always meant increased production within the Andes. (Reportedly, supply has remained static while prices increased.) Decision makers throughout the region should ensure that production increases to fill overseas demand, and that poor people get the maximum benefit from quinoa as both a food crop and a cash crop.

Other Developing Areas. Quinoa seems particularly promising for improving life and health in marginal upland areas. It probably could be cultivated in highland tropical regions, such as elevated parts of Ethiopia, the Himalayas, and Southeast Asia. The malted grains and flour hold promise as a weaning food for infants, and it is noteworthy that child malnutrition is common in many of these areas. Also, quinoa is one of the best leaf-protein-concentrate sources.[3]

[3] Information from R. Carlsson.

Ilave, Peru. Harvesting quinoa grain. Traveling through Colombia in the early 1800s, Alexander von Humboldt observed that quinoa was to the region what "wine was to the Greeks, wheat to the Romans, cotton to the Arabs." He was excited by the crop because at that time starvation was rampant all over the world, and he had gone to South America looking for new foods to combat it. Because of its high nutritive value, Thor Heyerdahl took quinoa grain on the raft *Kon Tiki*. (IAF/M. Sayago)

Industrialized Regions. In the United States, quinoa has found a market in restaurants, health food stores, and supermarkets. It sells at "gourmet prices" and in some stores is outselling wild rice. It should soon find similar demand in Europe, Japan, Australia, and other areas.

The plant's daylength requirements (for flowering) are, for now, likely to limit its successful cultivation in North America, Europe, Japan and other such industrialized areas to types that come from equivalent latitudes in the Andes (for example, from Chile). At present, these are not readily available.[4] On the other hand, tall, late-maturing, daylength-sensitive types could prove productive for forages, a use for which flowering is unnecessary.

Despite this limitation, the plant has already shown some promise in tests of farm-scale cultivation in high altitudes of Colorado and at near sea level in Washington and Oregon states as well as in England and Scandinavia.

---

[4] Most seed exported for food has been desaponized and is nonviable.

Before the Spanish Conquest, quinoa was apparently grown from southern Chile to northern Colombia (widely spaced dot pattern). Today, it is mainly restricted to Bolivia and Peru (dense dot pattern), where it is grown mainly in backyards, field margins, and as an intercrop. In a few areas (crosshatching) it is cultivated as a sole crop. (Map courtesy J. Risi C. and N.W. Galwey)

USES

Quinoa grains are traditionally toasted or ground into flour. They can also be boiled, added to soups, made into breakfast foods or pastas, and even fermented into beer.[5] When cooked, they have a nutlike flavor, and they remain separate, fluffy, and chewy. Quinoa flours, flakes, tortillas, pancakes, and puffed grains are produced commercially in Peru and Bolivia.

Quinoa has demonstrated value as a partial wheat-flour substitute for enriching unleavened bread, cakes, and cookies.[6] Blends of wheat flour containing up to 30 percent quinoa flour produce fully acceptable loaf breads.

Mixing quinoa with corn, wheat, barley, or potatoes produces foods that are both filling and nutritious. Malnourished children in Peru and Bolivia are now being fed such quinoa-fortified foods with good results.

The plant is sometimes grown as a green vegetable, and its leaves are eaten fresh or cooked. It is also used as an animal feed. The leaves and stalks are fed to llamas, alpacas, cattle, donkeys, sheep, and guinea pigs; the grain and leaves are excellent feeds for swine and poultry.

NUTRITION

Quinoa has an exceptionally nutritious balance of protein, fat, oil, and starch. The embryo takes up a greater proportion of the seeds than in normal cereals, so the protein content is high. Grains average 16 percent protein, but can contain up to 23 percent[7]—more than twice the level in common cereal grains.

Moreover, the protein is of unusually high quality, and is extremely close to the FAO standard for human nutrition. Quinoa's protein is high in the essential amino acids lysine, methionine, and cystine, making it complementary both to other grains (which are notably deficient in lysine), and to legumes such as beans (which are deficient in methionine and cystine).

As for carbohydrates, the seed contains 58–68 percent starch and 5 percent sugar. The starch granules are extremely small. They contain about 20 percent amylose, and gelatinize in the 55–65°C range. The

---

[5] A lightly fermented quinoa drink, one of the many types of *chicha*, was considered the "drink of the Incas."
[6] High-protein cookies and biscuits can be produced by mixing up to 60 percent quinoa flour with wheat flour. The nutritive value of wheat-flour noodles can also be considerably increased by using up to 40 percent quinoa flour, without affecting appearance or other characteristics of the end product. Information from E.J. Weber.
[7] Information from D. Cusack.

New Homes for Quinoa

Canada

Rossburn, Manitoba. In the western Canadian grain belt, quinoa has now been successfully tested on over 300 farms, some as far north as the 51st parallel. Farmers report yields of 1,300 kg per hectare and a good profit, even on small plots. They harvest the grain with haying equipment, allow it to dry, and then thresh it using small-grain machinery. The seed is processed using locally produced equipment. Much of the grain is consumed at home, but some finds its way into the local markets as well as into Canada's health-food trade. Although quinoa requires more labor than traditional grains, and is vulnerable to early weeds and late rains, the farmers expect—because of both their success and the widespread consumer acceptance—that it will become a viable crop for Canada. (G.A. Clarke)

Opposite:

Since the early 1980s quinoa has been cultivated on a trial basis at high altitudes in the Colorado Rockies, especially in the San Luis Valley. Conventional crops produce poorly here, but the trials have produced quinoa varieties that thrive. Commercial production began in the mid-1980s and has been rising steadily ever since. Quinoa is now an established crop for this challenging alpine environment. (J. McCamant)

United Kingdom

Near Cambridge, England, a Chilean variety, Foro, has thrived in trials conducted by N. Galwey (shown) during the past five years. As a result, in 1989 quinoa was grown commercially for the first time in Britain. (E. Long)

United States

fat content is 4–9 percent, of which about half is linoleic acid, an essential fatty acid for the human diet. Calcium and phosphorus—and iron in most varieties—also occur in higher amounts than in other grains.

In nutritional content, quinoa leaves compare favorably to other leafy vegetables—spinach, for example. When the fields are thinned, the offtake is used as edible greens. Leaf nutrient concentrates of quinoa have very low values of nitrate and oxalate, which are both antinutritive factors.[8]

AGRONOMY

In the Andes, quinoa is normally propagated by broadcasting seed over the land and raking it into the soil. Sometimes it is sown in narrow, shallow rows. The seedbed must be well prepared and well drained, for the seeds are easily killed by waterlogging. Seedling growth is extraordinarily fast. The growth period from planting to harvest ranges from 90 to 220 days, depending on variety and temperatures.[9]

Mechanized production has been successful in South America. Machinery used for grains or oilseeds (particularly rapeseed) can be used for quinoa with little or no modification.

HARVESTING AND HANDLING

Native quinoas have extremely variable periods of maturity, which increases the difficulty of mechanization. Thus, up to the present, harvesting has been done largely by hand and only rarely by machine. The grain yield often reaches 3,000 kg per hectare and sometimes goes as high as 5,000 kg per hectare, which is comparable to wheat yields in the Andean area.[10]

Handling involves threshing the seedheads, winnowing the seed to remove the husk, and drying the seed. (Seed must be especially dry when stored because it germinates so quickly.)

As noted, the seed of most quinoa varieties must be processed before use to remove the bitter saponins. In the normal household,

[8] Information from R. Carlsson.
[9] In southern Sweden (Scania), one adapted type (BP183) matures in 120–150 days. Information from R. Carlsson. Some varieties used in the U.S. Pacific Northwest mature in 90–100 days. This means that they can be grown without irrigation in cool, moist spring (sow in early April, harvest before July 4) before the onset of summer drought. Information from R. Valley.
[10] Reported for the variety Sajama under ideal conditions in Bolivia, and extrapolated for type Baer from small-plot trials in Cambridge, England. Information from J. Risi.

this is done by soaking, washing, and rubbing. On a commercial scale, mechanical milling, or a mixed washing and milling procedure, are the most common methods.[11]

The green leaves and stems are a nutritious feed, and the prospect of using them as forage is attractive. Apart from the grain, quinoa has been reported to produce 4 tons per hectare of dry matter containing 18 percent protein.[12] In test plots in southern Sweden, leaves have produced up to 1,000 kg per hectare of extractable protein after 70 days of growth.[13]

LIMITATIONS

Quinoa's agronomic limitations include problems with weeds, lodging, and the difficulty of harvest. If the harvest is not properly timed, shattering occurs with large loss of seed. Moreover, deciding on an exact harvesting time is difficult because panicles from the same plant mature at different times. Increasing the seeding rate encourages the plant to produce only a main panicle, which ripens more uniformly.

Losses to pests and diseases are generally low, but as quinoa becomes more intensively and widely planted, more serious problems are likely to emerge.[14]

RESEARCH NEEDS

The possibilities of improving quinoa genetically are most promising, especially within the Andean region. Various races contain many desirable characteristics, and as the genetics of quinoa become better understood, it seems likely that types will emerge that can compete on an equal or favorable footing with other grains, both in traditional farming and large-scale commercial production.

Germplasm collection should continue in Argentina, Bolivia, Chile, Colombia, Ecuador, and Peru, especially in remote areas where quinoa

[11] Combinations of milling and heat, as well as alternative traditional and modern methods, are also being explored in South America and the United States. Small-scale dehulling equipment has been developed and shows potential for village requirements. Information from IDRC.

[12] M.E. Tapia and J.N. Castro. 1968. Digestibilidad de la broza de quinua y cañihua por ovinos mejorados y ovinos no mejorados (chuscos), in *Primera Convención de Quenopodiáceas*, pp. 101–107. Universidad Nacional Técnica del Altiplano, Puno, Peru. Even higher amounts may be possible. Information from S. von Rütte.

[13] Information from R. Carlsson.

[14] More complete discussions of quinoa pests and diseases are contained in Consejo Internacional de Resursos Fitogenéticos (IBPGR), 1981; Tapia et al., 1979; and Risi and Galwey, 1984.

is a staple crop and where many relict cultivars are threatened with imminent extinction. Particular attention should be paid to harsh sites, such as the salt flats (*salares*), where quinoa's strengths are likely to be most obvious.

Further research is needed in the areas of adaptation, culture, and varietal improvement. Some varieties with distinctive flavors that may be preferred for certain uses, such as snacks and cereals, deserve additional consideration. The adaptability of different cultivars to salt stress merits special note.

Standard agronomic research should include investigations of weed control, plant density, crop-rotation sequences, and planting dates. The potential yield of quinoa is possibly even higher than it is for the true cereals, and the plant has an extraordinary response to fertilizer.[15]

For use in mechanized agriculture (notably outside the Andes), types might be developed that are short, unbranched, high in seed-to-stem ratio, and that carry their seedheads above the foliage so the two can be easily separated. Already some types are known that produce a single seedhead on top, and when grown at high density they are relatively branchless.

Processing The toxicity of saponins should be further defined, and their role in pest control and nutrition assessed. Large-seeded, high-saponin types might be the more pest-resistant and the most cost-effective to process. Methods of removing (and utilizing) saponins merit further study.[16]

Agronomy Advanced agricultural equipment used for other crops should be further adapted to quinoa. Agronomic and biological controls of pests (including traditional approaches) should be further evaluated, in conjunction with chemical practices. In particular, resistance to downy mildew needs to be bred into commercial cultivars.

Nutritional Improvement A protein fractionation study has shown that the proportions of albumin, globulin, prolamine, and glutelin vary for different species, and within the fractions, the amino-acid composition varies a lot.[17] Thus, it may be possible to select or breed a super-quality-protein quinoa, as has been done with corn.[18]

---

[15] Information from J. McCamant.
[16] Alcohol extraction is efficient but currently too costly to be practical. If its cost-effectiveness could be improved (for example, if saponins become commercially more valuable), the economic future of quinoa would be assured.
[17] Information from R. Carlsson.
[18] See companion report, *Quality-Protein Maize*. National Research Council. 1988. National Academy Press, Washington, D.C.

SPECIES INFORMATION

Botanical Name   *Chenopodium quinoa* Willdenow
Family   Chenopodiaceae
Common Names
  Quechua: kiuna, quinua, parca
  Aymara: supha, jopa, jupha, juira, aara, ccallapi, vocali
  Chibcha: suba, pasca
  Mapuche: quinhua
  Spanish: quínua, quínoa, quinqua, kinoa, trigrillo, trigo inca, arrocillo, arroz del Peru
  Portuguese: arroz miúdo do Perú, espinafre do Perú, quinoa
  English: quinoa, quinua, kinoa, sweet quinoa, white quinoa, Peruvian rice, Inca rice
  French: ansérine quinoa, riz de Pérou, petit riz de Pérou, quinoa
  Italian: quinua, chinua
  German: Reisspinat, peruanischer Reisspinat, Reismelde, Reis-Gerwacks

Origin. Quinoa was probably domesticated in several locations—perhaps in the Bolivian, Ecuadorian, and Peruvian Andes between 3,000 and 5,000 years ago. Quinoa and potatoes apparently were the staple foods of many ancient highland societies.

Description. Quinoa is an annual, broad-leaved, dicotyledonous herb usually standing about 1–2 m high. The woody central stem carries alternate leaves, generally pubescent, powdery, smooth (rarely) to lobed; it may be either branched or unbranched, depending on variety and sowing density, and may be green, red, or purple. The branching taproot, normally 20-25 cm long, forms a dense web of rootlets that penetrate to about the same depth as the height of the plant.

  The leafy flower clusters (panicles) arise predominantly from the top of the plant and also from leaf junctions (axils) on the stem. The panicles have a central axis from which a secondary axis emerges—either with flowers (amaranthiform), or bearing a tertiary axis carrying the flowers (glomeruliform). The small, clustered flowers have no petals. They are generally bisexual and self-fertilizing.[19]

  The dry, seedlike fruit is an achene about 2 mm in diameter (250–500 seeds per g), enclosed in the dryish, persistent calyx (perigonium) that is the same color as the plant. A hard, shiny, four-layered fruit

---

[19] Separate male and female flowers occur in a few cases. Information from J. Rea.

wall (pericarp) encloses each "seed," and contains 0–6 percent bitter saponins. The seed is usually somewhat flat, and is normally pale yellow, but may vary from almost white through pink, orange, or red to brown and black. The embryo can be up to 60 percent of the seed weight. It forms a ring around the endosperm that loosens when the seed is cooked.

Horticultural Varieties. The cultivated plant shows great variability, and there is an enormous range of diversity. It has thus far defied classification into botanical varieties and, as with corn, the various forms are termed "races" or strains. A classification based on ecotype recognizes five basic categories:[20]

• Valley type. Grown in Andean valleys from 2,000–3,600 m, these are tall, branched, and have long growth periods.
• Altiplano type. Found around Lake Titicaca, these are frost hardy, short, unbranched, and have short growth periods and compact seedheads.
• Salar type. Native to the salt flats in the Bolivian altiplano, these are hardy, adapted to salty, alkaline soil, and have bitter, high-protein seeds.
• Sea level type. Found in southern Chile (mid-height), these are mostly unbranched, long-day plants with yellow, bitter seeds.
• Subtropical type. Located in inter-Andean valleys of Bolivia, these are intense green plants that turn orange at maturity and have small, white or yellow-orange seeds.

The research of the past decades has produced several cultivars, selected and bred for their tolerance to heat and cold, resistance to disease, and for other desirable characteristics. Perhaps the oldest and most widespread of the new varieties are Kancolla and Blanca de Junín (selected in 1950 in Peru) and Sajama (selected in Bolivia in the 1960s). Sajama is particularly interesting as it has large, white seeds, no saponins, and under good conditions will yield 3,000 kg per hectare. In the early 1980s, a new sweet variety was obtained at Cuzco (from Colombian material) and named "Nariño."

Peru and Bolivia have the most extensive collections of these different races, each having over 2,000 ecotype samples. Other collections exist in Chile, Argentina, Ecuador, Colombia, the United States, England, and the Soviet Union.

[20] This is based largely on Tapia et al., 1979, and emphasizes agronomic differences. Gandarillas, emphasizing botanical differences, has recognized at least 17 races (based on inflorescence, plant form, leaf, and seed) whose names have become standard terminology. Other systems have been based on seed color, taste, or other end-product differences.

Environmental Requirements. This plant is highly variable. There is no *one* quinoa, and this rustic crop is more or less a complex of subspecies, varieties, and landraces. However, the following are its general environmental tolerances.

Daylength. Quinoa shows various photoperiod responses, from short-day requirements (for flowering) near the equator to no response in Chile.[21]

Rainfall. 300–1,000 mm.[22] Rainfall conditions vary greatly with variety and country of origin. Southern Chilean varieties get much rain, altiplano varieties get little. As with any grain crop, quinoa grows best with well-distributed rainfall during early growth and dry conditions during maturation and harvest. It can withstand excessive amounts of rainfall during early growth and development; on the other hand, it is notable for its drought tolerance, especially during late growth and seed maturation.[23]

Altitude. Quinoa ranges from sea level in Chile (36°S) and coastal Peru to over 4,000 m in the Andes near the equator. It is grown mainly, however, between 2,500 and 4,000 m.

Low Temperature. Quinoa tolerates a wide range of temperatures. The plant is normally unaffected by light frost (–1°C) at any stage of development, except during flowering. Quinoa flowers are sensitive to frost (the pollen is sterilized), so mid-summer frosts (which do happen in the high Andes) can destroy the crop. Although temperatures below – 1°C damage most types, some hardy types withstand even lower temperatures.

High Temperature. The plant tolerates but does not thrive in temperatures above 35°C.

Soil Type. Quinoa can grow in a wide range of soil acidities, from pH 6 to pH 8.5. It tolerates infertility, moderate salinity, and low base-saturation levels.

[21] Information from H. Wilson.
[22] Traditional Andean agricultural practices, such as microcatchments for water, sometimes allow worthwhile production in areas with as little as 100 mm of rain during the growing season. Information from D. Cusack.
[23] During the severe 1983 drought in Puno, Peru, quinoa was essentially the only grain crop that produced useful yields.

PART III

Legumes

For much of their protein supply, the Incas depended on the common bean, a plant of Central American and Mexican origin. But at high elevations the common bean performs badly—it grows slowly and its seeds take too long to cook because water boils at lower temperatures. As a result, ancient Andean peoples developed their own legumes: the large-seeded lima bean (see opposite), basul, nuñas, and tarwi.

It is important to give more research attention to all of these. While enormous resources have been expended in recent decades on grasses such as rice, wheat, corn, sorghum, and barley, in developing countries especially, the advancement of legumes has lagged. Yet the cultivation of legumes is the most practical and quickest way to augment the production of food proteins.

This section details the little-known Andean legumes basul, nuñas, and tarwi. The underexploited promise of the lima bean is detailed in a companion report.[1] Ahipa and pacay (ice-cream beans) are also legumes, but are handled in the roots and fruits sections of this report (see pages 39 and 277).

All of the Andean legumes described here have shown unusual promise. Basul and pacay are tree crops with exceptional potential for use in reforestation and the reclamation of wasteland. The nuñas are particularly interesting because the kernels burst upon heating, which makes them a bean counterpart of popcorn. This characteristic is especially useful since the grains become edible without the need for grinding or extensive cooking. And tarwi rivals soybean—the world's premier protein crop—in its composition and nutritive value.

---

[1] See companion report, *Tropical Legumes: Resources for the Future*. National Research Council. 1979. National Academy Press, Washington, D.C.

162

THE LIMA BEAN

To provide perspective on the possible future adoption of the under-exploited Andean legumes described in the following section, it seems instructive to consider the success of the one Andean legume that has already achieved worldwide renown.

The modern, large-seeded lima beans (*Phaseolus lunatus*) trace back to Peru. Indeed, they are named for the city where they were probably picked up and distributed throughout the world. Europeans first encountered the plant 400 years ago in the vicinity of what is now Lima.

Large lima beans were well known to the Incas and their predecessors, especially in the river valleys cutting across the coastal desert of Peru. Wild types are found in parts of the Andes from Peru to Argentina. However, like many other species, it may have first been domesticated on the eastern slope of the Andean highlands in warm, humid lands. But, if so, the people on the western side of the Andes learned to appreciate it at an early time. Limas have been found in excavations in coastal Peru dated at 6000–5000 B.C. A small-seeded (sieva) type is found in Central America and Mexico, but the earliest record of it is 500–300 B.C. Thus, there probably were two separate domestications of different lima bean strains; the small type in Central America, the large type in South America.

Exactly how lima beans left the Americas is not known, but since the time of Columbus they have become widely distributed, particularly in the tropics. In fact, they are one of the most widely cultivated pulse crops, both in temperate and subtropical regions. Spanish galleons took the small-seeded type across the Pacific to the Philippines, and from there it spread through Asia. It is now widely grown in Burma, for instance.

Slave traders took limas from Brazil to Africa. At an early date, they reached Madagascar. It is now the main pulse crop in the rain forests of tropical Africa. In many such areas, lima beans have escaped from cultivation and maintain themselves in a wild state. They are exported (under the name "white butter beans") from Madagascar.

The sieva type had already been spread from Mexico and Central America to New England by the time Europeans arrived. The large-seeded Peruvian lima types are known to have been carried by ship, perhaps as a curiosity, to be grown on a farm in New York State in 1824.

Today, both large-seeded lima beans and sievas are an important United States crop, grown for canning, freezing, and production of dried beans. They are also important as a fresh vegetable—commercial production running to more than 100,000 tons each year, chiefly in California.

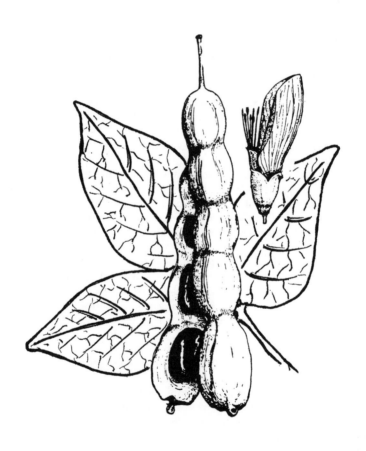

Basul

Agroforestry is gaining widespread recognition and research these days. In this agricultural system, shrubs and trees are grown together with food or plantation crops, and sometimes with livestock as a third partner. Such systems can be highly productive and resistant to perturbation, and they represent a major thrust in the programs emphasizing sustainable agriculture. Indeed, agroforestry is seen as one solution to the fact that the developing world will soon contain 500 million more persons than its current land resources can support.[1]

One of the least known but perhaps most promising candidates for inclusion in agroforestry is basul (*Erythrina edulis*). This smallish tree is native to the Andean region from western Venezuela to southern Bolivia. It is vigorous, fast-growing, and precocious, a pioneer species that colonizes newly cleared sites.

Basul (pronounced bah-*sool*) is not widely known even in the Andes. Yet it is found in many backyard gardens and along property boundaries as a beautiful "living fence." However, since pre-Columbian times it has been grown less for its beauty than for its large, edible seeds. Basul is one of the few trees that provides a basic foodstuff. It is a legume species and its seeds, like those of other legumes (also called beans, grain legumes, or pulses), are important sources of food both for humans and animals. It could be said that basul is the "tree bean of the Andes."

Basul is an important food crop because it grows in areas where seasonal food deficits occur often. In these "famine seasons," its dried seeds are an important nutritional safety net. They are used particularly in the months just before field crops are ready to harvest—a time when the previous year's harvest is often depleted and food is scarce. This tree-bean can then make the difference between health and malnutrition. Beans are rich in protein, and basul seeds complement the starch-rich cereals or root crops that make up the bulk of food consumed by the poor. Moreover, the amino acids in its protein complement those found in cereals and roots.

---

[1] This outcome was predicted to occur by 2000 A.D., assuming present levels of agricultural inputs, in a recent report of the UN Food and Agriculture Organization.

For all its importance as a food, this tree produces more than just beans. Indeed, it is one of the most versatile of all woody species. It supplies lumber for construction, poles for fencing, and wood for fuel. Its nitrogen-rich leaves are fed to animals and are used to mulch the garden. Its edible young flowers are used to decorate and season foods. And, although basul is mostly grown for home consumption, it is used as a cash crop as well—the pods being exchanged among families or sold in markets.

The living trees are also important themselves. Basul is one of the easiest trees to grow. Sections of stem—even large ones—take root readily and become living and long-lasting fence posts. A basul hedge requires little care and, once established, can live for decades. The tree can also be used for shading coffee, cocoa, and other sun-sensitive crops. Vines—such as pepper, betel, and grape—get double benefit because they also use the tree trunks as supports. As a leguminous tree, it supplies nitrogen that fertilizes the soil around it. Because of this it is sometimes called a "nurse tree."

Despite its extreme versatility, basul is so far little known outside the Andes. But—with the desperate need of developing countries for food, forage, firewood, paper, other wood products, reforestation, and erosion control—it is worth much more recognition than it now receives.

PROSPECTS

The Andes. Agronomic research aimed at regularizing the production of basul would greatly enhance its commercial use and might elevate it to one of the most prevalent and useful trees among the region's rural populations. Because of its "off-season" seed production, basul is potentially vital for health and survival in times of food scarcity. Owing to its robustness, it can be grown on unused patches of land as a source of "famine food," "famine feed," and other critical products. Given organized production it could also make some "wastelands" productive.

Other Developing Areas. Any crop that can be raised for home consumption to improve nutrition during times of scarcity is worthy of increased attention, not only in the Andes, but in other parts of the tropics as well. Versatile tree crops that yield food as well as other products are extremely useful in the lives of the rural poor, and offer important advantages to farmers in many developing countries.

Basul is a bean that grows on trees. Its extremely large seeds have a pleasant, slightly sweet flavor and are usually eaten like lima beans. They are also used in candies. (Wilson Popenoe © 1926 National Geographic Society.)

Because bacteria in its root nodules fix atmospheric nitrogen and because its fallen leaves enrich soil, basul is particularly promising for ameliorating the effects of soil impoverishment. It is also promising for use in reforestation, one of the most pressing of all environmental concerns.

Industrialized Regions. For regions outside Africa, Asia, and Latin America, basul is not a promising resource. The tree is killed by frost and has little or no potential for cultivation in North America, Europe, or other temperate regions, nor is there much need for its seeds there.

USES

The large, soft, succulent seeds are usually boiled in water with a little salt and are served as a side dish to corn, cassava, bread, or potatoes. They are sometimes mashed with cheese and are also fried. **They should not be eaten raw.**

The leaves and immature pods of this legume are used to feed cattle, pigs, sheep, guinea pigs, and chickens. In one agroforestry system, basul trees shade the animals while cover crops between the trees provide additional fodder. Meanwhile, the animal droppings help the trees remain productive.

As noted, the basul tree is widely used as a living fence. It is fast growing, stout spined, easily planted, and almost maintenance free.

NUTRITION

The seeds contain about 20 percent protein on a dry-weight basis.[2] Their amino-acid balance is similar to that of other legumes—rich in lysine, for example. Their limiting amino acids are methionine and tryptophan, both of which are also low in other legumes. They have a good balance of inorganic nutrients, particularly phosphorus.

AGRONOMY

Basul can be easily propagated by several methods. Its mature seeds germinate well and the young plants grow vigorously. Superior specimens can easily be propagated by stem cuttings, and, as noted,

[2] Perez et al., 1979.

branches stuck in the ground will sprout. Grafted trees begin producing seeds within 1–2 years of planting.

Small birds are the normal pollinators. The brilliant red flowers provide a rich and abundant nectar, an important food and water source for the birds at certain times of the year.

HARVESTING AND HANDLING

Although basul pods come from trees, they are harvested and processed much like beans. As with most legumes, the pods mature at slightly different times, and several pickings are necessary. However, unlike most plants, basul yields seasonal harvests twice a year. A single tree can yield as much as 200 kg of seed each year.

As the seeds dry out, they pull away from the pod wall and can be readily removed.

LIMITATIONS

Care must be taken to ensure that one is dealing with the right species. In some places the common name "basul" is applied to species other than *Erythrina edulis*. This could lead to a serious mistake, because the seeds of many other *Erythrina* species are poisonous even when cooked. (However, these are small and hard and are unlikely to be confused by knowledgeable people.) As already noted, they cannot be eaten raw.

The seeds are said to be rarely used in soups because they darken the soup and can give it a bitter flavor. (This is perhaps because of interaction with the metal pot.)

All *Erythrina* species are susceptible to insect borers that invade the heartwood. Proper care (and perhaps use of pesticides) can prevent the young trees from being destroyed. As the tree ages it seems to become resistant to the borers.

Basul is susceptible to extended droughts. (Grafting it onto *Erythrina falcata* rootstock greatly increases its drought tolerance.[3])

RESEARCH NEEDS

Among research topics for this species are the following.

Baseline Survey Traditional production methods should be surveyed and analyzed. Topics for investigation include planting density, pest control, pruning, harvesting, and tree maintenance.

[3] Information from I. Peralta V.

Genetic Improvement   Provenance evaluation for high yield, large seed size, pleasing taste, fast growth, and adaptability should be made. Most erythrinas are ' elf-sterile and require cross-pollination, an impediment for high fruit set. Whether basul requires this has not been reported and should be checked.

Nutritional Research   Nutritional trials could help demonstrate dietary importance. Details of amino acid and vitamin compositions are lacking. Toxicological analyses should be conducted on the seeds as a precaution.

Animal Production Trials   All woody plants that provide feed for livestock deserve greater recognition in animal production in tropical regions. Browse shrubs and trees complement (and often benefit) herbaceous pasture species and can be crucial to the nutrition—even the survival—of animals, especially during drought, when shallow-rooted species shrivel to straw. However, cattle are said to eat only limited quantities of erythrina leaves. Exploratory trials using basul are called for.

Basul trees fix nitrogen, and with their protein-rich foliage, pods, and seeds as well as their general robustness, they might enormously benefit developing country reforestation and soil-improvement programs in the future.

SPECIES INFORMATION

Botanical Name   *Erythrina edulis* Triana
Family   Leguminosae (Fabaceae)
Synonym   *Erythrina esculenta*
Common Names
   Spanish: basul, balú, antipurutu, baluy, chachafruto, chafruto, sachafruto, sachapuruto, calú, frísol calú, nopas (Colombia); pajuro (Peru); sachaporoto, sacha purutu (Argentina, Bolivia)

Origin.   Unknown, although the seeds are found in early burial sites. Basul is a semidomesticate, and wild forms are abundant at the transition zone between highland and forest.

Description.   Basul is a tree 8–10 m tall with trifoliate leaves. The trunk bears stout, conical spines, and the young branches are thorny. The two-petaled, red, fleshy flowers face upward, forming a large cup in which nectar gathers.

The 20–30 cm long seed pod is greenish purple, leathery, and spongy. It is smooth and nearly cylindrical, with constrictions between the 1–10 large, light-brown, glossy seeds (each 2.5–3.5 cm in diameter).

Horticultural Varieties. None.

Environmental Requirements

Daylength. Basul's requirements are unknown, but other members of the genus set seed to the limits of the subtropics.

Rainfall. 450–1,800 mm.

Altitude. Generally grown between 1,800 and 2,200 m in the central Andes; between 1,100 and 2,700 m in Colombia.

Low Temperature. Unknown, but probably about 5°C.

High Temperature. Unknown.

Soil Types. Apparently widely adaptable.

Nuñas (Popping Beans)

Like most present day Latin Americans, the Incas and their ancestors depended on boiled beans (*frijoles*) for much of their nourishment. Many lived so high in the mountains, however, that they couldn't cook ordinary dry beans—water boils at too low a temperature up there.[1] To circumvent this, they used a remarkable bean known as the nuña.

Nuñas (pronounced *noon*-yas) are a type of the common bean (*Phaseolus vulgaris*); they are roughly the bean counterpart of popcorn. Heated with a little oil, nuñas burst out of their seed coats. The effect is less dramatic than popping popcorn—nuñas don't fly in the air; they open like small butterflies spreading their wings.[2] The resulting product is soft and tastes somewhat like roasted peanuts.

Nuñas look much like common beans, but they are hard shelled. They come in many striking colors and patterns: white, red, and black-spotted, for instance. During cooking, the heat and moisture build up steam inside, and the hard shell and round shape mean that it can escape only by bursting out.

From Ecuador to southern Peru, nuñas are grown above 2,500 m altitude. They are produced mainly for home consumption and are much more common in houses than in markets. Nonetheless, they are often sold as part of a mixture of beans to be incorporated into soups.

That nuñas are unknown outside the Andes seems surprising. For industrialized nations, these popping beans could be a new and nutritious snack food. For developing nations, they could be a tasty source of high-quality protein. Toasting nuñas requires far less fuel than boiling beans, an important economic (and environmental) consideration in regions where fuel is scarce. In addition to having nutritional and energy-saving attributes, the plant is a nitrogen-fixing

[1] Common beans can be cooked at high elevation; it simply takes a long time and a lot of fuel, which is usually difficult to come by in the treeless uplands. Even at low altitudes, some varieties take hours to cook properly.
[2] Strictly speaking, they don't pop. They burst, and are more like toasted maize (cancha or chulpi), which is best known outside the Andes under the brand name "CornNuts."

Nuñas are normally cooked like popcorn. The seed shown was heated with hot air and exploded to double its size in under 2 minutes. This seed had been collected in Peru and stored in Pullman, Washington, for more than 10 years. The fact that nuñas burst normally is an indication that popping may not be restricted to fresh seeds nor to high altitudes. (S.C. Spaeth)

legume that benefits the soil in which it is grown and that is well suited to interplanting with other crops such as corn.

Although they have real potential, these popping beans are probably unknown elsewhere because they apparently have daylength requirements that at present seem to restrict their cultivation to equatorial latitudes. Research to overcome this could give the world a large and fascinating new crop.

PROSPECTS

Andean Region. Nuña cultivation is already widespread from at least northern Ecuador to northern Bolivia, although distribution is discontinuous and in many places the plant is unfamiliar. With research and promotion, this bean is likely to be more widely adopted throughout the mountain region, as well as in the lowlands. It is already widely used by some urban mestizos, and in both highland towns and some coastal cities it sells at prices similar to those obtained for the most favored bean cultivars.[3]

[3] Nuñas are most often found in "serrano" (Andean Indian) markets, such as the Mercado Mayorista in Lima, that are frequented by Indians from the highlands.

Other Developing Areas. For many Third World areas, the nuña's greatest inherent quality is its relatively low requirements for cooking fuel—a critical consideration where deforestation is acute or kerosene expensive.

This crop's close relationship to the common bean suggests that it could grow well and be readily adapted in regions far beyond its present range. So far, however, it seems best adapted to the montane tropics. Future research will likely give it broader adaptability.

Industrialized Regions. Nuñas could be a new and nutritious snack food with potential for North America, Europe, Japan, and other industrialized areas. This situation is analogous to the discovery of popcorn or roasted peanuts by the modern world. This remains to be tested, however. Some crops from very high elevations—and nuñas may be one—are highly restricted in their adaptive range. Although there is a possibility that they require the high light intensity and, perhaps, the high altitude to retain their popping quality, this is probably not the case (see later).

USES

In the central and northern Andes of Peru, nuñas are prepared in traditional ways much like popcorn. They are toasted for 5–10 minutes in a hot frying pan, which is usually coated with vegetable oil or animal fat.[4] The seed coat splits in two or more places, often between the cotyledons. The toasted product is served at main meals as a side dish, is eaten as a snack, and—in southern Peru—is often sold to tourists.[5]

NUTRITION

Nutrient levels are high and similar to those of the common bean. The protein content is about 22 percent.[6]

AGRONOMY

Agronomic techniques are the same as for common beans. Because of its viny habit, the nuña plant in the Andes is almost always interplanted with corn so it can climb on the stalks.

[4] They work well in microwave ovens, too. Information from L. Sumar.
[5] For instance, nuñas are common at the Pachar and Ollantaytambo stops of the Machu Picchu tourist train.
[6] Information from V. Ortiz.

Out-crossing contributes to a small gene flow, resulting in seeds with a gradient of "nuña-ness." Because of this, the capacity to pop can range from 30 to 90 percent of a given batch. To prevent cross-pollination, nuñas are grown in different parts of the field from common dry beans and string beans.

HARVESTING AND HANDLING

In the highlands, harvests occur 5–9 months after planting. However, at 25°C mean temperature, nuñas mature in about 80 days.[7] Yields appear to be similar to those of other traditional varieties of common bean.

LIMITATIONS

There has been little modern genetic improvement of this crop. Its yields can be erratic and, when compared with other bean varieties, it is relatively susceptible to pests and diseases.[8]

The presence of flatulence-producing factors could limit its acceptance. Severity, however, varies greatly with the strain, its preparation, and probably with the other foods with which it is eaten.

RESEARCH NEEDS

For a crop on which little recent research has been expended, almost everything about nuñas needs testing and assessment.[9] One basic need is ethnobotanical investigations of the plant's production and use in the Andes. Particular consideration should be given to its cultivation and agronomic characteristics, as well as to the ways in which it is prepared.

In addition, research is needed to determine the plant's range and ecological requirements, and to identify and protect different types of available germplasm, much of which has been lost already.

[7] Information from J. White.
[8] The Centro Internacional de Agricultura Tropical (CIAT) in Cali, Colombia, is now actively trying to introduce resistance to important diseases (such as bean common mosaic virus and anthracnose), a day-neutral photoperiod response, and a bush growth habit. Information from J. White.
[9] In the 1950s and 1960s, a program at the Universidad Nacional Agraria, La Molina, Peru, collected nuñas extensively in the Peruvian highlands and subjected the different types to tests. Research reports of this program, which was later dropped, should be republished.

It is important to understand what controls the popping qualities. The unique properties that enable toasting are now unknown. They may include seed shape, inelastic seed coat, or the quality or quantity of stored starch.

A test is needed that can determine whether a certain round bean is in fact a nuña. At the moment, there is no certain recognition, and developing such a test is necessary both for the national marketing of the crop and for understanding the genetic control of the nuña's intrinsic characteristics.

A major step in promoting nuñas is obtaining basic information on the popped product. Analysis of nutrients and biochemistry of the

NUÑAS IN THE UNITED STATES

In the past, the few researchers who knew about nuñas were uncertain whether these popping beans could succeed outside the high Andes. They thought that the plant might grow only at high altitudes or equatorial latitudes. They also feared that the popping character of the beans might disappear even if the plant could be grown outside its equatorial highland home. In addition, some expressed the opinion that the capacity to pop might occur only in newly harvested and sun-dried seeds. Old nuñas, it is rumored in the Andes, will not pop.

Now, however, it seems that these fears are unfounded. In 1978, the U.S. Department of Agriculture (to replenish germplasm in its seed bank) grew 14-year-old nuñas successfully in Pullman, Washington. The site is in the temperate zone at a high latitude (47°N) and low altitude (200 m).

In 1988, Stephen C. Spaeth, a USDA researcher, tested some of the seeds and found that even after 10 years of storage (at 4°C) they had lost none of their popping characteristics—they showed good "nuña-ness" by exploding to double their size after only 90 seconds on a hot-air gun (see page 174). Indeed, these 10-year-old "temperate-grown" beans popped just as well as nuñas newly harvested in Colombia.

Although the nuñas plant is now well known only in a relatively small region of the world, these results suggest that in future it may become a widely available, nutritious, tasty, and fuel-conserving food. Only its requirement for short daylength seems likely to initially limit its spread.

seeds is needed, including protein content, amino acid composition and nutritional availability, carbohydrate content and composition, fat content and fatty acid makeup, energy availability, and functional characteristics. Because of its radically different style of preparation from normal beans, it is important that the percentage of utilizable protein be assessed. (Antimetabolic factors may be less deactivated during a few minutes of toasting than in several hours of boiling, although, with the higher temperature, this seems improbable.)

Collections of seed should be made at the southernmost limits of nuña cultivation, and trials to identify types adapted for long daylength should be set up. In addition, any other methods for circumventing daylength limitations should be investigated.[10] Successful trials are likely to be the key to unlocking the nuña's global potential. However, breeding for other traits is also needed because the plants are currently late-maturing, tall and weak, and susceptible to diseases such as anthracnose.

SPECIES INFORMATION

Botanical Name   *Phaseolus vulgaris* Linnaeus
Family   Leguminosae (Fabaceae)
Common Names
   Quechua: ñuñas (Cajamarca, La Libertad, Trujillo, Lima); numia (Huanuco), nambia (Ancash), nudia and hudia (Cuzco), kopuro (Bolivia); chuvi, poroto, purutu, porotillo
   Spanish: nuñas
   English: nuñas, popping beans, popbeans

Origin.   Observations of ancient beans discovered at the Guitarrero Cave in Ancash, Peru, indicate that nuñas may have been available 11,000 years ago.[11] Thus, nuñas existed well before the Incas, and perhaps before the common (frijol) type, itself. Because of their ancient beginnings, nuñas have been called "a kind of witness of the first steps of plant domestication."

Description.   The morphology of the nuña plant is identical to that of the common bean. It is an indeterminate, climbing vine (2–3 m tall) that produces a large number of pods from abundant flowers that are primarily self-fertilized. Bush types may exist, but are unreported.

Like the common bean, each pod contains 5–7 seeds. Most seeds

[10] Work at CIAT has shown that photoperiod sensitivity may be overcome through use of 8-hour photoperiods. Information from J. White.
[11] For further information, see Kaplan and Kaplan, 1988.

are nearly spherical (occasionally oval), and range between 0.5 and 0.9 cm in diameter. Different strains have diverse coloration—white, yellow, gray, blue, purple, red, brown, black, and mixed.

Horticultural Strains. In the northern Peruvian Andes, where nuñas are almost a staple, the most common color forms marketed are gray and white speckled (nuña *pava*), light red (nuña *mani*), dark blue (nuña *azul*) and gray (nuña *ploma*). Within the vicinity of Cajabamba (near Cajamarca) and Huamachuco (La Libertad), there are scores of distinct types differing in seed size, shape, and color.

Although there are no discernible differences in taste between different types of nuñas, there is variation in the capacity to "pop." This quality is recognized by farmers and consumers: strains that pop the best are valued the most in the markets. The nuña *pava* (also called "*coneja*") is held in particularly high regard.[12] There is a white nuña at Cajabamba called "*huevo de paloma*" (pigeon egg), which is outstanding in popping ability, taste, and crunchiness. In Cajabamba there is also a red and white mottled nuña called "*parcollana.*"[13]

In the southern Andes of Peru, several strains of *Phaseolus vulgaris* appear to be closely related to the nuña. These are usually roasted (15–30 minutes) in either gypsum pebbles (*pachas*) or sand, rather than oil. They are called "*poroto de Puno*" and are found only in certain valleys, certain markets, and on special days. They do not pop, but the seed coat breaks open, and the shell comes off to leave a very dry, but tasty, product.

Environmental Requirements[14]

Daylength. As noted, certain nuña types are highly photoperiod sensitive. This sensitivity increases with growing temperature.

Rainfall. 500–1,300 mm throughout the growing season.

Altitude. 1,800–3,000 m in Peru

Low Temperature. 2–5°C; frost susceptible

High Temperature. About 25°C; may be intolerant of even moderately hot conditions.

Soil Type. As with modern beans. Nitrogen-fixing is more effective in light, well-drained soils because of better *Rhizobium* growth.

---

[12] The center of production of this nuña is the Citacocha district of southern Cajamarca.
[13] Information from J. Risi.
[14] Nuñas have seldom been grown outside the central Andes, and the outer limits of the plant's environmental requirements are unknown. The figures here are, therefore, indicative, but not definitive.

Tarwi

On the face of it, it is surprising that tarwi (*Lupinus mutabilis*) has not been developed as an international crop. Its seeds contain more than 40 percent protein—as much as or more than peas, beans, soybeans, and peanuts—the world's premier protein crops. In addition, its seeds contain almost 20 percent oil—as much as soybeans and several other oilseed crops. Tarwi[1] thus would appear to be a ready source of protein for food and feed as well as a good source of vegetable oil for cooking, margarine, and other processed food products.

One of the most beautiful food crops, tarwi (pronounced *tar*-wee) could also qualify as an ornamental. Its brilliant blue blossoms bespangle the upland fields of the Indians of Peru, Bolivia, and Ecuador. Indeed, corn, potato, quinoa, and tarwi together form the basis of the highland Indian's diet. In Cuzco, the former Inca capital, baskets of the usually bone-white tarwi seeds are a customary sight in the markets. The seeds are most often served in soups.

Tarwi seeds are outstandingly nutritious. The protein they contain is rich in lysine, the nutritionally vital amino acid. Mixing tarwi and cereals makes a food that, in its balance of amino acids, is almost ideal for humans. With its outstanding composition, tarwi might become another "soybean" in importance.[2] Because of this possibility, researchers in countries as far-flung as Peru, Chile, Mexico, England, the Soviet Union, Poland, East and West Germany, South Africa, and Australia have initiated tarwi research.

This "pioneer" species can be cultivated on marginal soils. Its strong taproot loosens soil and (because it is a legume) its surface roots collect nitrogen from the air. Both of these abilities benefit the land in which it is grown.

[1] Also widely known as "*chocho.*" At a 1986 conference of the International Lupin Association, the name "Andean lupin" was proposed for international use.
[2] This is not as improbable as it seems. Sixty years ago the soybean was hardly known outside Asia; today it is America's third largest crop and is a vital part of the economies of Brazil and several other non-Asian nations. Like tarwi it requires processing to rid its seeds of adverse components.

Despite all its qualities, tarwi is almost unknown as a crop outside the Andes. It has been held back mainly because its seeds are bitter. However, this can be overcome. The bitter principles (alkaloids) are water soluble, and they are traditionally removed by soaking the seeds for several days in running water. Technology now has modernized the process to the point where it can be done in a matter of hours. Also, in another approach to removing the bitterness, geneticists in several countries have created "sweet" varieties whose seeds are almost free of bitterness and need little or no washing. These efforts would appear to pave the way for a future role for tarwi in world agriculture.

PROSPECTS

Andean Region. Tarwi is found from Venezuela to northern Chile and Argentina, and in this area it is already getting modern attention. Engineers in Peru and Chile have developed machinery to debitter tarwi seeds. Indeed, small industrial installations are now operating at Cuzco and Huancayo in Peru.[3] Their product is being used to feed schoolchildren and to produce a line of cereals. Also, researchers in Chile and Bolivia have created varieties with only one-thousandth of the alkaloid levels found in bitter types.

Given these advances, tarwi cultivation and use should expand in the Andes. This is significant because the crop is an important contributor to the nutritional well-being of many campesinos for whom meat is a luxury. The highland diet is low in protein and calories, and the quality protein and high oil content of tarwi seed provide a double nutritional benefit.

Other Developing Areas. The lengthy process of washing the seeds has previously hindered tarwi's introduction to areas outside the Andes. However, the sweet types could make the plant into a major crop for tropical highlands and for a number of temperate regions. So far it is barely known outside of South America, but a hopeful sign is that in Chapingo, Mexico, tarwi has produced high yields of seed.[4]

---

[3] These pilot plants are capable of processing 7,000 tons of seed per year into vegetable oil and plant protein. The German government organization GTZ (Deutsche Gesellschaft für Technische Zusammenarbeit, Dag-Hammarskjold-Weg 1, D-6236 Eschborn 1) has spearheaded support for the investigation and development of tarwi.
[4] Information from J. Etchevers.

Field of tarwi high in the Peruvian Andes, near Chiara, Department of Cuzco. (D.W. Gade)

Industrialized Regions. This adaptable plant will flower both in the short days of the tropics and in the long summer days of the temperate zones. As noted, it has already been grown experimentally in Europe, South Africa, and Australia. Current types mature late in temperate latitudes, but a diligent search of the native germplasm in the Andes will likely turn up quick-maturing forms.

USES

As has been mentioned, tarwi appears to be a ready source of vegetable protein and vegetable oil for both humans and animals. It is also suitable for processed food products, high-protein meal for food and feed, and margarine. In the Andes, the cooked seeds are popular in soups, stews, and salads, or are eaten as snacks, like peanuts or popcorn. The soft seed coat makes for easy cooking.

Like other lupins (for example, the "lupini beans" of Italy and the white lupins of Eastern Europe), tarwi is an excellent green manure

Bowls of washed tarwi seeds are a common sight in markets throughout the Andes. Although unknown elsewhere, these seeds have a composition roughly comparable to that of soybeans, one of the world's premier crops. (H. Brücher)

crop, able to fix as much as 400 kg of nitrogen per hectare. Much of the nitrogenous product remains in the soil and becomes available to succeeding crops. With the high prices and shortages of this essential fertilizer element, tarwi could become increasingly important in crop-rotation systems.[5]

NUTRITION

As noted, the seeds are exceptionally nutritious. Protein and oil make up more than half their weight. In a survey of seed from more than 300 different genotypes, protein content varied from 41 to 51 percent (average 46 percent); oil content varied—in roughly inverse proportion—from 24 to 14 percent (average 20 percent).[6] Removing the seed coat and grinding the remaining kernel yields a flour that contains more than 50 percent protein.

Tarwi protein has adequate amounts of the essential amino acids lysine and cystine, but has only 25–30 percent of the methionine required to support optimal growth in animals. The protein digestibility and nutritional value are reportedly equivalent to those of soybean.[7]

Tarwi oil is light colored and acceptable for kitchen use. It is roughly equivalent to peanut oil, and is relatively rich in unsaturated fatty acids, including the nutritionally essential linoleic acid.

The seeds' fiber content is not excessive, and they are thought to be good sources of nutritionally important minerals.[8]

AGRONOMY

Like most Andean crops, tarwi is hardy and adaptable. It is easily planted and tolerates frost, drought, a wide range of soils, and many pests. Its soft-skinned seeds germinate rapidly, producing vigorous, rapid-growing seedlings.[9] Robust vegetative growth continues throughout the growing season and the plants become masses of foliage topped by showy purplish-blue flowers.[10] Most fields, however, remain under 1 m tall and at the end of the season bear many tiers of pods, held high above the leaves. Each pod contains the beanlike seeds that are white, speckled, mottled, or black.

---

[5] For centuries, tarwi has been an important component in the traditional crop rotations of the Andes. It is likely that tarwi's alkaloids are a factor in controlling potato nematodes when the two crops are rotated. Information from M. Tapia.
[6] Information from R. Gross.
[7] Ortiz et al., 1975.
[8] Pakendorf et al., 1973.
[9] However, the young plant often remains in a state of suspended growth for several weeks, during which time it is very susceptible to pests.
[10] On Taquile Island in Lake Titicaca, there is a variety with rose-colored flowers. Information from T. Plowman.

Reportedly, the green-matter production considerably exceeds that of the commercial European lupin species (*Lupinus albus* and *L. luteus*). In experiments in the Soviet Union, the plant produced 50 tons per hectare of green matter containing 1.75 tons of protein per hectare.[11]

Many forms of the plant resist the lupinosis fungus, which sometimes kills livestock that feed on the foliage of other lupin species.[12] Also, many ecotypes are resistant to lupin mildews and rots.[13]

HARVESTING AND HANDLING

Unlike many lupin species, tarwi pods do not split and shed their seed on the ground. Ancient Indian cultivators probably selected plants that held their seed until they could be harvested.

As noted, the water-soluble alkaloids must be washed out before the seeds can be eaten.

LIMITATIONS

As already stated, tarwi's most serious known liability is the alkaloids in its seeds. Apart from that, its long vegetative cycle during and after flowering is a major limitation.

Because there has been little agronomic improvement of tarwi, the available cultivars are primitive. Most have indeterminate growth and produce multiple tiers of flowers. Unless drought or cold causes the plants to "dry down," they keep on flowering endlessly. Continuous flowering boosts yields and can be a good thing for a small farmer because it means a continuing source of food. However, it also means that pods ripen at different times, and for a large-scale farmer this greatly hinders harvesting, especially if machinery is used.

Also, current tarwi cultivars require a growing season as long as 5–11 months to fully ripen their seeds. In temperate zones, this means that the seeds do not always ripen before the onset of winter. In the Andean highlands, the drought season starts as the plants begin ripening, thus bringing the flowering to a natural end.

Compared with other commercial lupin crops (for example, *Lupinus angustifolius* or *L. luteus*), tarwi has a lower leaf to stem ratio, so that it produces fewer seeds than would be expected from such a mass of vegetation.

---

[11] Brücher, 1968.
[12] Van Jaarsveld and Knox-Davies, 1974.
[13] These ecotypes are not yet in intensive cultivation, however. Information from K.W. Pakendorf.

Tarwi cross-pollinates so readily (the percentage of outcrossing may exceed 10 percent) that to preserve specific cultigens, such as low-alkaloid types, may require a special system of seed production and distribution. This factor particularly limits the increase of "sweet" tarwi in the Andes because of the recessive inheritance of the genes for low-alkaloid types and because bitter types are always nearby. Elsewhere (for example, in parts of North America), wild lupines[14] may be a source of pollen pollution.

Current types are particularly sensitive to alternaria, a fungal disease that destroyed a large area of the crop in Peru in one recent year.

RESEARCH NEEDS

Eliminating problems caused by the bitter alkaloids would help tarwi advance as a world crop.[15] Research is needed into improving the technology of debittering the seeds on a large scale. Also, the nonbitter varieties should be advanced to commercialization.

From the research already completed, it seems that strains with almost no alkaloids are available in nature or can be created artificially. The challenge now is to make them stable, so that the low alkaloid content is inherited uniformly by succeeding generations.[16] Also, the initial low-alkaloid strains have been proven highly susceptible to insect attack. Breeding programs should seek plants that have alkaloids in the leaves but not in the seeds.[17]

Research to boost yield is also needed. Today, many flowers fail to set seed. Studies of pollination and fertility could point the way to helping the plant to approach its potential. (The failure to set seed is a characteristic of all lupins, and research on other species may also benefit tarwi.)

In any breeding program, high priority should be given to selecting early-maturing varieties for areas where growing seasons are short (for example, temperate latitudes and semiarid areas with short rainy seasons). Early-maturing types may also suffer less damage from pests

[14] "Lupine" is the standard American spelling; "lupin" is standard elswhere. We suggest that the former be reserved for undomesticated species and the latter for the domesticated species.
[15] However, it may not be vital. Several staple foods require elaborate processing before they are safe to eat. Cassava is an example.
[16] This has been achieved by E. von Baer (see Research Contacts). The stable, sweet line, called "inti," has been created through successive breeding that has reduced the alkaloid content to a level below 0.003 percent while maintaining 51 percent protein and 16 percent oil. The seeds are small, however, and the yields lower than normal. Both problems seem likely to be overcome soon.
[17] A method that uses specially prepared reagent paper to screen for the presence of alkaloids is already available. Information from E. Nowacki.

and diseases. In addition, selection for mutants with a single short flowering period is needed. (Such determinate types have been located in southern Peru and Bolivia and probably can also be found elsewhere in the Andes.[18]) Synchronous ripening would be particularly useful in many locations dependent on mechanical harvesting.

Adaptability trials should be conducted in different parts of the world. Other cultivated lupins are fairly specific in their temperature and soil requirements; tarwi, too, might prove to have limited adaptability.

Practical tests of the seeds in food products should be undertaken. This is likely to generate a demand that will stimulate commercial production, especially in the high Andes.

SPECIES INFORMATION

Botanical Name   *Lupinus mutabilis* Sweet
Family   Leguminosae (Fabaceae)
Common Names
   Quechua: tarwi
   Aymara: tauri
   Spanish: altramuz (Spain), chocho (Ecuador and northern Peru), tarhui (southern Peru and Bolivia), chuchus muti (Bolivia)
   English: tarwi, pearl lupin, Andean lupin

Origin.   Pre-Inca people domesticated this lupin more than 1,500 years ago, and it became a significant protein contributor to the region's food supply. It provides a common motif on both ancient and modern ceramics and weavings.

Description.   Tarwi is an erect annual, growing 1–2.5 m tall, with a hollow, highly branched stem and short taproot. The showy, multicolored purple to blue flowers (each with a yellowish spot) are held high above the digitate leaves. To attract pollinating insects, the flowers exude a honeylike aroma.

The hairy, 5–10 cm long pods are flattened, about 2 cm across, and contain 2–6 (or more) ovoid seeds 0.6–1.0 cm across.

Horticultural Varieties.   Many ecotypes and landraces exist throughout the central Andes. As mentioned, South American researchers have begun selecting for higher, more uniform yields and less bitter seeds. However, standardized varieties are not yet available.

---

[18] Information from E. von Baer.

Germplasm collections are maintained at the Universidad del Cuzco in Peru; at the Estación Gorbea in Chile; at the Institute of Plant Genetics in Poznan, Poland; in East Germany; and in the U.S.S.R.

Environmental Requirements. The plant is native to tropical latitudes (from 1°N to 22°S) but occurs mainly in cool valleys and basins at high altitudes. Thus it is a crop for cool climates (tropical highlands and temperate regions), not for the humid or arid tropics.

Daylength. Apparently neutral. Tarwi will flower and set seed both in the short (12-hour) tropical days and in the longer summer days in temperate zones.

Rainfall. The limits are unknown, but tarwi withstands exceptional levels of drought.

Altitude. From Colombia to Bolivia this species grows in the Andes at altitudes from 800 m to well over 3,000 m. In Australia, Europe, and California, it has been grown at or near sea level.

Low Temperature. Tarwi is semihardy. Mature plants are frost resistant; young plants are frost sensitive.

High Temperature. Unknown.

Soil Type. The plant is tolerant of sandy and acid soils, but in acid soils the production of rhizobium is very poor.

Related Species. The genus *Lupinus* is very diverse, with more than 100 different species in the New World and a smaller number in the Mediterranean region. Other than tarwi, all agriculturally important lupins derive from Mediterranean species. These, too, have important global promise.

The seed of Mediterranean lupins were rendered free of toxic alkaloids in the late 1920s and 1930s by the German researcher R. von Sengbusch, who isolated low-alkaloid ("sweet") strains. These are now used as feed and fodder in Europe (especially the Soviet Union and Poland), the United States, Australia, and South Africa.

Of particular note is the narrowleaf lupin (*Lupinus angustifolius*). Nonbitter types with soft seed coats were discovered in Germany in the 1920s. Through 20 years of dedicated selection, the Australian scientist John S. Gladstones developed an early-maturing, sweet-seeded, nonshattering type. It is now widely planted in Western Australia as livestock feed, and its grain is exported to Europe.

PART IV

Vegetables

Fresh vegetables are increasingly popular today owing to a growing appreciation of their importance in human nutrition. They supply vitamins, minerals, trace elements, dietary fiber, and some protein. They tend to be particularly rich in vitamins A and C. However, people are also consuming more because vegetables provide the diet with variety, flavor, and zest.

Nevertheless, millions of people, particularly in the tropics, do not get enough. Indeed, the lack of fresh vegetables is often so serious in the tropical diet that it contributes to malnutrition. Vegetables are especially important for providing much-needed vitamins and minerals to malnourished children in particular, and the current lack of them is a serious concern. For example, lack of vitamin A—an abundant ingredient in many colored vegetables—is the world's major nutritional deficiency and the leading cause of blindness in children in Africa and elsewhere.

Vegetables are good for both subsistence and commercial use. They produce the most food per area planted and they grow quickly. There is every indication that their production will increase in importance, especially for a large proportion of the world's neediest people.

This section describes two groups of promising, but little-known, Andean native vegetables: peppers and squashes and their relatives. These provided the Incas and their predecessors with food, and they show outstanding promise to improve the nutrition of peoples in many parts of the world today. Both squashes and peppers are popular throughout the world, but the species described here are unknown outside the Andes.[1] They offer good nutrition, many new tastes, and potential new crops for scores of countries.

---

[1] Botanically speaking, peppers and squashes are fruits, but they are not normally used as dessert fruits, and in this book we have collected them in this section on vegetables.

HOW THE TOMATO SUCCEEDED

To provide perspective on the possible future for the vegetables described in the following section, it is helpful to consider the unusual development of another vegetable with an Andean origin, the tomato.

The tomato derives from a genus of weedy Andean plants with red, orange, or green berries of currant to cherry size. But although ancient graves have yielded remnants of dozens of different native Andean food crops, nothing indicates that tomatoes were ever cultivated for food in their ancient homeland. No samples or pottery depictions have been found.

However, although it was not a food of the Incas, by the time of the Spanish Conquest the cherry tomato had reached Mexico and apparently was being cultivated and eaten there, at least in a small way. Indeed, the plant's common name derives from the Mexican (actually Nahuatl) word *"tomatl."*

The tomato apparently reached Europe in 1523. However, for at least another century, it remained largely unappreciated. Although by 1600 it had spread throughout Europe, almost everywhere it was regarded as toxic and as a mere curiosity, the *"pomme d'amour"* or "love apple."*

The first tomato seeds to cross the Atlantic undoubtedly went to Spain and were of yellow-fruited varieties.** It is thought that they were quickly passed on to Italy, probably through the kingdom of Naples, which had come under Spanish rule in 1522. Italians were the first people anywhere to show real enthusiasm for the tomato as a food. It was in Italy that the large-fruited tomatoes of commerce first gained acceptance. Eventually, this "nonfood of the Incas" became synonymous with Italian cuisine—the base for sauces to go on pastas from lasagna to linguine.

Although Italians eagerly accepted the plant, northern Europeans stubbornly resisted. They started consuming tomatoes on a large scale only in the second half of the *last* century. To them, the smell of the plant's foliage was said to be as revolting as the thought that southern Europeans would eat the fruit.

European voyagers spread the tomato to Southeast Asia before 1650 and to North America by about the time of the American Revolution. Thomas Jefferson had tomatoes in his garden by 1781. For another

---

* "The whole plant is of a ranke and stinking savour," said John Gerard (*The Herball or Generall Historie of Plants*, 1597; reprinted 1984, Apt. Bks., Inc., New York) under the heading "Apples of Love." Pierandrea Mattioli described it as *mala insansa*, "unhealthy apple."
** In France, Olivier de Serres, agronomist under Henry IV, wrote that "love apples are marvelous and golden." His enthusiasm, however, was not necessarily for their taste, for, he continued, "they serve commonly to cover outhouses and arbors."

century, however, Americans regarded the plant with suspicion. Even into the 1900s there was much doubt about possible toxicity and adverse health effects. To eat a raw tomato was generally believed to be suicidal. Only by prolonged cooking, it was widely reported, could the tomato's venom be neutralized. In 1860, the bible of the American housewife, *Godey's Lady's Book*, cautioned that tomatoes should "always be cooked for three hours."

Only relatively recently has the tomato become a major world food. It was in commercial production late last century in the United States but came into intensive production only about the time of World War I. Even today, its acceptance in tropical Asia and Africa is not great, but the tomato is still continuing its diaspora. In just the last 20 years, it has taken hold so strongly in China that it is now the "number one" vegetable there. Despite conservative culinary traditions and a wealth of native vegetable crops, the tomato is becoming increasingly important in the Chinese diet. (A special kind of tomato sauce has long been known the world over by the name "ketchup," derived from the Chinese word *koechiap*, meaning brine of pickled fish.)

Regardless of its slow beginnings, the tomato is now one of the top 30 food crops of the world. Nearly 20 million tons are produced annually, most in Europe and North America. The United States now not only grows a big proportion of the world's tomatoes, it is the leading consumer, followed by Italy, Spain, the Arab nations, Brazil, Japan, and Mexico.

The Modern Tomato's Mixed Parentage

Today's tomato is more than just the one species *Lycopersicon esculentum*. Plant breeders have engineered it to meet modern requirements by incorporating genes from many of its wild relatives. For this purpose, plant explorers have gathered wild tomato seeds, especially in the Andes. One such collection was made in 1962 when two young botanists, Hugh Iltis and Donald Ugent, were studying the wild potatoes of the dry valleys near Abancay, Peru. Eating lunch on a rocky mountain slope, they picked the fruits from a scraggly wild tomato plant growing nearby. The fruits were green and only the size of marbles, but they helped make a tasty meal. Although not involved in tomato studies, the two plant taxonomists saved the seed and later mailed it to renowned tomato breeder Charles Rick.

In his University of California plots, Rick planted the seeds and discovered that the plant was new to science. It was named *Lycopersicon chmielewskii* in honor of a deceased Polish scientist and fellow tomato breeder. Rick soon noticed that the tiny fruits of this new tomato had a very high sugar content (11.5 percent)—almost twice the normal level. During almost a decade of cross-breeding, he transferred genes for high sugar content into horticultural lines of the common tomato. The result was large, red tomatoes with unusual sweetness and flavor. The content

The Inca's sacred valley, the Vilcanota. Typically, Andean crops have been grown on mountainsides, and many varieties have evolved to fit the diverse environments occurring between the broad valley floors and the tiny terraces at the topmost heights. (N. Vietmeyer)

Inca crops in a market in Ipiales, Colombia. These staples of a time-honored diet include arracacha (in foreground), ulluco (pink vegetables), potatoes, and oca (white crop in rear). (C. Sperling)

Pacay. One of the most unusual trees, the pacay and its botanical relatives produce giant pods filled with a white pulp that is smooth and sweet. For this reason, these pods are sometimes called "ice-cream beans" (see page 277). (W.H. Hodge)

Goldenberries. Also widely known as cape gooseberries, these are common wayside fruits of the Andes. A few countries have begun commercially producing these tangy fruits that come in their own "paper" husks (see page 241). (Turners and Growers)

Peppers. Although one Andean pepper, the chile, dominates the cuisines of many lands, the rocoto (background) and Andean ají (foreground) are two peppers that still remain to be discovered by the rest of the world (see page 195). (N. Vietmeyer)

Nuñas, Common beans are among the world's major foods, but the special variety called "nuñas" remains unexploited outside the Andes. When heated, these nuñas pop, somewhat like popcorn. They are a tasty, nutritious, quick-cooking food with much future promise (see page 173). (J. Kucharski, U.S. Department of Agriculture)

Cherimoya. Enjoyed by all who taste it, the cherimoya (see page 229), given research, could become a major fruit around the world. (A. Rokach)

Naranjilla. This yellow relative of the tomato produces a green juice that is one of the culinary delights of Ecuador and the northern Andes (see page 267). With fruit juices in rising demand, this could become a major international product, but the crop first needs research attention. (W.H. Hodge)

Tamarillo. Another tomato relative, the tamarillo grows on trees. It has a sharp tangy flavor, quite unlike its well-known cousin. Tamarillos are beginning to enter international commerce, and, given research, this crop could have a bright future in a number of nations (see page 307). (D. Greenberg)

of soluble solids (mainly sugars) was elevated from an average of about 5 percent to about 7 percent—a potential benefit of great economic value to California's tomato industry.

This demonstrates how important even the least-known wild relatives of crops can be, especially for increasing yield and for introducing disease- and pest-resistance. For reasons of space, this report cannot deal with them, but it should be understood that the wild relatives are so important that, without them, many modern crops simply would not exist.

Following are some of the other wild Andean ancestors and relatives that contribute genes to the modern tomato.

• *L. peruvianum*, a widespread Peruvian species, contributes resistance to pests and increases vitamin C content.

• *L. pennellii*, from dry hill slopes of western Peru, contributes drought resistance and high levels of vitamins A and C and sugar.

• *L. esculentum* var. *cerasiforme*, which grows on the warm eastern slopes of the Andes, helps plants tolerate high temperature and humidity and resist certain fungal diseases.

• *L. chilense*, native to coastal deserts of northern Chile and southern Peru, contributes genes for drought resistance.

• *L. cheesmanii*, from the Galapagos Islands, provides salt tolerance, high soluble solids, and the jointless fruit stalks that help tomatoes break off cleanly during mechanical harvesting.

• *L. hirsutum*, which grows in high altitudes of Ecuador and Peru, offers genes that resist numerous insects and mites and code for cold tolerance.

Andean aji

Peppers

Peppers have become the number one spice ingredient in the world. Red, yellow, green, or brown; hot, mild, or in-between—more are now consumed than any other. In almost every country of Europe, Asia, Africa, the Caribbean, North America, and Latin America, they are the most popular condiment, employed to enliven rice, beans, cassava, corn, and myriad other staples.

What is more, peppers have a large—and growing—following in countries that have not traditionally used them. In the United States, for example, produce markets carry fresh peppers in rainbow colors from white to purple, sizes from a gram to a half kilo, and shapes from flat to spherical. Grocery shelves display dozens of concoctions to fire the taste buds. Restaurants serve everything from chile relleno to Korean beef. Even some cocktails come spiced with pepper.

All this would have seemed unbelievable to the South American Indians who were probably the first to use peppers—extremely hot, pea-sized fruits they found growing around them—perhaps more than 7,000 years ago.[1] Such pungent foods should have limited appeal, but the history of peppers is one of enthusiastic acceptance wherever they were taken. By the time of Columbus, peppers were a principal seasoning of the Incas and the Aztecs. Montezuma received them as tribute. Columbus came to the New World looking for the black pepper of Asia and stumbled upon this even more piquant spice. (Believing he had reached the Indies, he named the people "Indians" and the spice "pepper," thereby creating endless subsequent confusion.)

After Columbus, peppers quickly spread around the world. The plants adapted to new environments and became so thoroughly entrenched in many cultures that the little green and red fruits have

---

[1] The ancestors of all peppers are believed to have originated in an area of Bolivia, but peppers spread quickly and reached Central America and Mexico in very early times. Information from W.H. Eshbaugh.

195

become an even bigger spice than the black one the Admiral had been looking for. Chili powder, cayenne, Tabasco, pimientos, and paprika all derive from peppers. Today, one can hardly imagine what many national diets must have been like without them. The foods of India, Hunan and Szechuan (China), Thailand, Indonesia, Ethiopia, West Africa, and others became synonymous with highly spiced foods. And Hungary and Spain became known for paprika and pimientos.

These developments are most often based on one species, *Capsicum annuum*. Two others, *C. frutescens* and *C. chinense*,[2] are also used in a few tropical areas. But in the Andes, the probable homeland of peppers, there remain other promising species that have scarcely spread outside the region. These are grown mainly in rural home gardens, but several are still wild plants. Examples of both cultivated and wild types are highlighted below.

SPECIES

Rocoto.[3] The rocoto (*Capsicum pubescens*) is widely cultivated in the high Andes. Its purple and white flowers, fuzzy leaves, and black, wrinkled seeds make it easy to recognize. It produces fruits sometimes almost as large as bell peppers, but instead of being mild in flavor, they are pungent like hot chiles. When ripe, these beautiful, thick-fleshed fruits are brilliantly colored—shiny red, orange, yellow, or brown—and they come in a variety of shapes and sizes. They are often eaten stuffed with meat.

The plant is the most cold tolerant of the cultivated peppers. It grows at higher altitudes than other species, generally from 1,500 to 2,900 m, but cannot tolerate the heat of the lowland tropics. It is a perennial that grows for 10 or more years and is sometimes called the "tree chile."

The Incas prized rocoto for its special flavor, and 450 years later it is still mainly confined to the Andean area formerly occupied by the Incas. However, it is also cultivated a little in the highlands of Costa Rica (particularly for producing yellow food coloring), Guatemala (where it is called "*siete caldos*"), and southern Mexico (where it is known as "apple chile" or "horse chile"). It is virtually unknown

---

[2] *Capsicum* taxonomy is not clear-cut; these two species may be one and the same. Their best-known use in North America is as the prime ingredient of Tabasco Sauce.
[3] The name "rocoto" (also spelled rokkoto) is used in Peru, Chile, and Ecuador; locoto (or lokoto) is used in Peru (Puno) and Bolivia. Other names are "chile manzano" (Mexico) and "panameño" (Costa Rica).

Peppers in the market at Cliza, Bolivia. Foreground, ají; background, rocoto. (W.H. Eshbaugh)

anywhere else, and its introduction north of Colombia is almost certainly post-Columbian, perhaps even twentieth century.

Rocoto occurs only in cultivation; its wild ancestors have not been defined, although genetically it is closely allied to the ulupicas (see below).

Andean Ají. The common cultivated pepper of the southern part of the Andean area is the brilliantly colored *Capsicum baccatum*.[4] Cultivated forms seem to have been domesticated from wild, weedy plants in a large band of territory stretching from southern Peru eastwards through Bolivia and Paraguay to southwestern Brazil. The center of origin is probably Bolivia. Some peppers unearthed from archeological sites resemble this species, and this has led people to

[4] More properly, *C. baccatum* var. *pendulum*. The wild progenitor of this crop, *C. baccatum* var. *baccatum*, grows at even higher elevations (up to 1,600 m) in Bolivia. Its local names are "arivivi" and "cumbai," and it has a narrow distribution from central Peru, through Bolivia, to northern Argentina and southern Brazil. It, too, deserves research and testing. Information from W.H. Eshbaugh.

Peppers appear to have originated in what is today central Bolivia. Here can be found the greatest wealth of *Capsicum* species and varieties. From this small but very diverse location (it includes temperate, subtropical, and tropical zones), one species spread throughout much of the world to become the chili and sweet peppers that liven foods on every continent. The other species, including two domesticated and perhaps a dozen wild ones, remain to be exploited beyond the Andes. (J. Andrews)

believe that it has been in cultivation for perhaps 4,000 years. Today, the Andean ají[5] (pronounced ah-*hee*) is cultivated in Bolivia, Peru, Ecuador, Argentina, and Brazil, as well as in Costa Rica, where it is called "*cuerno de oro*" (golden horn).

This is primarily a lowland species, but it is found today up to around 1,100 m elevation. It is a sprawling shrub, distinguished by the (yellow- or green-) spotted corollas of its flowers and by its long, conical fruits. Although described as a distinct species more than 150 years ago, it was later regarded as a variant of the common *C. annuum*, some types of which resemble it in everything but the spotted corollas. It was "restored" to the status of a separate species only in 1951.

Andean ají fruits are most commonly shiny orange and red, but rare yellow and brown forms are also known. They are very hot and are made into sauces, some of which are bottled commercially with herbs and onions. Typically, these sauces are used on cassava (yuca) and in marinated, uncooked fish (ceviche). A few are sweet types. Only in this species and the common pepper are nonpungent cultivars known.

---

[5] Ají is a name often used in Spanish for any pepper, including the common pepper (*C. annuum*), more generally called "pimiento." In the central Andes, however, ají ordinarily refers to *C. baccatum*. For this reason we have coined the common name Andean ají.

CAPSAICIN

A pepper's pungency is caused by capsaicin (pronounced cap-*say*-i-sin), a chemical that is odorless, colorless, and flavorless, but that irritates any tissue it contacts. Biting into a pepper stimulates nerve receptors in the mouth to signal "pain," and the brain in turn induces sweating, salivation, and increased gastric flow in an attempt to rid the body of the irritation.

Capsaicin is related in structure to vanilla (structurally, it is the vanillyl amide of isodecylanic acid), but it is very acrid. A single drop diluted in 100,000 drops of water will produce a persistent burning of the tongue. Diluted in 1 million drops of water, it still produces a perceptible warmth.

Capsaicin is concentrated in the pepper's placenta, the inner part that supports the seeds. There is a rough correlation between the amount of capsaicin and the amount of carotenoid pigment. Thus, the stronger the flavor, the deeper the color of the fruits.

Capsaicin has many uses of its own. Applied in concentrated form to the skin, it induces a feeling of warmth (actually an irritation), and because of this it is used in sore-muscle remedies. It is also the ingredient that gives the "bite" to commercial ginger ale and ginger beer. It is so powerful an irritant that it is used to make antidog and antimugger sprays. Also, it is used in concoctions to deter deer and rabbits from devouring vegetable crops.

Recently, it has been found that the brain probably releases painkillers when nerve receptors send it the "capsaicin signal." It has been tested in topical treatment for relief of pain caused by shingles, psoriasis, and other skin conditions.

In another modern twist, capsaicin and peppers are being touted for the diet conscious. Peppers, it is said, perk up the taste of food without adding fat. Indeed, it is thought that they may burn more calories than they provide.

Wild Andean Peppers.[6]   In the peppers' probable "homeland" in Bolivia are two essentially undomesticated species, known as "ulupicas," both of which are greatly appreciated by the local peoples.[7] They are aromatic, tasty, and much hotter than rocoto or other common peppers. As the Indian name implies, they are closely related to one another, and perhaps to rocoto. Both have purple flowers or, occasionally, white flowers.

[6] Information in this section from W.H. Eshbaugh. There are about 20 other wild species of *Capsicum*. Those are less closely related to the domesticated peppers, but some of them will cross with the domesticated species.
[7] It is not unusual for wild peppers to be popular and pricey. In northern Mexico and the southwestern United States, some wild peppers (called "chiltepins," one of the bird peppers) are in such high demand that they sell for up to 10 times the price of cultivated bell peppers. The chiltepins are often smuggled across borders, and harvesting pressures are so heavy that in many places these wild plants have become scarce.

The more widespread ulupica is *Capsicum eximium*, found between about 1,400 and 2,800 m in the drier, cooler parts of Bolivia, northern Argentina, and parts of Paraguay. It may reach 2 m in height, and its small (6 mm diameter) fruits are round, red, and fiery. They are frequently bottled or pickled. The natives like this ulupica so much that they often encourage it to grow, even though it is basically a weed.

The other ulupica (*C. cardenasii*)[8] is known only from the Andean sierra of Bolivia. It can be found at elevations between about 2,600 and 2,900 m, and may be cultivated in certain places. The exceptionally pungent fruits are found in markets in La Paz.[9] They are sometimes boiled and diluted with water (to reduce their bite), and then preserved in oil and vinegar and used as pickles. Also, the boiled fruits are dried and ground with tomatoes to make a very popular, aromatic condiment.

Two other wild species, *C. chacoense* of northern Argentina and Paraguay (locally called "covincho") and *C. tovarii* of Peru (called "mukúru") are even less well known, but they, too, are used locally. *C. chacoense* seems to have the advantage of being extremely drought tolerant. It is found between 1,450 and 2,200 m elevation.

Although the wild peppers seem to be in the process of domestication, they still lack certain qualities for widespread commercial success. For example, the fruits are small and tend to fall from the plants if touched or jarred. Also, they ripen rapidly and become soft soon after being harvested.[10]

PROSPECTS

The Andes. Peppers are already common in the Andean diet and their use is widespread, but there is nonetheless ample opportunity to select better growing and more diverse varieties. Moreover, the use of peppers (particularly the pungent types) in prepared foods could increase with expanded industrialization and export markets.

The five cultivated species are all highly variable in plant type, fruit type, pungency, and degree of adaptation. Wild and primitive cultivars undoubtedly contain useful sources of resistance to viral, bacterial, and fungal diseases, as well as nematodes, in addition to possessing desirable culinary qualities. Also, genes for greater environmental

---

[8] This scientific name, by P.G. Smith and C.B. Heiser, honors Martín Cárdenas, one of the foremost authorities on Andean food crops and botany (see dedication to this report). This ulupica is little known—the original scientific specimen was bought by Cárdenas in a La Paz market in 1958.

[9] The pungency in peppers is due to capsaicin (see sidebar), and research has shown that ulupicas have 4,000–5,000 units of capsaicin, a very high number. Information from G. Veliz.

[10] Information from G. Veliz.

adaptation surely exist. Collection of all species should continue and accessions be carefully preserved.

The Andean germplasm is a potentially vital source of resistance to diseases that afflict common peppers, such as fungal root diseases that sometimes kill a high proportion of the plants before they can be harvested. Also, it could be a significant resource for increasing the pungency and color of peppers. Both these qualities are of huge economic importance.

Other Developing Areas. Peppers are now perhaps the most widespread commercial vegetable in the tropics. Millions of the poor and destitute live on rice, beans, cassava, dahl, or another bland staple set off by a little dash of peppers. India's ubiquitous chutney and curry, for example, are unimaginable without the "heat" of peppers, whose origin is actually the Andes.

Peppers, therefore, constitute a potentially key intervention for the improved health of the Third World. Their nutritional content is relatively high, and they are good sources of vitamins, particularly vitamin C, and in the dried pungent types, vitamin A. Fortunately, these nutrients are not lost during the varied types of processing.

So far, selection has concentrated almost entirely on *Capsicum annuum*, but the other species deserve exploration. The screening and evaluation of the "lost" Andean species could provide useful characters to common peppers. Unfortunately, the species fall into a number of distinct genetic groups that do not hybridize freely with one another, thus limiting the usefulness of genes from other species.[11]

Industrialized Regions. The common pepper is increasingly popular in the United States and other nations that have had no tradition of eating spicy food. The related species of the Andes therefore are potentially valuable future resources, even in such countries. Rocoto, for instance, is thick fleshed like a bell pepper, but spicy like a chile. That combination could encourage specialty uses that might project this now unknown pepper into a place in the cuisines of the world. For use outside the tropics, however, varieties that are daylength neutral and adapted to cultivation outside the Andes will have to be located. Thanks to the foresight of biologists, a number of germplasm collections are in place. What now needs to be done is to identify and create superior types suitable for commercial production.

[11] *C. annuum*, *C. frutescens*, and *C. chinense* hybridize among themselves easily. *C. baccatum* hybridizes with each of them with some difficulty. Unfortunately, however, *C. pubescens* is genetically isolated and will not hybridize with any of the others. Information from W.H. Eshbaugh. Perhaps advances in biotechnology will change this picture.

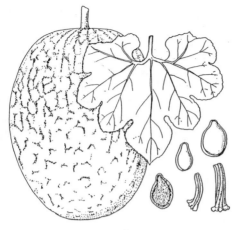

zambo

zapallo

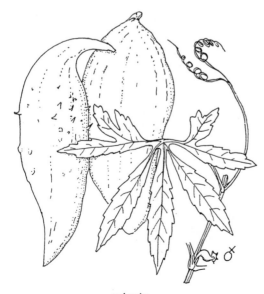

achocha

Squashes and Their Relatives

Cucurbits (*Cucurbita* species) are a collection of botanically related food crops that includes what are variously called squashes, pumpkins, vegetable marrows, and gourds. Among the first plants used by mankind,[1] they have long been among the most widely distributed. Most are extremely versatile, being used as fruits, vegetables, edible seeds, and oilseeds, as well as sources of fodder and fiber.

Traditionally, cucurbits have been particularly important in the Americas. Together with corn and beans, they were a nutritional mainstay of pre-Columbian civilizations such as the Incas, Mayas, and Aztecs. Since Columbus' time, however, they have become popular throughout most of the world. Today, they are eaten by millions of people, but almost nowhere are they major crops. Moreover, for all their value to people, cucurbits are (at least by comparison with the major grain crops) much neglected by scientists.

This is unfortunate, because these plants, which typically are trailing vines with extensive roots and harsh (often prickly) leaves and stems, are well suited to the peasant or individual gardener. They have wide adaptability and are easily cultivated. Their needs are usually satisfied by moderate soil moisture, and once vigorous growth starts, they seldom need weeding. They are little bothered by insect pests or heat. When judged by nutritional yield and labor required per hectare, they are among the most efficient of all crops.

Fruits are the major cucurbit product. Immature fruits are eaten as green vegetables. Mature fruits are boiled or baked and are important sources of starchy and sugary foods. The excellent keeping qualities of the ripe fruits of some species allows them to be stored for months— even years—without special care. And, if cut in strips and dried in the sun or over coals, the flesh of others will also keep for years.

Nutritionally, these fruits are excellent sources of vitamin A, vitamin C, iron, and potassium. They are low in sodium. The young leaves

---

[1] The archeological record indicates that some *Cucurbita* species entered into agriculture by at least 6000 B.C.

203

and the flowers (especially the surplus male flowers) of several species are sometimes eaten, and they, too, are sources of vitamins and minerals. In some species, the seeds are roasted and consumed as a snack and are often more prized than the flesh that surrounds them. The seeds can have protein *and* oil contents of 30–40 percent.

Five richly flavored Andean cucurbits are discussed below.

SPECIES

Zapallo (Winter Squash). The squash[2] (*Cucurbita maxima*), called "zapallo" (pronounced za-*pie*-oh) in the Andes, is of exclusively South American origin. Its center of diversity lies in northern Argentina, Bolivia, southern Peru, and northern Chile, but by the 1400s it had been spread northward throughout the warmer parts of the Inca realm.[3] At the time of Columbus, it was still confined to South America, but today it is widely grown throughout the world, particularly in Europe, India, the Philippines, and the United States. It is a winter-type[4] squash and includes the table vegetable most often called "pumpkin," as well as many common vegetables called "squash."

More tolerant to cool temperatures than other squashes, this species is grown as far south as the limits of agriculture in Chile. Using this species, Chile and Peru have developed the most gigantic form of all commercial "pumpkins." Fruits of 20–40 kg are commonly found in markets.

This squash is noted for its rich diversity—some authorities claim it has more forms than any other cultivated plant. In the main, the fruits are cylindrical, often bulbous, and have a central cavity filled with fibers and seeds. Some brightly colored, highly attractive varieties have become extremely popular specialty vegetables in the United States in recent years. Chilean varieties that have become common foods in the United States include Acorn, Banana, Boston Marrow, Buttercup, Golden Delicious, and Hubbard.

Crookneck. This species (*Cucurbita moschata*) is apparently Mexican or Central American in origin. However, it must have been spread widely in prehistoric times because its center of diversity extends as

---

[2] The common names of cucurbits are a muddle. Names such as "pumpkin" and "squash" are used for different species in different countries. There are no internationally recognized common names for *Cucurbita maxima, C. moschata,* and *C. ficifolia.*
[3] A recent excavation in northern Argentina has disclosed a wealth of well-preserved specimens, suggesting that it was a common cultivated plant in northern Argentina at least as long ago as 500 B.C.
[4] The name refers to an ability to be stored through the winter, not an ability to grow in the cold.

Zapallo can reach giant size. (H. Popenoe)

far south as northern Colombia and Venezuela. Apparently, it was introduced to Peru as early as 3000 B.C.

At the time of the Spanish colonization, the crookneck was abundant in northern South America and Central America.[5] Today, it is grown extensively in other parts of the world, especially in tropical Asia and Japan. Highly esteemed varieties in the United States include such cultivars as Butternut and Cushaw. It is the chief canning "pumpkin" of the midwestern United States, eaten each year by millions of families in Thanksgiving pie.

It, too, is a winter-type squash. However, it is well adapted to the tropical lowlands where high temperatures and high humidity prevail. It is notably resistant to the pesky squash-vine borer.

The plant yields five different products: mature fruits, which are baked, steamed, or made into pie; young fruits, which are boiled; male flowers, which are dipped in batter and fried as fritters (*buñuelos*); seeds that are roasted; and young tips of the vines, which are eaten boiled. The seeds have a delightful, nutty flavor, and were probably the product for which this plant was initially domesticated.

[5] It had also been carried (probably via Mexico) to Florida, where the Indians grew its vines on girdled oak trees. Early Florida settlers adopted it and called it the "Seminole pumpkin." Common names used in Latin America include ayote (Central America), lacayote (Peru), joko (Bolivia), and auyama (Colombia, Venezuela).

Zambo. Indians in the Andes commonly grow this "import" from Mexico. In fact, this squash (*Cucurbita ficifolia*)[6] has become so popular in the Andes that it is grown more frequently there than in its native land.[7] Today, it occurs from central Mexico through the high plateaus of Central America and along the highlands of the Andes as far south as central Chile. So far, it is little known elsewhere.

This species is another cool-climate (but not frost-tolerant) member of the genus *Cucurbita* and is the only perennial among commercial cucurbits. It is pest resistant and short-day flowering. In some places, the rampant, irrepressible vine runs wild, climbing trees and shrouding shrubs with its figlike leaves. Its elongated or globe-shaped fruits may weigh 11 kg (even when not grown under forcing conditions) and are white, green, or white and green striped. It has white flesh and is the only squash with black seeds (a white-seeded race also exists).

Cultivated extensively in the Andean highlands—mostly at 1,000–2,000 m elevation[8]—the young fruits are used like zucchini. The mature fruits are prized especially for desserts, usually cooked and served in sweet syrup. They are also fed to domestic animals (horses, cattle, and sheep) during the dry season.

No fruit anywhere keeps as well as these. Mature, they are commonly stored (kept dry, but without any other special care) for two years, and yet their flesh remains fresh and actually gets sweeter with age. They are eaten boiled or in preserves. Immature ones can pass for zucchini in looks and in recipes. Especially delicious and nutritious is a pudding made by simmering this squash with milk and cinnamon.

The seeds are baked and eaten like peanuts and are greatly appreciated. They have an unusually high concentration of oleic acid, the prime ingredient in olive oil.

Achocha. Achocha[9] (*Cyclanthera pedata*) is not a true squash, but it belongs to the same family, Cucurbitaceae. It, too, is common in the Andes. The fruits are small "gourds" 6–15 cm long, with flattened sides and soft spines. Pale green with darker green veins, they have a spongy interior containing up to a dozen seeds.

Some immature achochas look and taste like tiny cucumbers, for which they are fair substitutes in many culinary uses. (They are never

[6] This species is known by several names in the Andes—for instance, zambo (Ecuador), Vitoria (Colombia), and lacayote (Peru).
[7] Where it is called "chilacayote" or "tzilacayote." In Costa Rica and Honduras, its name is "chiverre"; in New Zealand, "pie-melon."
[8] It is common, for example, in the coffee belt of Colombia, its immature fruits selling for good prices in Bogotá.
[9] A Quechua word. In the Andes it is also widely called "caihua." Elsewhere, it is known as pepino de rellenar (Colombia), pepino andino (Venezuela), and variations on "achoca" and "caihua."

Achocha. (N. Vietmeyer)

crunchy, however.) Others are covered in soft green spines and have a curious shaggy appearance. In the immature form (that is, before the seed becomes black and hard), they can be eaten raw or cooked. When mature, they are better cooked, and the hard, black seeds must be removed. Filled with mincemeat or vegetables and baked, mature achochas make a tasty dish, not unlike stuffed peppers, with a flavor that has been likened to artichoke.

Achocha is undoubtedly of South American origin—probably including the Caribbean—but it is found in Mexico as well. In fact, the crop is cultivated from Mexico to Bolivia and grows prolifically in mountainous valleys up to 2,000 m elevation.

Achocha has been tested in cultivation outside the Americas and seems to have widespread promise. It fruits well in subtropical climates, such as northern New Zealand.[10] In South Florida and southern Taiwan, it has grown and set fruit well. In Nepal, it is occasionally cultivated at about 2,000 m elevation and has escaped in places. In England, it has fruited in a greenhouse.

In several parts of the Andes, a wild relative, *Cyclanthera explodens*, is used.[11] The fruits of this species are eaten by peasants, boiled or as a salad. Like achocha, this is a "poor-people's plant." It seems to

---

[10] Information from D. Endt.
[11] Information from H. Cortes B. and J. León. This plant is so named because when the fruit matures, it throws the seeds explosively.

Casabanana. (E. Sarmiento G.)

tolerate more cold than other cucurbits. It grows at 2,600 m elevation near Bogotá.

Casabanana. Another cucurbit, this species (*Sicana odorifera*) is found growing around houses in the foothills and lowlands of the Andes. A fascinating and useful plant, its fruits look like long, cylindrical, red-colored squashes. They are edible only when young, at which time they can be eaten both raw and cooked. It is the mature fruits, however, that are most prized. Although inedible, they exude a strong, pleasant fragrance reminiscent of a blend of ripe melon and peach. They are used as air fresheners to perfume kitchens, closets, clothes, and Christmas crèches. In Nicaragua, they are used to flavor *frescos*, especially a drink known as "*cojombro.*"

The casabanana is known only in cultivation (or as an escape from cultivation); its origins are therefore uncertain. It is probably not of Andean origin, although it was originally described from Peru. It may have been brought from the eastern part of South America—Paraguay or Brazil perhaps.

The plant is well known in Mexico and Central America and has been introduced, as a curiosity mostly, to France and possibly to other European countries.[12] The young fruits are eaten cooked in soups, but

[12] In Spanish it goes by many common names, including secana (Peru), pavi (Bolivia), cagua (Colombia), pepino de olor, melocotón, and melón calabaza.

the main product in Europe, as in Latin America, is this gourd's pleasing and penetrating fragrance that will perfume a whole house.

PROSPECTS

The Andes. In the Andes, as in some other parts of the world, squashes are considered to be food for the poor. Unfortunately, this means that they have not received the scientific recognition and research funding they deserve. This should not continue. Because they are so easy to grow and so well liked, efforts to introduce pest-resistant strains and improved modes of cultivation could bring big benefits to some of the neediest people in the hemisphere.

Other Developing Areas. "Pumpkins" and "squashes" have vast potential in subsistence farming. They are exceptionally attractive to peoples lacking ready means of food preservation. And they are outstanding as multipurpose plants. As noted, the young fruits, mature fruits, seeds, and even flowers can serve as food.

The germplasm of the Andes—home to many cucurbits for thousands of years—is especially important for the entire developing world. Currently, many Third World countries (Ethiopia, for example) have only one or at best a small number of squashes, and even those have almost no genetic variation. Thus, by and large, people outside Latin America are unaware of the wealth of types available.

Industrialized Regions. Cucurbits now grow throughout the temperate world and contribute a wide variety of products ranging from the Halloween pumpkin of the United States to the glasshouse cucumber of England. The important cultivated species are major market crops in North America, southern Europe, and temperate Asia. In addition, there is large commercial production of cucumbers in a number of more northern countries.

However, the types that remain in the Andes are an important unexploited resource. The squashes on the dinner tables of the future could be far more colorful and tasty than those of today. Cucurbits are excellent food for those who require acid-free diets. Most are noted for their keeping qualities.[13]

Moreover, the casabanana, with its penetrating fragrance, and the achocha, with its eye-catching shaggy appearance, could both make unusual specialty-produce items in many wealthy countries.

[13] Recently in Florida, for example, the crookneck has been found superior to the Cuban calabaza for shipping to distant northern markets. Information from J. Morton.

PART V

Fruits

Although the Incas feasted on roots such as potatoes, oca, and ulluco; grains such as quinoa and kiwicha; and legumes such as tarwi and nuñas, they also had a wealth of fruits and nuts. This section describes Andean berries, capuli cherry, cherimoya, goldenberry, highland papayas, lucuma, naranjilla (lulo), pacay (ice-cream beans), passionfruits, pepino, and tamarillo (tree tomato).

Many of these fruits exist exactly as they were at the time of discovery by the Spanish. Few are cultivated anywhere on a large commercial scale as yet. Most are dooryard plants, whose cultivation is primitive by modern standards: varieties are unselected, soil requirements are unknown, propagation techniques have not been perfected.

Despite this, the native Andean fruits are not inferior to those of other areas. Each is prized in one part of the Andes or another. They have limited use only because of lack of awareness of their possibilities, not because they taste bad. They are a unique and very diverse set of resources for the future.

It is important to develop these crops. Fruits in general are good sources of vitamins and are probably a dietary necessity. Vitamin A and vitamin C content can be notably high. In addition, the high contents of calcium, phosphorus, and iron in some varieties are of special value to growing children. These qualities make them exceptionally valuable for daily use in tropical villages and towns.

Specimens of any undeveloped fruit tend to vary greatly in taste, size, appearance, and texture, but careful selection, clonal propagation, and appropriate horticultural manipulations can bring huge improvements almost overnight. One of the most vital and rewarding activities is to collect and sort through such varieties, seeking the individual specimen with outstanding qualities.[1]

[1] In this, they are no different from more conventional fruits. For instance, all named apples (Golden Delicious, for example) as well as all seedless oranges, come from single mutant trees (sometimes only a branch) that someone noticed and propagated vegetatively.

210

DOMESTICATION OF THE STRAWBERRY

To see what might happen in the future with the little-known fruits described in the following chapters, consider the case of the strawberry.

All the cultivated strawberries we enjoy today are at least partly Andean. They are the offspring of a union between two American species brought independently to Europe: *Fragaria chiloensis* (native to Chile)* and *F. virginiana* (native to eastern North America). The first hybrid seems to have been an accident, arising in a garden near Amsterdam around 1750. The way it occurred has been described by Ruth Epstein, editor of *The American Festival*:

> *In 1712 a French naval captain named André Frezier was sent to South America to report Chilean coastal defenses, and when he came home, he bore armloads of the big-berried, pineapple-flavored Chilean strawberry plants. In his enthusiasm, he had selected only the most beautiful, the most vigorous, the most flower-filled plants to transport back to France. He had unwittingly selected only females. The plants transplanted happily enough—they bloomed profusely—but for 30 years they bore no fruit. And then, by chance and by mistake, some foundlings of the Virginia variety were set in amongst them. The South American spinsters mixed and mingled with Virginia's dandies and a union was consummated in the beds of France.*

* Sadly, this species, which gave us the modern strawberry, is now becoming rare, dying out because farmers would rather plant its more vigorous hybrid "child."

Fruits have particular merits for peasants and small farmers because they can be homegrown and provide both food and a cash crop. However, with today's transportation, refrigeration, and food processing, industries based on rare fruit are also becoming increasingly important. Exotic Andean fruits, for instance, are potentially profitable exports to affluent buyers in wealthy countries. Diets in North America and Europe are becoming extremely cosmopolitan, with increasing demand for exotic foods. Indeed, for two decades New Zealand has been sending its fruits to North America, Japan, and Europe, where they sell briskly at premium prices. Some—such as goldenberry,[2] cherimoya, naranjilla, pepino, passionfruit, and tamarillo—are Andean natives described below. It is a measure of their delicious taste that consumers pay the enormous costs of fresh fruit that have been airfreighted from the other side of the globe.

[2] Cape gooseberry.

mora de Castilla

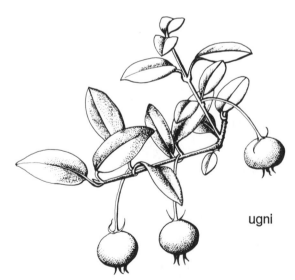

ugni

Berries

Scattered throughout the highlands of tropical America from Mexico to Peru are dozens of species of native berries. Their fruits are common both in the countryside and in the markets of Bogotá, Quito, and other large cities. Some are said to be superior in flavor and size to their well-known cousins, the commercial raspberries, blackberries, and blueberries. In Latin countries they are important fresh fruits as well as ingredients in jellies, jams, juices, thick syrups (*jarabes*, from which are made refreshing confections), and even wines.

Despite their popularity, these native berries have not been widely studied. Most are still gathered from the wild, and only a few are seen in regular cultivation. Because the plants have received little or no horticultural attention, their fruits exhibit widely variant size, color, and quality, and the flavors in any batch may range from extremely acidic to cloyingly sweet.

This chapter highlights several berries that are found mainly at elevations between 1,300 and 3,000 m in the Andean region. These seem to show promise as new cash crops for farms and backyards both in the Andes and elsewhere.

SPECIES

Mora de Castilla. This blackberry[1] (*Rubus glaucus*) is native to the broad area from the northern Andes to the southern highlands of Mexico. Although common in the wild, it is also abundant in the gardens of hundreds of towns and villages, especially in Ecuador and Colombia. In two Ecuadorian towns, Ambato and Otavalo, nearly every garden has the plants, and mora de Castilla (pronounced mor-a dey cast-*ee*-ya) fruits appear in the markets most of the year. In Colombia, the mora de Castilla has become an increasingly important cash crop. During recent years, its cultivation has increased because

---

[1] By accepted definition it is a blackberry because, when picked, the floral receptacle detaches from the plant and remains on the fruit. Other than that, it resembles a raspberry in the appearance of its leaves and the flavor of its fruits. Like other species in the genus, it exhibits a wide variability.

Although many species of wild berries are found in the Andes, the mora de Castilla is the most famous and popular. This Andean counterpart of the loganberry could have a big international future. Test samples of its high-quality, deep-red juice have been well received at a large U.S. fruit-drink corporation. This product might prove valuable for giving pallid juices (such as grapefruit) a rich ruby red color. (Wilson Popenoe © 1926 National Geographic Society)

RASPBERRIES AND BLACKBERRIES

The status of the Andean berries today is not markedly different from that of the commercial raspberries and blackberries in other parts of the world, even in relatively recent times.

Worldwide, there are more than 3,500 species of the genus *Rubus*. These are brambly, wild bushes that carry edible fruit. Two (subgenera *Eubatus*, the blackberries, and *Ideobatus*, the raspberries) have enormous commercial significance in many countries throughout the world; most of the rest are little known. Even the commercial species were neglected until the last century. Blackberries (there are several different species) were first cultivated in the United States in the 1800s, becoming common about 1850. In Europe, although blackberries had been gathered for centuries, their cultivation as a domestic crop is perhaps even more recent than in the United States. The red raspberry (*Rubus idaeus*), a wild European bramble, was first domesticated in Greece and Italy in about 1600, and the black raspberry (*R. occidentalis* and *R. leucodermis*) was domesticated in the United States within the last 150 years.

it is profitable, and because its fruits are now exported to the United States. More than 2,500 hectares are planted in it, and near Bogotá 1,300 hectares are in commercial production. Three commercial varieties have been selected and are under cultivation in Colombia.

Although it is often the most common blackberry in the Andean markets, the mora de Castilla is barely known elsewhere. However, it has flourished in Haiti and is being grown in a small way in Guatemala and El Salvador. This could be an indication of its future spread.

Mora de Castilla is one of the Andean berries that is said to be superior in flavor and quality to most cultivated blackberries and raspberries. Its fruits are large (up to 3 cm long). When fully ripe, they range from dark red to nearly black in color. Their seeds are small and hard, with little flesh adhering to them. In flavor, they are rich and rather tart, much like loganberry, making them well suited for eating fresh as well as for use in juices, jams, and preserves. They are exceptionally juicy (the juice has a striking, purple-red hue) and make excellent jam, which tastes like jam made from black raspberries.

The plant is a vigorous shrub of luxuriant growth that, climate permitting, produces fruit year-round. Its canes, 3–4 m long, are armed with small, but very annoying, hooked prickles.[2] They have a whitish,

---

[2] The hooked prickles on canes and stems, and especially those on backs of leaves, make working on the plants disagreeable. They adhere to skin of the backs of hands, as well as to most gloves. Smooth gloves made of heavy plastic or hard leather are the only answer.

waxy surface, which is characteristic of the species.

This apomictic species can be grown from seeds, but is normally propagated vegetatively (using tip layers or stem pieces) because it yields sooner. It grows well on many types of soil—reportedly thriving in almost anything from heavy clays to loose volcanic sands.[3] Nonetheless, it does best on moist, organic soils.

Although the species has not had the benefit of much modern horticultural attention, in well-tended plantings its annual yields are said to reach 20 tons per hectare. Improved cultural methods are needed, such as growing the plants on trellises (already done successfully in El Salvador) as well as means of controlling pests and diseases. These problems are being addressed in programs in Colombia and Venezuela.

Like other *Rubus* species, it exhibits wide variability because of segregation. For this reason, selection of outstanding plants and vegetative propagation may be an easy way to establish superior cultivars. Successful crosses have been made between mora de Castilla and a number of other *Rubus* species. So far, however, most of the hybrids have been infertile and lacking in hardiness.[4]

Giant Colombian Blackberry.   The giant Colombian blackberry (*Rubus macrocarpus*) is native to a narrow, rather inaccessible zone in the higher areas of Colombia (2,600–3,400 m elevation). Its canes, leaves, and flowers resemble those of blackberries, while its light red fruits resemble raspberries in appearance and loganberries in taste. Cultivated fruits are huge—up to 5 cm long and 2.5 cm wide—several times larger than today's commercial berries. Well-grown fruits are said to be as big as a hen's egg. A number of attempts to grow this species outside its natural region of dispersal have failed, and to date, no hybrids between this and other *Rubus* species have been produced.

The fruit is marketed in Colombia but is usually classed as a "*zarzamora*" (wild bush berry), to distinguish it from the mora de Castilla. When ripe, it is wine red, with compact pulp and slight acidity.

Mora de Rocota.   At the end of summer, peasants gather various wild berries, collectively called "*zarzamoras*," and sell them in practically all Andean markets. Housewives buy them to make delicious jams and drinks. Chile exports some to Europe.

The mora de rocota (*Rubus roseus*) is one of the three leading wild *zarzamoras* of Bolivia, Peru, and Ecuador. When ripe, the fruits are

---

[3] Popenoe, 1924.
[4] Information from H.K. Hall.

The giant Colombian blackberry, one of the biggest berries in the world, is almost too large to be taken in a single mouthful. (Wilson Popenoe © 1926 National Geographic Society)

crimson to nearly black in color, acid to sweet in flavor, and similar to cultivated raspberries in size and taste. They are eaten fresh, as juice, or made into preserves, wine, and aguardiente.

The plant[5] grows wild at an elevation of 2,800 m in Bolivia, 3,000–3,700 m in Ecuador. So far, it has not been cultivated.

Mora Común.   Most of the berry fruits in the highlands of tropical America are produced by *Rubus adenotrichus*, the most common species from Mexico to Ecuador. It is seldom cultivated, but the fruits of wild plants are collected and sold in the markets for the preparation of jellies, refreshments, and even wine. The plant is characterized by the long, reddish, glandular hairs that cover the branches. The fruits are red, conic, compact, and up to 2 cm long; in quality they are inferior to *R. glauca*, but the plant yields more and is more resistant and adaptable to different conditions. Because of its wide variability, it offers the possibility that major improvements can be made merely through the selection of superior clones.

[5] Also known as huagra mora, kari-kari, cjari-cjari, or chilifruta.

Ugni. Chile is marketing this fruit internationally under the name "myrtle berry." Although now one of the least-known fruits in the world, it is meeting with a good reception, especially in Japan, and could have a splendid future. (ProChile)

Mortiño. Throughout most of the Andean sierra at elevations between 2,800 and 4,000 m, the mortiño[6] (*Vaccinium floribundum*) is abundant. This "blueberry of the Andes" remains undomesticated, but given research it, too, could have a future in cultivation and widespread commerce.

The plant is especially profuse in the northern Andes—in Colombia, Bolivia, and Venezuela—where it occurs mainly at elevations from 1,800 to 3,800 m. It is not cultivated, but its fruits are gathered from wild bushes and sold in village and city markets. In Ecuador it is eaten raw, made into preserves, and used in a special dish with molasses, spices, and other chopped fruits on "The Day of the Dead" (November 2, All Souls' Day). In some areas its ripening season is the occasion for picnics, the people going together into the countryside to pick and eat the fruit.

The mortiño is a slender shrub. Some specimens grow 2–3 m high, others are dwarf and prostrate. Pink flowers and deep green foliage give it a handsome appearance. Its round berry is blue to nearly black, very glaucous (covered in a whitish bloom like grapes), and up to

[6] Also known as macha-macha (Bolivia and Peru), congama (Peru), and mortiño falso and chivacu (Venezuela).

about 8 mm in diameter. Because the plant has been given no selection, its fruits are variable in quality; they are sometimes pleasant and juicy, and at other times are barely edible. They contain numerous, though hardly detectable, small seeds.

Mortiño fruits closely resemble the blueberries of the United States, and superior types could probably be developed into commercial crops for temperate climates and tropical highlands.

Andean Blueberry. This blueberry[7] (*Vaccinium meridionale*) grows between 2,400 and 4,000 m elevation in the cold, windy highlands (*páramos*) of Colombia and also between 1,000 and 2,000 m in the mountains of Jamaica.

It is a shrub 1–4 m tall, or sometimes a tree growing up to 13 m high, with reddish, flaking bark. The berries are black, nearly round, and about 1 cm in diameter (larger than most blueberries). They are sweet and juicy and are borne in clusters of 10–15. The skin is somewhat tough and may be difficult to digest. The fruits are marketed in Bogotá and are popular in preserves, pastries, frozen desserts, and wines.

Ugni. Outside of Chile, this plant (*Myrtus ugni*[8]) is one of the least known commercial fruits; almost nothing about it can be found in the international research literature. In Chile, however, ugni[9] is not only cultivated, but the processed fruits are being exported to fill a growing demand in Japan.

The fruits are oblate, up to 1.5 cm in diameter, and purplish to deep cranberry in color. They are said to fill the air with the fragrance of strawberries and have a pleasant wild-strawberry taste.[10] Like the cranberry of North America,[11] they have a "spritely" flavor and make piquant drinks, desserts, jams, and jellies.[12]

The slow-growing evergreen shrub reaches about 2 m in height and flowers in 3–5 years. It is drought resistant and tolerates some frost. With its profuse, pink-tinged blossoms, it makes a showy ornamental. In Chile, most ugni is found growing wild in mountainous forest

---

[7] Locally called "agraz," this fruit is not to be confused with the "agraz," or bejuco de agua, *Vitis tiliaefolia*, that prospers only below 1,800 m from Colombia to Mexico as well as in the West Indies.

[8] Synonyms are *Ugni molinae* and *Eugenia ugni*.

[9] Other common names include uñi, murta, murtilla, strawberry myrtle, Chilean cranberry, and Chilean guava.

[10] Botanically speaking, ugni is a myrtle and not related to the cranberry (which the small red berries so resemble in appearance) or to the other berries in this chapter.

[11] *Vaccinium macrocarpon*.

[12] These jellies were a favorite of Queen Victoria.

clearings south of Temuco.[13] It is also highly prized in city gardens and is sometimes used as a border hedge. The sweeping branches establish roots when they contact the ground, and ugni can be easily propagated from such offsets as well as from cuttings and seed. In addition, researchers at the University of Concepción, Chile, have developed techniques for large-scale production of selected types using tissue culture.

The plant reportedly bears well on the coast of California, is commonly grown as an ornamental in the southern United States, and is also found in New Zealand.[14]

PROSPECTS

The Andes. At present, little or no effort is being exerted to cultivate or domesticate most of these Andean berries. Research is now needed to determine their range, types, and distribution. Additional information on the phenology and morphology should be gathered and an evaluation made of the commercial potential of each species, including the technical problems to be overcome. These results could then be used to direct attempts at domestication and improved use. Germplasm collections will almost certainly turn up the high-quality specimens needed for large-scale success. The development of basic cultivation practices will enable plantations to be established and management strategies to be formulated.

It is likely that domestication of the wild species will produce a cultivated fruit larger and sweeter than today's. However, there is also promise for the wild berries, as many people prefer their tartness. The existing stands of wild and semidomesticated plants can be made more productive by pruning to produce fruiting laterals, instead of leaving them to develop unchecked, as is now the case.

Efforts to exploit these berries will probably be successful—as is being demonstrated in the departments of Cundinamarca and Boyaca in Colombia. There, small-scale growers have organized into farmers' cooperatives, and have created a prosperous agroindustry based on the mora de Castilla.

Emerging technologies could make these fruits even more valuable commercially. For example, in North America, the drying or freezing of fruits to extend their marketing season has not been as effective with blueberries as it has been with other fruits. However, explosion

---

[13] It is sometimes confused with murtilla blanca (*Ugni candollei*), which is restricted to the Chilean coast, and which has larger leaves and larger berries.
[14] W.R. Sykes writes that "it is a fairly common garden plant in many cooler parts of New Zealand because it tolerates -10°C without seeming to suffer any damage."

puffing—a process used for drying many fruits and vegetables—has recently been used successfully. This may greatly expand many markets for these popular fruits and could perhaps be applicable to the South American blueberries also.

North American growers have trouble supplying enough blueberries to consumers at home and abroad, given that the harvest season is only 6 weeks in the spring. Thus, the success of new technologies for preserving berries would open the possibility of countries in the Southern Hemisphere supplying northern markets in the off-season.

Other Developing Areas.   Because of their extraordinary size and flavor, the mora de Castilla and the giant Colombian berry deserve trials in upland areas of the tropics. However, because their growth habits are not well understood and their genotypes largely uncollected, substantial commercial efforts should await the results of development trials in South America. Also, because of their vigor, thorniness, and easy dispersal by birds, no *Rubus* species should be introduced to new areas without extreme care.

Industrialized Regions.   The main value of these Andean berries for horticulture in Europe, North America, and other temperate zones is as sources of genes. Because of its vigor and the size and quality of its fruit, the mora de Castilla, in particular, could prove an excellent subject for crossing with northern raspberries. In addition, the unusually large size of the Colombian berry is a valuable characteristic that might be combined, by means of hybridization, with cultivated raspberries. However, previous trials have shown this plant to be susceptible to some North American raspberry diseases, and the process of capitalizing on its genetic characteristics may be slow and difficult.[15]

Ugni suffers no such limitations, and it deserves trial plantings and development in many parts of the temperate zones. To ease its introduction into markets in English-speaking regions, the fruit would benefit from a new name. "Murtilla" (pronounced "mur-*tee*-ya"), a common name for it in Chile, is one possibility. Perhaps the best, however, is "myrtle berry," a name now being used by Chilean exporters.

[15] Information from G. Galleta.

Capuli Cherry

Around highland villages from Venezuela to southern Peru, the capuli[1] (*Prunus capuli*[2]) is one of the most common trees. Easily identifiable, it has been said to characterize the Andean region much as the coconut palm typifies tropical coasts. Yet it is probably not an Andean native; capuli (pronounced ka-poo-*lee*) is an Aztec word, and most botanists believe that Spaniards introduced the tree from Mexico or Central America in Colonial times. Whatever its origin, this attractive tree has become so popular that it is seen from one end of the Andes to the other, especially around Indian settlements. In fact, it is now cultivated much more in the Andes than in its probable northern homeland, and the fruit is often much larger and more flavorful.

At harvest, capuli fruits are abundantly available in Andean markets. Capuli is a cousin of the commercial black and bing cherry, which it usually resembles both in appearance and taste. However, fruits are carried on short stalks and in bunches almost like grapes, and some taste like plums. They are round and glossy and are maroon, purple, or black in color. Their flesh is pale green and meaty, and most are juicy. The skin is thin, but sufficiently firm for the fruits to be handled easily without bruising. Although mostly eaten as fresh fruit, they can also be stewed, preserved, or made into jam or wine.

This fruit could become popular throughout much of the world. Although it grows in the Andes at tropical latitudes, it thrives there only in cool upland areas (2,200–3,100 m at the equator; fruit set occurs between 10–22°C).[3] It is therefore a plant for subtropical or warm temperate regions. Some newly introduced specimens are growing particularly well in northern areas of New Zealand, where little or no frost occurs.

Despite its promise, the fruit also has a down side. The pit is rather

[1] Locally, the fruit is often just termed "cherry" (cereza or guinda). Little-used Quechua names are "murmuntu" and "ussum." In English it is sometimes called "American cherry."

[2] The name *Prunus capuli* is used extensively in agricultural and horticultural publications. Research indicates that the capuli is actually a large-fruited subspecies of the North American black cherry, formally designated as *Prunus serotina* subsp. *capuli*. Information from R. McVaugh.

[3] Much of the technical data in this chapter is from R. Castillo.

large in proportion to the size of the fruit. Also, there is usually a trace of bitterness in the skin. However, in the best varieties it is so slight as to be unobjectionable and the fruits compete well with imported cherries.

It is curious that this fruit doesn't have more negative features because it has scarcely benefited from concentrated horticultural improvement and so far has been propagated primarily by seed.[4] This is not because of any inherent difficulty: both grafting and budding are easy and successful, and the plant also roots easily from softwood cuttings.[5]

The tree is extremely vigorous. It sets flowers and fruits heavily in its third—or even in its second—year of growth. It eventually reaches 10 m or more in height. Apparently, it is not exacting in its soil requirements and grows well on any reasonably fertile site. It can thrive in poor ground, even clays, and seems to prefer dry sandy soils. Although resistant to damping-off, powdery mildew, and other seedling diseases, it is susceptible to the common black-knot fungus[6] and does not thrive in wet areas (areas receiving 300–1,800 mm are said to be best in Ecuador).

Apart from bearing fruit, this is a useful, fast-growing timber and reforestation species (because it produces in poor soils, cost of production is also lowered). A few years after planting, its wood is suitable for tool handles, posts, firewood, and charcoal. After 6–8 years it yields an excellent reddish lumber for guitars, furniture, coffins, and other premium products. The wood is hard, is resistant to insect and fungus damage, and sells at high prices.[7] Young branches are supple and strong, like willow canes, and the prunings are often used to make baskets.

Capuli seems particularly suitable for agroforestry systems. Its deep roots help prevent erosion, and it may not dry the soil. It can be interplanted with field crops such as alfalfa, corn, and potatoes. It is a good plant for wind protection and it acts like a biological barrier— the birds enjoy its fruits so much they leave nearby crops alone.[8]

PROSPECTS

The Andes. Horticulturists in the Andes should investigate this species soon. Because of its fine woodworking and fuel qualities, trees are disappearing from some areas, with consequent loss of valuable

[4] In the late 1980s, grafted trees are just coming into use in Ecuador. Information from B. Eraso and R. Castillo.
[5] Information from P. Del Tredici.
[6] Information from P. Del Tredici.
[7] Information from R. Castillo.
[8] Information from B. Eraso.

Ambato, Ecuador. Although the capuli is known throughout Latin America, and probably originated in Mexico, it is in the Andes where the best types in size, color, and flavor are found. Great quantities appear in native markets, especially in Ecuador, where the capuli is an important food, not only of the Indians, but of all the inhabitants. After the Conquest, it was at times a mainstay of the invading Spaniards. The best capuli cherries (right) have dark maroon skins, firm texture, juicy greenish-brown flesh, and a flavor resembling that of the common cherry. (D. Endt)

germplasm. There are excellent possibilities for selecting better tasting, fleshier fruits. With vegetative propagation, selected horticultural varieties could become increasingly important and widely popular in many parts of Latin America where only the inferior seedling forms now grow.[9] Some trees already produce large and fleshy fruits, and it is important to select and propagate them by budding or grafting.[10] Flavor differences are particularly important. The fruits of many trees are so bitter as to be disagreeable, whereas others (often called "capuli chaucha") are sweet, pleasant, and delicious. This type, which compares favorably with imported cherries, should be selected and propagated. (The region around Ambato, Ecuador, is said to be a good source.)

As demand increases, processing facilities should become available to smooth out the price swings caused by glut then scarcity.

Other Developing Areas.   Although this tree is known throughout most of the Americas, the best fruits are found in the Andes. Andean capulis deserve serious attention in other Latin American countries as well. It is an excellent street tree for urban areas, adding shade, beautification, and even a little nutrition. (In downtown Quito, for example, it is not uncommon to see schoolchildren clambering for fruit in the capuli trees by the bus stops.)

It also has promise outside the Americas. It can be cultivated in many regions, including some where European cherries are not successful. It may be of value in many parts of Asia Minor, northern India, and other regions with similar climate. (Strict quarantine procedures must, of course, be followed in the Old World homelands of so many *Prunus* species.)

Industrialized Regions.   Although unequal to the cultivated European-derived cherries—produced by generations of selection and vegetative propagation—the capuli fruit is of good quality and has much potential for improvement.

The plant seems photoperiod insensitive and sufficiently hardy to permit successful cultivation as far north as California, Florida, and

[9] Because seeds have nearly 100 percent germination, they are generally sown where the tree is to grow. Information from R. Castillo.
[10] A collection of almost 100 accessions has been planted at Santa Catalina, near Quito. Information from R. Castillo.

the Gulf states of the United States. Indeed, it has already grown and borne fruit in Massachusetts (42°N).[11]

As noted, this South American cherry has recently been introduced to New Zealand and also deserves greater attention in certain parts of southern Europe, including the shores of the Mediterranean, and perhaps South Africa and Australia. Trials are already under way in Sicily, where the tree is growing well.[12]

[11] Where it holds its leaves into January and has survived intermittent temperatures of less than -20°C. Information from P. Del Tredici.
[12] Information from A. Raimondo.

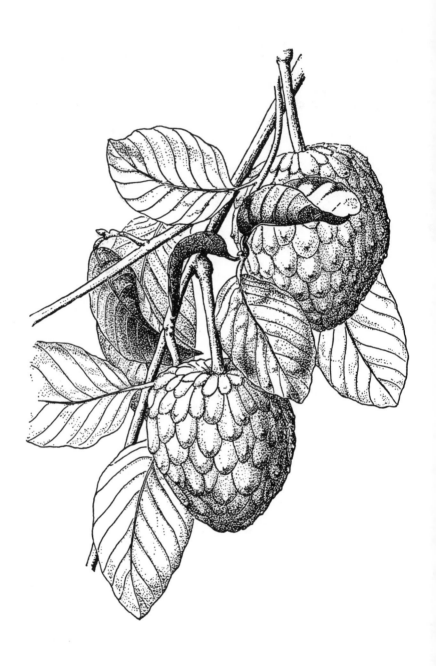

Cherimoya

Universally regarded as a premium fruit, the cherimoya (*Annona cherimola*) has been called the "pearl of the Andes," and the "queen of subtropical fruits." Mark Twain declared it to be "deliciousness itself!"

In the past, cherimoya (usually pronounced chair-i-*moy*-a in English) could only be eaten in South America or Spain. The easily bruised, soft fruits could not be transported any distance. But a combination of new selections, advanced horticulture, and modern transportation methods has removed the limitations. Cushioned by foam plastic, chilled to precise temperatures, and protected by special cartons, cherimoyas are now being shipped thousands of kilometers. They are even entering international trade. Already, they can be found in supermarkets in many parts of the United States, Japan, and Europe (mainly France, England, Portugal, and Spain).

Native to the Ecuadorian Andes, the cherimoya is an important backyard crop throughout much of Ecuador, Colombia, Venezuela, Bolivia, and Peru. Chileans consider the cherimoya to be their "national fruit" and produce it (notably in the Aconcagua Basin) on a considerable commercial scale. In some cooler regions of Central America and Mexico, the plant is naturalized and the fruit is common in several locales. In the United States, the plant produces well along small sections of the Southern California coast where commercial production has begun.

Outside the New World, a scattering of cherimoya trees can be found in South Africa, South Asia, Australasia, and around the Mediterranean. However, only in Spain and Portugal is there sizable production. In fruit markets there, cherimoyas are sometimes piled as high as apples and oranges.

A good cherimoya certainly has few equals. Cutting this large, green, heart-shaped fruit in half reveals white flesh with black seeds. The flesh has a soft, creamy texture. Chilled, it is like a tropical sherbet—indeed, cherimoya has often been described as "ice-cream fruit." In Chile, it is a favorite filling for ice-cream wafers and cookies. In Peru, it is popular in ice cream and yogurt.

World demand is strong. In North America and Japan, people pay more for cherimoya than for almost any other fruit on the market. At present, premium cherimoyas (which can weigh up to 1 kg each) are selling for up to $20 per kg in the United States and more than $40 per kg in Japan. Despite such enormous prices, sales are expanding. In four years, the main U.S. supplier's weekly sales have increased from less than 50 kg a week to more than 5,000 kg a week.

Today, the crop is far from reaching its potential peak. Modern research is only now being applied—and in only a few places, principally Chile, Argentina, Spain, the Canary Islands, and California. Nonetheless, even limited research has produced a handful of improved cultivars that produce fruit of good market size (300–600 g), smooth skin, round shape, good flavor, juiciness, low seed ratio, resistance to bruising, and good storage qualities. With these attributes, larger future production and expanded trade seem inevitable.

But growing cherimoyas for commercial consumption is a daunting horticultural challenge. In order to produce large, uniform fruit with an unbroken skin and a large proportion of pulp, the grower must attend his trees constantly from planting to harvest; each tree must be pruned, propped, and—at least in some countries—each flower must be pollinated by hand.

Nevertheless, the expanding markets made possible by new cultivars and greater world interest in exotic produce now justify the work necessary to produce quality cherimoya fruits on a large scale. Eventually, production could become a fair-sized industry in several dozen countries.

PROSPECTS

The Andes. Although cherimoyas are found in markets throughout the Andean region, there has been little organized evaluation of the different types, the horticultural methods used, or the problems growers encounter. Given such attention, as well as improved quality control, the cherimoya could become a much bigger cash crop for rural villages. With suitable packaging increasingly available, a large and lucrative trade with even distant cities seems likely. Moreover, increased production will allow processed products—such as cherimoya concentrate for flavoring ice cream—to be produced both for local consumption and for export.

Other Developing Areas. Everywhere these fruits are grown, they are immediately accepted as delicacies. Thus, the cherimoya promises to become a major commercial crop for many subtropical

Cherimoya has been grown for centuries in the highlands of Peru and Ecuador, where it was highly prized by the Incas. Today, this subtropical delight is gaining an excellent reputation in premium markets in the United States, Europe, Japan, and elsewhere. (T. Brown)

areas. For example, it is likely to become valuable to Brazil and its neighbors in South America's "southern cone," to the highlands of Central America and Mexico, as well as to North Africa, southern and eastern Africa, and subtropical areas of Asia.

Industrialized Regions. The climatic conditions required by the cherimoya are found in pockets of southern Europe (for example, Spain and Italy), the eastern Mediterranean (Israel), the western United States, coastal Australia, and northern New Zealand. In these areas, the fruit could become a valuable crop. In Australia and South Africa, the cherimoya hybrid known as atemoya is already commonly cultivated (see sidebar).

The cherimoya could have an impact on international fruit markets. The United Kingdom is already a substantial importer, and, as superior cultivars and improved packing become commonplace, cherimoyas could become as familiar as bananas.

USES

The cherimoya is essentially a dessert fruit. It is most often broken or cut open, held in the hand, and the flesh scooped out with a spoon. It can also be pureed and used in sauces to be poured over ice creams, mousses, and custards. In Chile, cherimoya ice cream is said to be the most profitable use. It is also processed into nectars and fruit salad mixes, and the juice makes a delicious wine.

NUTRITION

Cherimoya is basically a sweet fruit: sugar content is high; acids, low. It has moderate amounts of calcium and phosphorus (34 and 35 mg per 100 g). Its vitamin A content is modest, but it is a good source of thiamine, riboflavin, and niacin.[1]

HORTICULTURE

Because seedling trees usually bear fruits of varying quality, most commercial cherimoyas are propagated by budding or grafting clonal stock onto vigorous rootstock. However, a few forms come true from seed, and in some areas seed propagation is used exclusively.

The trees are usually pruned during their brief deciduous period (in the spring) to keep them low and easy to manage. The branches are also pruned selectively after the fruit has set—for example, to prevent them rubbing against the fruits or to encourage them to shade the fruits. (Too much direct sunlight overheats the fruits, cracking them open.)

Under favorable conditions, the trees begin bearing 3–4 years after planting. However, certain cultivars bear in 2–3 years, others in 5–6 years. Many growers prop or support the branches, which can get so heavily laden they break off.

[1] Information from S. Dawes.

Pollination can be irregular and unreliable. The flowers have such a narrow opening to the stigmas and ovaries that it effectively bars most pollen-carrying insects. Honeybees, for instance, are ineffective.[2] In South America, tiny beetles provide pollination,[3] but in some other places (California, for instance) no reliable pollinators have been found. There, hand pollination is needed to ensure a high proportion of commercial-quality fruit.[4]

HARVESTING AND HANDLING

The fruits are harvested by hand when the skin becomes shiny and turns a lighter shade of green (about a week before full maturity). A heavy crop can produce over 11,000 kg of quality fruits per hectare.

LIMITATIONS

A cherimoya plantation is far from simple to manage. The trees are vulnerable to climatic adversity: heat and frost injure them, low humidity prevents pollination, and winds break off fruit-laden branches. They are also subject to some serious pests and diseases. Several types of scale insects, leaf miners, and mealy bugs can infect the trees, and wasps and fruit flies attack the fruits.

Pollination is perhaps the cherimoya's biggest technical difficulty. Not only are reliable pollinators missing in some locations, but low humidity, especially when combined with high temperatures, causes pollination failure; these conditions dry out the sticky stigmas, and the heavy pollen falls off before it can germinate.

Hand pollination is costly and time consuming. However, it improves fruit set of all cultivars under nearly all conditions. It enhances fruit size and shape. It allows the grower to extend or shorten the season (by holding off on pollination) as well as to simplify the harvesting (by pollinating only flowers that are easy to reach).[5]

[2] The male and female organs of a flower are fertile at different times. Honeybees visit male-phase flowers but not female-phase flowers, which offer no nectar or pollen.
[3] Reviewer G.E. Schatz writes: "Pollination is undoubtedly effected by small beetles, most likely *Nitidulidae*. They are attracted to the flowers by the odor emitted during the female stage, a fruity odor that mimics their normal mating and ovipositing substrate, rotting fruit. There is no other "reward" per se, and hence it is a case of deception. The beetles often will stay in a flower 24 hours—the flower offers a sheltered mating site, safe from predators during daylight hours. Studies on odor could lead to improved pollination."
[4] California growers use artists' paint brushes with cut-down bristles to collect pollen in late afternoon. The next morning they apply it to receptive female flowers.
[5] Schroeder, 1988.

The fruits are particularly vulnerable to climatic adversity: if caught by cold weather before maturity, they ripen imperfectly; if rains are heavy or sun excessive, the large ones crack open; and if humidity is high, they rot before they can be picked.

The fruits must be picked by hand, and, because they mature at different times, each tree may have to be harvested as many as 10 times. In addition, the picked fruits are difficult to handle. Even when undamaged, they have short storage lives (for example, 3 weeks at 10°C) unless handled with extreme care.

The fruit has a culinary drawback: the large black seeds annoy many consumers. However, fruits with a low number of seeds exist,[6] and there are unconfirmed reports of seedless types. So far, however, neither type has been produced on a large scale.

RESEARCH NEEDS

The following are six important areas for research and development.

Germplasm   The danger of losing unique and potentially valuable types is high. A fundamental step, therefore, is to make an inventory of cherimoya germplasm and to collect genetic material from the natural populations as well as from gardens and orchards, especially throughout the Andes.

Selection   Future commercialization will depend on the selection of cultivars that dependably produce large numbers of well-shaped fruit with few seeds and good flavor. Selection criteria could include: resistance to diseases and pests, regular heavy yields of uniform fruit with smooth green skin, juicy flesh of pleasant flavor, few or no seeds, resistance to bruising, and good keeping qualities.

Pollination   The whole process of pollination should be studied and its impediments clarified. Currently few, if any, specific insects have been definitely associated with cherimoya pollination.[7] The insects that now pollinate it in South America should be identified. Spain, where good natural fruit set is common in most orchards, might also teach much.[8] Selecting genotypes that naturally produce symmetrical, full-sized fruits may reduce or eliminate the need to hand pollinate, bringing the cherimoya a giant step forward in several countries.

Cultural Practices   Horticulturists have not learned enough to clearly understand the plant's behavior and requirements. Knowledge of the effects of pruning, soils, fertilization, and other cultural details is as

---

[6] Flesh: seed weight ratios from 8:1 to 30:1 have been reported.
[7] Schroeder, 1988.
[8] Information from J. Farré.

yet insufficient. The current complexity of management should be simplified. Evaluation of the plants in the Andes, and the ways in which farmers handle them, could provide guidance for mastering the species' horticulture. Also, there is a need for practical trials to identify more precisely the limits of the tree's environmental and management requirements.

Intensive cultural methods, such as trellising and espaliering,[9] may help achieve maximum production of high-quality fruits. These growing systems facilitate operations such as hand pollination; they also provide support for heavy crops.

Breeding Ongoing testing of superior cultivars is needed. Low seed count, good keeping quality, and good flavor have yet to coincide in a cultivar that also has superior horticultural qualities. In addition, it is advisable to grow populations of seedling cherimoyas in all areas where this crop is adapted. From these variable seedling plants, selections based on local environmental conditions can be made. Elite seedling selections can be multiplied by budding or grafting. Mass propagation of superior genotypes by tissue culture could also provide large numbers of quality plants.

Improved cherimoyas might be developed by controlled crosses and, perhaps, by making sterile, seedless triploids. Breeding for large flowers that can be more easily pollinated might even be possible.

Hybridization Members of the genus *Annona* hybridize readily with each other (see sidebar), so there is considerable potential for producing new cherimoyalike fruits (perhaps seedless or pink-fleshed types) that have valuable commercial and agronomic traits.

Handling Improved techniques for handling, shipping, and storing delicate fruits would go a long way to helping the cherimoya fulfill its potential. Ways to reduce the effects of ethylene should be explored. Cherimoyas produce this gas prodigiously, and in closed containers it causes them to ripen extremely fast.

SPECIES INFORMATION

Botanical Name   *Annona cherimola* Miller
Family   Annonaceae (annona family)
Common Names
 Quechua: chirimuya
 Aymara: yuructira

---

[9] It has been reported that on Madeira, trees were espaliered so successfully that in some locations they have replaced grapes, the main crop of the island. The branches were trained so that fruit ripened in shade.

ATEMOYA

Queensland Department of Primary Industries

Like the cherimoya, the atemoya has promise for widespread cultivation. This hybrid of the cherimoya and the sugar apple was developed in 1907 by P.J. Wester, a U.S. Department of Agriculture employee in Florida. (Similar crosses also appeared naturally in Australia in 1850 and in Palestine in 1930.) The best atemoya varieties combine the qualities of both cherimoya and sugar apple. However, the fruits are smaller and the plant is more sensitive to cold.

The atemoya has been introduced into many places and is commercially grown in Australia, Central America, Florida, India, Israel, the Philippines, South Africa, and South America. In eastern Australia, for at least half a century, the fruit has been widely sold under the name "custard apple."

The atemoya grows on short trees—seldom more than 4 m high. The yellowish green fruit has pulp that is white, juicy, smooth, and subacid. It usually weighs about 0.5 kg, grows easily at sea level, and apparently has no pollination difficulties.

The fruit may be harvested when mature but still firm, after which it will ripen to excellent eating quality. It finds a ready market because most people like the flavor at first trial. It is superb for fresh consumption, but the pulp can also be used in sherbets, ice creams, and yogurt.

Seedling progeny are extremely variable, and possibilities for further variety improvement are very good. So far, however, little work has been done to select and propagate superior varieties.

Spanish: chirimoya, cherimoya, cherimalla, cherimoyales, anona del Perú, chirimoyo del Perú, cachimán de la China, catuche, momona, girimoya, masa
Portuguese: cherimólia, anona do Chile, fruta do conde, cabeça de negro
English: cherimoya, cherimoyer, annona
French: chérimolier, anone
Italian: cerimolia
Dutch: cherimolia
German: Chirimoyabaum, Cherimoyer, Cherimolia, peruanischer Flaschenbaum, Flachsbaum

Origin. The cherimoya is apparently an ancient domesticate. Seeds have been found in Peruvian archeological sites hundreds of kilometers from its native habitat, and the fruit is depicted on pottery of pre-Inca peoples. The wild trees occur particularly in the Loja area of southwestern Ecuador, where extensive groves are present in sparsely inhabited areas.

Description. A small, erect, or sometimes spreading tree, the cherimoya rarely reaches more than 8 m in height. It often divides at the ground into several main stems. The light-green, three-petaled, perfect flowers are about 2.5 cm long. The fruit is an aggregate, composed of many fused carpels. Depending on degree of pollination, the fruits are heart-shaped, conical, oval, or irregular in shape. They normally weigh about 0.5 kg, with some weighing up to 3 kg. Moss green in color, they have either a thin or thick skin; the surface can be nearly smooth, but usually bears scalelike impressions or prominent protuberances.

Horticultural Varieties. A number of cultivars have been developed. Nearly every valley in Ecuador has a local favorite, as do most areas where the fruit has been introduced. Named commercial varieties include Booth, White, Pierce, Knight, Bonito, Chaffey, Ott, Whaley, and Oxhart. These exhibit great variation in climatic and soil requirements.
In Spain, 200 cultivars from 10 countries are under observation.[10]

Environmental Requirements

Daylength. Apparently neutral. In its flower-bud formation, this plant does not respond to changes in photoperiod as most fruit species do.

[10] Information from J. Farré.

Rainfall. The plant does not tolerate drought well. For good production, it needs a fairly constant source of water. In Latin America, the tree thrives under more than 1,200 mm rainfall during the growing season. As noted, high humidity assists pollen set, and a dry period during harvesting prevents water-induced damage to fruit. Also, water stress just before flowering may increase flower (and hence fruit) production.

Altitude. The cherimoya does best in relatively cool (but not cold) regions, and is unsuited to the lowland tropics. (In equatorial regions it produces well only at altitudes above 1,500 m.)

Low Temperature. The plant is frost sensitive and is even less hardy than avocados or oranges. Young specimens are hurt by temperatures of –2°C.

High Temperature. The upper limits of its heat tolerance are uncertain, but is is said that the tree will not set fruit when temperatures exceed 30°C.

Soil Type. Cherimoya can be grown on soils of many types. The optimum acidity is said to be pH 6.5–7.5. On the other hand, the tree seems particularly adapted to high-calcium soils, on which it bears abundant fruits of superior flavor. Because of sensitivity to root rot, the tree does not tolerate poorly drained sites.

Related Species. The genus *Annona*, composed of perhaps 100 species mostly native to tropical America, includes some of the most delectable fruits in the tropics. Most are similar to the cherimoya in their structure. Examples are:

• Sugar apple, or sweetsop (*Annona squamosa*). Subtropical and tropical. The fruit is 0.5–1 kg, and yellowish green or bluish. It splits when ripe. The white, custardlike pulp has a sweet, delicious flavor.
• Soursop, or guanabana (*A. muricata*). This evergreen tree is the most tropical of the annonas. The yellow-green fruit—one of the best in the world—is the largest of the annonas, sometimes weighing up to 7 kg. The flesh resembles that of the cherimoya, but it is pure white, more fibrous, and the flavor, with its acidic tang, is "crisper."
• Custard apple, or annona (*A. reticulata*). This beige to brownish red fruit often weighs more than 1 kg. Its creamy white flesh is sweet but is sometimes granular and is generally considered inferior to the other commonly cultivated annonas. However, this plant is the most vigorous of all, and types that produce seedless fruits are known.

---

[10] Information from J. Farré.

• Ilama (*A. diversifolia*). This fruit has a thick rind; its white or pinkish flesh has a subacid to sweet flavor and many seeds. It is inferior to the cherimoya in quality and flavor, but it is adapted to tropical lowlands where cherimoya cannot grow.

• *A. longipes*. This species is closely related to cherimoya and is known from only three localities in Veracruz, Mexico, where it occurs at near sea level. Its traits would probably complement cherimoya's if the two species were hybridized to create a new, man-made fruit.[11]

[11] Information from G.E. Schatz.

Goldenberry (Cape Gooseberry)

The goldenberry[1] (*Physalis peruviana*) has long been a minor fruit of the Andes and is found in markets from Venezuela to Chile. It has also been grown in Hawaii, California, South Africa, East Africa, India, New Zealand, Australia, and Great Britain. So far, however, it has nowhere become a major crop. Nonetheless, this interesting and unusual botanical relative of potatoes and tomatoes has commercial promise for many regions.[2]

Goldenberry fruits are succulent golden spheres the size of marbles with a pleasing taste. They are protected by papery husks resembling Chinese lanterns. The attractive and symmetrical husk with its edible yellow fruit inside gives it an eye-catching appearance and potential market appeal.

In most places where they are grown, goldenberries are now considered fruits only for backyard gardens or for children to pluck and eat. Given research, however, they could become commercial fruits of particular interest to the world's up-scale restaurants and bakeries. This is the strategy that established markets for kiwifruit in the 1960s and led to a multimillion dollar annual crop. Goldenberries already carry prestige in some international markets. Europeans, for example, often pay premium prices to dip them in chocolate[3] or decorate cakes and tortes.

In addition to having a future as fresh fruits, goldenberries make excellent jam; in fact, in India, they are known commonly as "jam fruit." In the United States they are best known as preserves marketed under the Hawaiian name "poha."

---

[1] The plant is more commonly called "cape gooseberry." It is, however, not a true gooseberry. The name was adopted by Australians, who received their first plants from the Cape of Good Hope. A less cumbersome and more mellifluous name could help change its consumer image (as did the change from "Chinese gooseberry" to "kiwifruit"). The panel suggests the descriptive name "goldenberry," which is sometimes used in Great Britain and South Africa and has been recommended as the commercial name (see Legge, 1974). Moreover, the name "goldenberry" is reminiscent of other berry fruits, such as blackberry and blueberry.

[2] We recommend this species to tomato breeders seeking a challenge.

[3] The husks are peeled back and make an attractive, eye-catching—and tasty—dessert.

Although not yet well known internationally, the goldenberry is an established commercial crop in several places: India, South Africa, New Zealand, and Hawaii among them. The rather wild and weedy plant can be outstandingly productive. (D.W. Watt, courtesy *Sunset* magazine)

This plant has few severe horticultural problems. It tolerates poor soils and is forgiving of neglect. It is easy to grow from seeds or cuttings and matures quickly. Although harvesting the fruit is labor intensive, the yields are high.

All these features would seem to make the goldenberry an ideal addition to horticulture. However, current types do not travel well, some have a slightly bitter taste that invokes either intense like or dislike, and they can be sticky and mottled in appearance. Currently, the crop is probably best suited for local use as a fresh fruit or for jams. Eventually, however, selections and techniques will be found that deliver top-quality fruits to market. Then, the goldenberry is likely to become a well-known international commodity.

PROSPECTS

The Andes. Many Andeans are familiar with the plant and its requirements. The germplasm to create a global crop is in their gardens. Other parts of the world use vast quantities of blueberries and blackberries in baked goods and cooking; Andean nations now have the opportunity to create an equal demand for goldenberries. Some countries—Colombia and Chile, for example—could add the golden-berry to their fresh-fruit exports.

Other Developing Areas. The goldenberry seems destined to become a popular cash crop for smallholders far beyond its native lands. The plant seems to succeed wherever tomatoes are grown, and it is both cultivated and naturalized in Malaysia, China, India, southern and eastern Africa, and the Caribbean. It is likely that local and export potential for both fresh and processed fruits will develop. (This is already occurring in Kenya. See sidebar.)

The goldenberry plant is one of the first to pioneer newly created waste areas. Its robustness and adaptability could lead to cultivation in many now-unused marginal areas.

Industrialized Regions. With rising consumer awareness and demand, goldenberries will become more common in farmers' markets and urban groceries in North America, Europe, and other industrialized areas. They could become standard produce items, meriting intensive, mechanized agriculture with advanced cultivars. In addition, golden-berry jams and baked goods will become better known.

USES

Generally, goldenberry fruits are eaten fresh—many of them picked casually off the weedlike plants. Chilled, the fruits provide a crisp, tartly sweet addition to both vegetable and fruit salads. Sometimes they are sweetened by pricking the skin and rolling them in sugar, which they absorb.

Goldenberries are also used in sauces and glazes for meats and seafood, and they add an intriguing flavor to desserts. Their tangy taste combines well with meats, vegetables, and other fruits.

The processed fruits are commonly used in sauces and preserves such as jams and jellies.[4] They are also canned whole in syrup and exported from South Africa. Sun dried, they form what has been called "a very agreeable raisin." Goldenberry ice cream, although now unknown, seems a promising vehicle for introducing the fruit.

The goldenberry plant is a good ground cover for protecting land from erosion. Its spreading robust growth rapidly covers erosion-prone sites.

NUTRITION

The goldenberry is an excellent source of provitamin A (3,000 I.U. of carotene per 100 g) and vitamin C, as well as some of the vitamin B-complex (thiamine, niacin, and vitamin B_{12}). The protein and phosphorus contents are exceptionally high for a fruit, but calcium levels are low.[5]

AGRONOMY

This crop is easy to plant and maintain. Propagation is generally by seed. However, rooted cuttings flower earlier, yield more, and grow true to type. The crop is normally treated as an annual, even in the tropics.

The plant forms huge, straggly bushes. In New Zealand, at least, it is grown on poor, droughty soil just to limit the size of the bush. High soil fertility fosters useless vegetative growth, while low fertility induces

---

[4] The juice of the ripe berry contains so much pectin and pectinase that jams and preserves are self-jelling and require no additional pectin.

[5] One sample of the fruit contained 78.9 percent water, 0.3 percent protein, 0.2 percent fat, 19.6 percent total carbohydrate, 4.9 percent fiber, 1.0 percent ash. Per 100 g, there were 73 calories, 8 mg calcium, 55 mg phosphorus, 1.2 mg iron, about 1 mg sodium, 320 mg potassium, 1,460 mg B-carotene equivalent, 0.10 mg thiamine, 0.03 mg riboflavin, 1.70 mg niacin, and 43 mg ascorbic acid. Information from J. Duke.

Inside its Chinese-lantern-like husks, the goldenberry has bright yellow succulent fruits the size of marbles. (N. Vietmeyer)

fruit production.[6]

Goldenberry can be grown along the margins of fields, ditches, and roadways, or interplanted with other crops.[7] Although it thrives in full sun, some shelter from strong wind is desirable.

Most growers prune plantings back sharply after the first harvest to reduce pest infestations and to allow fruit to form on new growth. Flowering occurs 65–75 days after planting,[8] and harvest may commence 85–100 days after that.[9] Ripening occurs over a period of several months and the plants normally require more than one harvest.

The papery husk helps protect the berries from some birds, insects, and disease organisms.

[6] "If you're nice to it, you get a jungle," notes David Klinac, a horticultural researcher who has worked with the crop. "You get vast numbers of fruit per bush, but the effort of cutting your way in is enormous. It's best to be as hard on the plants as possible."
[7] In the Andes, it is often grown among root crops, beans, or corn. Plants growing in waste places yield a favorite wayside treat for children and hikers, and it is commonly an encouraged weed as well as a tended crop.
[8] Goldenberry has a long flowering period. The first fruits to form are often sacrificed to ensure the establishment of strong, healthy plants. They flower year-round in frost-free areas.
[9] In colder regions, seedlings can be started under glass.

HARVESTING AND HANDLING

Mature fruit will drop and can lie on dry ground for several days without ill effect, other than possible soiling of the husk. Shaking the plants and gathering the fallen fruit helps yield more uniformly mature berries.[10]

Harvesting is currently not mechanized because the dried husks (which should be intact for commercial fresh-fruit markets) shatter easily. Fruits (including husk and stem) are easily removed from the plant with a simple twist. For processed products, machinery has been developed to husk the fruit.[11]

Plants give their maximum yield the first season. They will produce for 2 or 3 years but the fruits get smaller after the first.[12] Yield is highly variable, depending on environment and intensity of cultivation. Untended plants may yield as little as 3 tons per hectare; carefully tended plants can provide 20 tons per hectare; yields of more than 33 tons per hectare have been reported.[13]

The berries continue to ripen after picking. Over a period of 2–3 weeks most achieve a uniform, bright golden yellow. Soluble solids increase from 11 percent to as much as 16 percent.

If carefully harvested when nearly mature and kept inside the husk, fruits may keep for several months in a dry container. Storage life is influenced by handling, wetness of husk at harvest, and fruit size. (Big fruits tend to split.) Removing the husk causes the fruits to deteriorate, irrespective of storage conditions. (Even after cold storage, fungal infection [*Penicillium* and *Botrytis*] is generally evident.) Keeping the husks intact minimizes handling damage and contains most infections. Fruits in the husk have been stored (below 2°C) for 4–5 months before marked shrinkage and collapse occur. Especially good results have been achieved by drying the fruits in the sun or in warm air (30°C) until husks are crisp to the touch. The fruit also freezes well.

LIMITATIONS

In this half-wild plant there is much variation in size, shape, and flavor of the fruits, time of maturity, and plant form. It has been so

[10] Sheets of polyethylene or other material can be spread on the ground. Shaking the plants then drops the fruit on the sheet, which can be pulled out and the fruit dumped into containers.
[11] A mechanical husker has been developed at the University of Hawaii. (Jaw-Kai Wang, 1966.)
[12] Information from H.Y. Nakasone.
[13] Information via J. Duke.

neglected that uniform horticultural varieties and superior cultivars remain to be identified. Currently, it is difficult to get lines that perform consistently. Growers in New Zealand, at least, select their own favorite plants from their own fields and replicate them by planting cuttings.

The plant's susceptibility to pests and diseases is not well understood, but when plantings are small and separated, pests and diseases are seldom major problems. In fact, once established, the crop requires little attention. However, potential threats (especially in large plantings) include birds, which prey upon the fruits; tobacco mosaic virus and bacterial leaf-spot, which infect the plants; and a number of insects that attack the foliage.

Palates accustomed to sweet fruits may find the slightly bitter aftertaste of some goldenberries unappealing. Even so, the uniquely fruity flavor is almost universally enjoyed.

Despite the ease of production, this may be an expensive crop to produce on a large scale. The plant probably needs staking for ease of handling in commercial production. Moreover, harvesting and manually husking each fruit is labor-intensive and is time-consuming. (This should not be more of a concern than with other berry crops, however.)

The husk is an asset, but can also be a liability. It conceals the fact that the fruit inside may have split or otherwise been damaged. (The experience of finding a split and rotten berry inside can be irritating.)

This robust plant could become a weed when introduced to new locations. And it may be a threat to foraging animals. Its leaves and stems are suspected of having caused the erosion of intestinal membranes (diptheresis) in cattle.[14]

The berry is sometimes covered with a sticky coating. If the weather is damp, a black mold often develops on this coating, although no harm seems to come to the fruit and the discoloration is easily washed off. The consumer, however, is likely to reject the whole batch after finding a few sooty berries. Of course, this is hidden by the papery shell and comes as a surprise to the unwary shopper.

Only ripe fruits should be eaten. Although unreported, there may be toxic glucosides in the unripe fruit.

RESEARCH NEEDS

The following are among this crop's general research needs.

Genetic Selection Since the 1940s and 1950s, goldenberry has been

---

[14] Information from J. Duke.

given almost no serious horticultural attention. Researchers should now scour the Andes as well as New Zealand, South Africa, Hawaii, and other likely places for the sweetest and most pleasant types. Superior strains should be documented. The greatest variation is likely to be found along the Andean cordillera. Selection for growth habit, stable characteristics, and superior fruit types can probably best be accomplished in this region. In particular, a determinate cultivar, whose fruit mature all at once, would be valuable for intensive agriculture. Selecting plants whose shapes are amenable to mechanical harvesting would also be a major advance.

Fruit Quality Much could be done with producing color variations. Fruits that are striped, variegated, and oddly shaped have often been observed, but so far have not been developed.

It is easy to select for larger berries, but these tend to split during transport. Overcoming splitting would be an important factor in advancing the crop. The causes of splitting should be investigated.

Postharvest handling and improved methods of shipping, harvesting, and storing goldenberries would allow a greater return to the grower and a more satisfied consumer. (The techniques used for handling tomatoes might apply here. For example, the use of ethylene might be employed to force uniform ripening.)

Agronomic Research Techniques of pruning that adapt the plants to mechanical harvesting could increase yields while lowering the costs of production in countries where labor costs are high. Also, modern trellising methods (as used with pepino, for example) should be tested.

There is little practical understanding of pests and diseases. Baseline studies are needed.

Applying to the goldenberry the tissue-culture techniques developed for tomatoes and potatoes may open up avenues for the rapid propagation of desirable cultivars with predictable performance.

The use of hormones to induce fruit set and increase fruit size should be explored.

End-Use Research Additional research is needed on the vitamin and mineral content as compared with other berry fruit. It seems likely that nutritional quality could become a valuable tool in promoting the goldenberry.

Research is also needed on new uses for the processed fruits, especially in products that ship and store well—for example, marmalades, preserves, and jams—but also in wine, ice cream, milk shakes, yogurt, and toppings for desserts and cakes.

GOLDENBERRY IN KENYA

In the highland areas of East Africa, the goldenberry (locally called "cape gooseberry") is popular, especially among the urban population. The fruit is a common item in greengrocers' shops. It is also widely available packed in cans or as jam or preserves.

In the early 1980s, Kenya Orchards Limited (Mua Hills, Machako District) was canning 10–15,000 kg per year. Fruit was purchased from small producers in the immediate area. More fruit would have been processed had it been available. The cannery considered exporting but could not meet the local demand, which at that time was steadily increasing. In addition to pure preserves, the cannery produced tasty mixtures of goldenberry with mulberries, guavas, and peaches. The cannery was experimenting with canning goldenberry juice, but problems were encountered with brewing, precipitins, and separation after heating in the sterilization process.

Fruit could be held, if necessary, at ambient temperatures (in the 20°C range) for 3–7 days, if not too ripe.

The cannery was realizing a large profit—that is, the finished products were selling at 12 times the purchase price of the fruit.

Farmers in the area were giving little or no attention to the young plants once their seedlings or cuttings were established. However, in the higher, cooler areas near Nairobi, at least one large vegetable producer was growing a selection from South Africa under improved agronomic conditions. All fruit was sold at premium prices in the better greengrocers' shops. It was especially popular among the local Europeans and the large expatriate population, who consumed it both raw and stewed.

B. H. Waite

SPECIES INFORMATION

Botanical Name   *Physalis peruviana* Linnaeus
Family   Solanaceae (nightshade family)
Synonym   *Physalis edulis* Sims
Common Names
   Quechua: topotopo
   Aymara: uchuba, cuchuva
   Spanish: uvilla, capulí, uchuva (Colombia), aguaymanto, amor en bolsa, cereza del Perú, cuchuva, lengua de vaca, motojobobo embolsado, sacabuche, tomate silvestre, topotopo, yuyo de ojas
   English: golden berry (South Africa, U.K.), cape gooseberry, giant

groundcherry, Peruvian groundcherry, Peruvian cherry (U.S.), poha
(Hawaii), jam fruit (India), physalis
  German: Ananaskirsche, essbare Judaskirsche, Kap-Stachelbeere,
peruanische Schlutte, judenkirche
  Dutch: lampion
  French: coquerelle, coqueret, coqueret du Pérou, alkékénge du
Pérou
  Italian: fisalis
  Portuguese: batetesta, camapú, camapum, groselha do Perú, herva
noiva do Perú, tomate inglês, tomateiro inglês
  Hindi: teparee, makowi
  Sinhalese: thol thakkali
  Malawi: jamu, Peruvian cherry
  Arabic: habwa (Sudan)

Origin. The goldenberry was known to the Incas, but its origin is
obscure. It grows wild in many parts of the Andes (for instance, in
Colombian forests above 2,200 m elevation), but whether these are
wild ancestral plants or just cultivated plants run wild is not clear.

Description. The goldenberry is a branched, shrubby herb normally growing to about 1 m, with velvety, heart-shaped leaves. Before
reaching full height, side branches develop and soon grow larger than
the main stem, causing the plant to straggle sideways. If staked,
pruned, and given good care, height may reach 2 m.
  The yellow, bell-shaped flowers are easily pollinated by insects and
wind. (Insect pollinators, such as bees, generally appear to help fruit
set.) The calyx at the base of the flower forms a "bladder" around
the fruit as it begins to form, eventually enclosing it fully. This husk
becomes straw-colored and parchmentlike on maturity. In warmer
climates, the plant can flower and fruit year-round.
  The fruit measures 1.25–2 cm in diameter and contains many flat
seeds—it is somewhat like a miniature tomato in internal structure.
When fully mature, the husk and fruit drop to the ground together.

Horticultural Varieties. Although goldenberry has been commercially cultivated in some areas for more than 200 years and local
genotypes are common, selected strains for commercial use are not
widely available. A vigorous, large-fruited type (more than 2 cm in
diameter) has recently been developed in Oregon (United States).[15]

---

[15] Information from Peace Seeds.

Also, several plant nurseries in the United States and United Kingdom sell named types aimed primarily at home gardeners.

Environmental Requirements

Daylength. The plant is apparently not greatly restricted by day-length because it yields fruit well both near the equator and at high latitudes (in New Zealand, for example).

Rainfall. At least 800 mm of moisture is necessary during the growing season. Greater amounts (up to 4,300 mm have been reported if soil drainage is good) increase yield, although excessive moisture can promote diseases as well as hamper fruit set (probably because it decreases pollination).

Altitude. Apparently unimportant. The fruit is grown from sea level in New Zealand, for instance, to 2,600 m near the equator in the Andes.

Low Temperature. Some tolerance to light frost has been noted, but plantings will not prosper when night temperatures are consistently lower than about 10°C.

High Temperature. Heat apparently does not inhibit fruit setting. In Hawaii, the plant produces fruit where day temperatures are in the range of 27–30°C.

Soil Type. The plant is fairly adaptable to a wide variety of soils (pH 4.5–8.2), most notably highly weathered tropical latosols.[16] Fertile, well-drained, sandy soil is preferred, although vegetative growth can overwhelm fruit production if soils are too rich.

[16] Information from J. Duke.

Highland Papayas

The tropical papaya probably originated in the lowlands of Central America, but since the time of Columbus it has spread throughout the tropics and become well established. Languishing in the highlands of South America are several intriguing relatives that should be more widely cultivated. They, too, may have worldwide potential.

These "highland papayas"[1] are particularly common in upland valleys of Ecuador and Colombia, but they can be found from Venezuela well into the southern cone countries. Like the common papaya, they are *Carica* species, but compared with their well-known tropical cousin, the highland papayas tend to be smaller, less succulent, and quite different in taste. The vast range of their diversity has not yet been collected or codified to any extent, and surprising discoveries quite likely await explorers, researchers, and entrepreneurs.

These mountain papayas incorporate a wide array of flavors and qualities. Although some specimens are tasteless and many have to be cooked with sugar to make them palatable, a few specimens are highly appealing as fresh fruits—having mild, fresh, melonlike textures and flavors. Extremely fragrant, they add an alluring scent to meals or special occasions. The types used for cooking are appealing in their own way. They are commonly added to soups and stews, to which they lend rich, fruity flavors.

Generally speaking, highland papaya plants resemble the tropical papaya plant in appearance and have similar cultivation requirements. All are "herbaceous trees." All can have enormous yields: often within a year or two of planting, the palmlike trunks are stacked with 60 mature fruits; some can have as many as 200. The fruits look somewhat like papayas, and all contain the enzyme papain—at least when immature.

There are four important potentials for these crops:

- Direct use. The types that produce tasty, high-quality fruits could

---

[1] We have elected to use "highland papayas" as a collective name for these Andean species. They have an utterly confusing variety of local names, the same word often being applied to different fruits in different areas.

be propagated and commercialized. One, the babaco (see later), is already entering international trade.

• Creating new fruits. Highland papayas are fascinating "raw materials" from which new fruits can be created. Given their great variability and the fact that many are interfertile, the opportunities for generating new taste combinations are immense. Horticulturists in several South American countries, as well as in New Zealand, Australia, Taiwan, Singapore, Italy, and Israel, are now exploring such crosses.

• Extending the range of papaya cultivation. These Andean cousins come from subtropical areas with elevations up to 3,000 m. Their genetic endowment for cold resistance could be of great significance: adding a few degrees of cold adaptability could expand enormously the world's production of, and appreciation for, common papayas.[2] They might, for instance, result in papayas that are suited to subtropical regions (such as Southern California and the shores of the Mediterranean) where commercial papaya cultivation is now impossible.[3]

• Improving papaya production. The tropical papaya is plagued by pests. Genes from highland papayas have already been effectively employed in creating cultivars for regions where diseases (especially viruses) and pests (such as fruit flies) now restrict papaya cultivation, but more genetic benefits remain to be tapped.

The following pages highlight six promising Andean highland papayas (four species and two hybrids).

SPECIES[4]

Chamburo. From Panama to Chile and Argentina, the chamburo[5] (*Carica pubescens*[6]) is commonly found around mountain villages. It

[2] The highland species are difficult to cross with common papaya using traditional breeding techniques, but newer methods seem likely to make the process routinely successful. For example, tissue culture techniques to propagate plants from fertile but nonviable seeds add an important tool to the quest for new papaya combinations. Information from J. Martineau, R. Litz, and H.Y. Nakasone.

[3] Highland papayas will not withstand heavy frost, but they yield under cooler climatic conditions than normal papaya plants. Their flowers and fruits are less affected by cool weather. Thus, they produce ripe fruit at cool temperatures where normal papaya fruits remain immature and insipid until they rot.

[4] For the sake of simplicity, throughout the chapter we have used the scientific names of Badillo (1971) for species, and those of Heilborn (1921) for hybrids. Only the most common of the multitudinous synonyms are listed.

[5] Other common names include chambur, chamburu, chambura, papaya de olor, papaya de montaña, papaya de altura, papaya de tierra fria, col de monte, papayuella, papayo, siglalón, chihaulcan, chiehuacan, bonete (Mexico), and mountain papaya or mountain pawpaw. Some of these names are also used for other highland papayas. In southern Ecuador as well as outside the Andes (notably New Zealand), the name "chamburo" commonly refers to *Carica stipulata* (see below).

[6] Strictly speaking, the botanical name is *Carica pubescens* Lenné & Koch. Synonyms include *C. cundinamarcensis* and *C. candamarcensis*.

is by far the most common and widespread species of highland papaya. In eastern Peru it occurs in almost every backyard. In northern Chile it is grown commercially in plantations.

The five-sided yellow fruits have firm flesh and a fragrance both pleasing and penetrating. They are 15–20 cm long, weigh about 130 g, and have a medium papain content. The interior cavity contains many spiky seeds.

Fruits vary greatly in sweetness: some can be eaten fresh, but most must be cooked. They are suitable for stuffing with fruits, vegetables, or other fillings because their firm flesh holds its shape during cooking. They yield a clear juice and are excellent in pies, ice cream, marmalades, or sweets. Canned preserves are marketed in Chile.

The plant is distinguished by the coating of hairs on the underside of the leaves. It grows in areas ranging from dry, windy, open plateaus to humid, shaded forests. It is found at altitudes up to 3,000 m, and withstands –3°C without serious injury. Most plants have flowers of a single sex (dioecious); a few have flowers of both sexes (monoecious).

Chamburo is generally propagated by seeds. It grows vigorously and bears fruit in its second year. It is fairly tolerant of nematodes and is perhaps resistant to papaya ring spot virus, the most devastating disease of the common papaya.

Siglalón.   This species (*Carica stipulata*[7]) is limited to a small region of southern Ecuador,[8] usually at elevations of 1,600–2,500 m. Again, there is much variability among individual plants. In comparison with common papayas, the fruits are quite small, and most are cooked with sugar and eaten as sweets. They are said to be the best highland papayas for jams and sauces. They are also boiled, strained, sweetened, and made into fruit drinks, normally in blends with other juices. Only a few plants yield fruits that can be eaten raw, but some of those are delicious.

The yellow fruit has 10 or 11 ridges, weighs 40–150 g, and possesses a strong, pleasant aroma. The creamy flesh contains many seeds— some smooth, some corky. Many fruits are soft skinned and long; others are firm and squat. They are easier to peel than those of chamburo (*C. pubescens*), but the immature fruits have perhaps the highest papain content of any papaya and, even in ripe ones, the raw juice tends to irritate the skin, notably the corners of the mouth.

The spiny, occasionally branched, tree is robust and grows vigorously

[7] This species was first distinguished taxonomically only in 1966 as *Carica stipulata* Badillo.
[8] In Azuay and Loja provinces, where it is native, it is also known as siglalón silvestre, paronchi, toronche, toronchi del campo, and chamburo, the name often used in popular literature.

its first year to as much as 2–3 m. (It may eventually reach 8 m.) Male and female flowers are borne on separate plants. Fruits develop in the second year, and some plants have borne fruits annually for more than 20 years. The seeds come fairly true, but vegetatively propagated plants bear fruit at an earlier age.

It is possible that this (and some of the other highland species) might be grown as a source of papain. This enzyme is a valuable international commodity used, among other things, to clarify beer.

Siglalón has fairly good resistance to papaya ring spot virus.

Col de Monte.   The col de monte[9] species (*Carica monoica*[10]) is a "dwarf" papaya, most commonly found in cultivation in Ecuador and eastern Peru between 600–1,700 m elevation. There, the people like the fruit so much that they protect and nurture even wild specimens.

The smooth, hard-skinned fruit has faint ridges and an orange, sometimes brilliant, skin. Typically, it has firm, orange or deep-yellow flesh with only a slight taste, large-horned seeds, and a high papain content.

Some of these small, fragrant fruits are eaten raw—usually mixed with other fruits. Cooked with lemon and sugar they have been likened to stewed apricots. Their firmness makes them suitable for drying and candying, and they freeze well. The young seedlings and mature leaves are cooked as greens (hence the common name "col de monte," which, translated, means "mountain cabbage").

The plant grows vigorously, but reaches only 1–3 m in height. It is found mostly in areas of high rainfall and mild temperatures. It is generally monoecious, commonly having male and female flowers together on the same inflorescence. The flowers are usually self-fertilized and the plant comes fairly true from seed. It crosses easily with *C. pubescens*, often yielding hybrids that bear much fruit of good flavor.

It is susceptible to papaya ring spot virus.

Papayuelo.[11]   As with the other highland papayas, this small, hardy Colombian species (*Carica goudotiana*[12]) is highly variable. Some of its fruits are quite sweet with an attractive taste somewhat like apples. Others are astringent and barely edible, even cooked.

The fruits are usually five-angled, pale yellow, with occasional

---

[9] Other common names include col de montaña, papaya de selva, tomate de monte, peladera, peladua, dwarf papaya, orange paw paw, and Peruvian cooking papaya. The name "col de monte" is often loosely used for any short-statured highland papaya.
[10] One synonym is *Carica boliviana*.
[11] Sometimes called "col de monte" or "payuello." Both these common names are used in other areas for other highland papayas.
[12] One synonym is *Papaya gracilis*.

shades of purple, red, or orange. They grow up to 20 cm long and weigh up to 200 g.

The plant occurs in humid forests between 1,500–2,300 m elevation. It can, however, tolerate some dryness. It often branches at the base and has an upright habit with a gray trunk. Generally small, it can reach heights of 8 m. It is mostly dioecious (both male and female flowers are sometimes intensely red or purple, with red or green petioles), bears heavily, and its fruits transport well if harvested before full maturity.

HYBRIDS

Several of the more popular highland papayas are natural hybrids. Found most commonly in southern Ecuador, these present a great deal of variation in fruit size and quality. Some produce a few viable seeds whereas others produce no seeds at all. To maintain clonal uniformity, the plants are most often propagated vegetatively.

Two of these hybrids are highlighted here.

Babaco. The babaco[13] is the most commercially advanced highland papaya. Its fruits are fragrant and flavorful, with quite a different scent and taste from the common papaya. Its distinctive flavor has been likened to the taste of strawberry with a hint of pineapple. Sweet to some palates, it is pleasantly subacid—unique and refreshing. The large, normally seedless, "zeppelinlike" fruits, some weighing 2 kg, can be consumed fresh or stewed.

When picked, the babaco ("ba-*bah*-co") is green to yellow; when mature, it is bright yellow. The pulp is white to cream in color, juicy, and melting in consistency. The fruit is easy to prepare because it can be eaten skin and all, and there are no seeds to remove.

Babacos are popular and common over a wide area of the cool highlands of Ecuador between elevations of 1,400 and 2,500 m. Most are found in small holdings, but some larger plantations have been established. Until recently it was unknown in neighboring countries, but plantings have now been started in Colombia. It is also being grown successfully in New Zealand, Australia, Israel, Italy, Guernsey, and California. It is probable that babaco could grow well in many other locations.[14] It is especially successful as a greenhouse crop because the plant is small and needs no pollinators.

[13] In the terminology of Heilborn, who treated the plant as a species, this plant's botanic name is *Carica pentagona* Heilborn. The full binomial that indicates its hybrid status is *Carica x heilbornii* Badillo n.m. *pentagona* (Heilborn) pro species. It seems to be one of the possible crosses between *Carica pubescens* and *Carica stipulata*.

[14] Kenya, Greece, and Brazil are reportedly starting commerical production.

Auckland, New Zealand. Babaco has a phenomenal capacity for high yield. Yields of 100 tons per hectare have been recorded merely a year or two after the trees were planted. Shown here are the fruits of just three babaco trees. (D. Endt)

Babaco has potential for export as a fresh fruit. New Zealand growers are already shipping it to Japan and the United States. When properly harvested and carefully packed, it remains in good condition for a month or more at 6°C. The cut fruit also keeps well and does not turn brown (oxidize) over time.

Babaco is a small plant (1–2 m), only occasionally branched, but it coppices well. Total yields are better than those of a good papaya plantation, and have exceeded 100 ton per hectare.[15] Propagation is by cuttings or tissue culture.

Toronchi. Toronchi[16] is a natural hybrid[17] found scattered around houses and villages in southern Ecuador. It is a frost-resistant, vigorous plant that may grow to 5 m, although most selected types are smaller, reaching only 2.5 m. It is found up to about 2,500 m elevation, and can withstand temperatures down to 1°C. There is much variation, but all specimens produce attractive, quality fruits with a delightful fragrance. They are eaten fresh or processed and are readily accepted, even by those tasting them for the first time. Cooked, they are useful for sauces, jams, pie fillings, and pickles, as well as for adding to cheesecake and dairy products such as yogurt.

The mature fruit is five-sided, generally 10–15 cm long, green to lemon-yellow in color, and weighs up to 0.5 kg. It is extremely juicy with a soft, creamy-white pulp and is one of the least seedy of the highland papayas. Smooth skinned, it can be eaten without peeling, and has a medium papain content. A New Zealand variety called "Lemon Creme" is a vigorous cultivar that produces an abundant crop of sweet, lemon-scented fruits.

Fast growing, these plants bear within 12 months. Picked ripe, the fresh, raw fruits are superior in flavor to all other highland papayas. However, they must be handled more delicately than the babaco; even when moderately ripe, they are poor shippers.

Although the plant functions as a sterile hybrid, pollination seems to increase production, and the fruits often contain viable seeds. Toronchi hybridizes readily with siglalón (*C. stipulata*), resulting in many intermediate forms with different flavors and varying amounts of seed.

PROSPECTS

The Andes. All of these fruits are consumed locally throughout much of the Andean highlands, but currently only Colombia, Ecuador,

---

[15] Information from F. Cossio. See also Cossio, 1988, p. 50.
[16] Common names include toronche, toronche de Castilla, poronchi, chamburo, and chamburo de Castilla.
[17] *Carica chrysopetala* Heilborn. This hybrid is now commonly known as *Carica* x *heilbornii* Badillo n.m *chrysopetala* (Heilborn) pro species.

Auckland, New Zealand. As with many little-known Andean fruits, innovative and courageous private researchers in New Zealand have pioneered the production and international export of babaco. Shown here on January 7, 1982, is part of the first-ever export shipment, ready for loading into the temperature-controlled cargo hold of a 747 aircraft bound for Frankfurt, West Germany. (D. Endt)

Chile, and Venezuela are seriously exploiting their commercial potential.[18]

These plants deserve much more horticultural attention. They could add appreciably to fresh-fruit production in the Andean highlands, and they are generally well suited to the smallholder. The benefits to child nutrition, family income, and general welfare are likely to be significant.

Given adequate quality control, it might be possible to develop both a fresh-fruit export business and an Andean papain-extraction industry. (This enzyme is a low-volume, high-value item with export potential.)

Because highland papayas are so variable and adaptable, it seems likely that types superior to those of today will be discovered or developed. With their great adaptability and high yield these "new" papayas—given genetic selection and improvement—should be a bonanza.

[18] In Colonia Tovar, Venezuela (near Caracas), babaco production has become established since two scientists introduced cultivars 20 years ago.

Other Developing Areas. For cooler parts of the developing world, highland papayas could provide a range of future fruits. Indeed, with their extreme variability, they could become a veritable backyard fruit bazaar for upland villages from Morocco to Papua New Guinea. Like the common papaya in the lowlands, these Andean species and hybrids could provide masses of tasty, nutritious fruits in the highlands. It is likely they would enter local trade and, in time, develop into profitable small-scale or even large-scale domestic and export operations. Before their commercial value can be exploited, however, there is a need to select and develop types that have good and uniform qualities and that can be propagated on a large scale.

Industrialized Regions. Highland papayas may be potential fruits for subtropical and warm-temperate areas of North America, Europe, Japan, and Australasia. However, they require climates that are free from both frosts and excessive summer heat.

These plants can be susceptible to certain pests and diseases, particularly to nematodes, mites, root rot, and viruses. Their climatic requirements and cultural practices for maximum production and best flavor are little known. Thus, considerable horticultural attention is needed before they can be exploited with confidence. Postharvest handling studies would greatly increase their market potential.

Although the fragrance of many of these fruits is instantly appealing, the taste is unusual, often must be acquired, and can be disappointing to consumers expecting a papayalike flavor. Although for simplicity we have called these "highland papayas," an important strategy in the wider marketing of these species is to create or adopt new names that avoid the papaya image and that allow the fruits to be judged on their own merits.

To create cold-tolerant papayas has been the plant breeder's goal for almost a century. With modern technology, better germplasm collections of Andean species, and old-fashioned persistence, it seems probable that this goal can be achieved.

Lucuma

In the temperate highlands of Colombia, Ecuador, Peru, and northern Chile, lucuma[1] (*Pouteria lucuma*) is common. Well known to the Incas, it is an unusual fruit with smooth, bronze-yellow skin, and somewhat resembles a persimmon, but there the similarity ends. Its bright yellow or orange flesh is usually blended into other foods.

Lucuma (pronounced *luke*-mah) pulp is popular in drinks, puddings, pies, cookies, and cakes. It tastes and smells like maple syrup. Added to milk or ice cream, it contributes both color and flavor. It is a frequent component of milk shakes, typically made with lucuma but without ice cream.

For all that, lucuma is little known outside its homeland—which is strange because it is rich, nutritious, and satisfying; is versatile; and possesses a distinctive flavor. It is enjoyed largely for its flavor, but in some parts of Peru and Ecuador it plays a significant part in the basic diet of the poor. Lucuma fruits can weigh 1 kg, they are very filling, and one tree can produce as many as 500 fruits during a year—enough to feed whole families. And at times when field crops are out of season or stressed by drought, lucuma, with its year-round production and deep roots, literally becomes the tree of life.

Unlike most sweet fruits, the lucuma is high in solids and is a good source of carbohydrate and calories. When the fruit falls from the tree, it is still unripe. It is stored in hay or other dry material until soft. Even fully ripened, the pulp is firm and almost pumpkinlike in texture. Low in acid, it is a good source of minerals, particularly iron, as well as of vitamins, especially carotene (provitamin A) and niacin (vitamin B_3).

An unusual advantage is that the fruit, when ripe, can be dried and milled into a mealy flour. The flour can be shipped long distances, stored for years in airtight containers, and (in Peru at least) is found in markets year-round. Fresh, undried lucuma pulp can also be frozen and stored safely for long periods.

Besides feeding people, its fruits are said to make a good feed for chickens, promoting both growth and eggs with bright-yellow yolks. In addition to its fruits, the lucuma tree is valued for its dense, durable timber.

---

[1] Also spelled lucmo. The botanical name is often also given as *Pouteria obovata*, or *Lucuma obovata*.

Not enough is known about this fruit to fill in the horticulture, harvesting, nutrition, research needs, and other details as given in previous chapters. However, the tree has the following environmental requirements.

Daylength. Fruits are set in latitudes from the equator to 33°S in Chile, so daylength seems unimportant.

Rainfall. The plant grows well in areas subjected to occasional dryness and tolerates seasonal rains well, but not waterlogging or extended humid weather.

Altitude. Although most common in inter-Andean valleys between 1,500 and 3,000 m elevation, lucuma grows well and produces fruits of high quality in the Peruvian lowlands and at sea level in Chile.

Temperature. Although it thrives in cool highlands, lucuma seems to require frost-free climates and is killed by −5°C temperatures. Its climatic requirements are roughly comparable to those of lemons.

Soil Type. Lucuma appears adapted to sandy and rocky sites and needs well-drained soils. It tolerates moderate salinity, calcareous soils, and trace element deficiencies (particularly iron) that often restrict other fruit trees. However, it yields best in deep alluvial soils high in organic matter.

Lucuma is highly variable in fruit size and quality, but has received little horticultural or botanical attention.[2] Fruit quality seems highly dependent on seedling type, climate, and horticultural practice. Commercial orchards would be more feasible if elite types were selected and propagated vegetatively.[3]

PROSPECTS

The Andes. Lucuma is best known and enjoyed in Chile, Peru, and southern Ecuador. In spite of local popularity, it has suffered elsewhere in the Andes because the types tried there produced fruits that were too dry or of poor flavor. If superior types are selected and propagated, this fruit has a much greater future throughout the region.

Peru and Chile have recently established named lucuma cultivars from selected grafts and seedlings. However, many valuable types remain in orchards and backyards—still to be "discovered." Given the availability of superior cultivars suited to different climates, markets could be stimulated from Venezuela to Chile and Argentina.

[2] Traditional cultivars are Seda and Palo. More recent selections have been made at the Universidad de Chile and at La Molina University.
[3] Grafting is difficult to achieve, but tissue culture is showing some promise. Information from M. Morán-Robles.

Lucuma fruits, unlike most fruits, are rich in starch and relatively dry. They are often used as a basic food and can be dried into flour that is easily stored and can provide a tasty treat even years later. (ProChile)

The dried, ground pulp is prepared in small factories in Chile and Peru, but this easily transportable powder should be suitable both for expanded home processing and for increased commercial use. Export markets could develop in the United States, Japan, and other affluent societies looking for new flavors for dessert foods. Already, Chile is shipping lucuma to Switzerland, where the fruit is used to flavor ice cream. It is said that the flavor cannot be reproduced artificially.[4]

Other Developing Areas. Except for plantations in Costa Rica, the species is virtually unknown in commercial production outside South America. Plantings should be tried in other dry and frost-free highland areas of the tropics and subtropics. It seems likely to become useful in parts of Mexico, Central America, Brazil, and Central and southern Africa.

Industrialized Regions. Lucuma has been tested on a backyard scale outside Latin America. It yielded satisfactorily in Hawaii, but so far has produced only poor-quality fruit in Florida. Some trees in California did well at first, but were eventually frozen out. Trials are now under way in Queensland, Australia, and the plant has shown early promise in sheltered frost-free sites in northern New Zealand.

[4] Information from A. Endt.

Naranjilla (Lulo)

For centuries the naranjilla (*Solanum quitoense*) has been an immensely popular fruit of Colombia and Ecuador. Writers have described it as "the golden fruit of the Andes" and "the nectar of the gods."

Orange-yellow on the outside,[1] the fruits look somewhat like tomatoes on the inside, but their pulp is green. Their juice, considered the best in the region, is used to flavor drinks.[2] In fact, many even prefer it to orange juice.

Although little known to the outside world, naranjilla (usually pronounced na-ran-*hee*-ya in English) appears likely to produce a new taste for the world's tables. It promises to become a new tropical flavor with a potential at least as great as the increasingly popular passionfruit (see page 287).

However, producing naranjilla is a scientific challenge; before it can achieve its potential, it needs intensive research. Despite its overwhelming popularity in the northern Andes, it has been given little serious commercial development. In fact, owing to several factors, naranjilla fruits have become scarce and expensive in Ecuadorian and other Andean markets.[3] Through misfortune and lack of funds, efforts to check the devastation of nematode pests have failed so that production in some areas is declining. On the other hand, demand is higher than ever, owing to naranjilla's local popularity and the increasing export of both fresh fruits and canned products.

Given attention, problems such as these should be entirely avoidable, but even when such operational difficulties are overcome, naranjilla will still be a challenge to produce. It has little genetic diversity and, consequently, is probably restricted to a narrow range of habitats. It almost certainly requires a cool, moist environment—a type that is of limited occurrence. It may also require a specialized pollinator.

[1] Although "naranjilla" is Spanish for "little orange," the fruit is not a citrus, but is a relative of the tomato and potato. In many areas it is called "lulo," a pre-Columbian word, possibly of Quechua origin.
[2] In the 1760s, the Majorcan missionary Juan de Santa Gertrudis Serra wrote of the naranjilla: "The fruit is very fresh [and diluted] in water with sugar, makes a refreshing drink of which I may say that it is the most delicious that I have tasted in the world."
[3] In the last decade, prices have increased more than tenfold (even accounting for inflation).

Nonetheless, with study, these problems can probably be overcome, or at least mitigated. Then the taste of naranjilla should become known to millions.

PROSPECTS

The Andes. Although now in decline, the naranjilla could become one of the major horticultural products of the region and an important market crop for small-scale producers. The fruit or juice (canned, frozen, or concentrated) has considerable export potential.[4] What is needed is a coordinated effort to fully understand the crop's status and difficulties.

Nematocides and biological controls are currently available to forestall the devastation caused by root-knot nematodes. In addition, at least two closely related species, apparently highly resistant to the root-knot nematode, seem promising as rootstocks. They may also be sources of genetic resistance, for they form fertile hybrids with naranjilla.

Other Developing Areas. With the increasing international demand for exotic fruits, this is a budding crop for the uplands of Central America and for other areas of similar climate. Already, naranjilla has been established as a small-scale crop in Panama, Costa Rica, and Guatemala. Both there and in other frost-free, subtropical sites, it promises to become a substantial resource. However, because of the plant's restrictive climatic and agronomic requirements, success will not be achieved easily. Establishing naranjilla in commercial production will require much work and dedication.

Industrialized Regions. Naranjilla can provide the basis for a new fruit-drink flavor that could become popular in North America, Japan, Europe, and other such areas. In a test at Cornell University several years ago, blindfolded panelists unfamiliar with the fruit chose naranjilla juice over apple juice by three to one, and a blend of naranjilla and apple juice over apple juice alone by nine to one. In the 1970s, a major U.S. soup manufacturer created a fruit drink based on naranjilla for nationwide sale, but it reluctantly abandoned the project because of problems in producing a large and reliable supply of fruit.

---

[4] Colombia is already exporting small amounts to Central America and the United States.

Pasto, Colombia. Naranjilla is among the most popular fruits in the northern Andes. (R.E. Schultes)

USES

The naranjilla is versatile. It can be eaten raw or cooked or used to make juice. It is also cooked in fruit pies and confections, and is used to make jellies, jams, and other preserves. In Venezuela, Panama, Costa Rica, and Guatemala, unstrained pulp is used for toppings on cheesecakes, sponges, ice cream, yoghurt, and fruit salads. The fresh juice is also processed into frozen concentrate and can be fermented to make wine.

Despite its versatility, naranjilla is mainly used at present to flavor drinks. In Ecuador and Colombia, naranjilla *sorbete* is something of a national drink, often served in hotels and restaurants. It is made like lemonade: the freshly extracted juice is beaten with sugar into a foamy liquid that is green, heavy-bodied, and sweet-sour in flavor. (Most tasters express surprise that it is not a blend of several fruits.)

Naranjillas are eaten only when fully ripe, at which time they yield to a soft squeeze and their rather leathery skin is bright orange or yellow (though sometimes still marbled with green). On average, they are about the size of golf balls. The slightly acid flavor is more pronounced if the fruit is not completely ripe. However, even some ripe fruits are too acid to be eaten raw, and the pulp must be sweetened to be palatable.

Naranjilla fruit on a plant growing near Versailles, Colombia. (J. Morton)

NUTRITION

The naranjilla is rich in vitamins, proteins, and minerals.[5] It is said to contain pepsin, the stomach enzyme that aids digestion of proteins.

AGRONOMY

The plant is propagated by seeds, cuttings, or grafts onto the rootstock of other species. Seeds germinate freely. Cuttings root easily, especially when parts of older, slightly woody stems are used. Like many members of the Solanaceae, it can also be regenerated in tissue culture from pieces of leaf or stem tissue.

The plant grows rapidly. Seedlings begin bearing in 6–12 months; grafted plants mature even faster, flowering at 3–4 months of age and maturing fruits at 6 months. In principle, this perennial could continue bearing for years, but in the Andes and Central America plantings usually succumb to root-knot nematodes after about 4 years.

[5] The composition per 100 g edible portion: calories, 23; water, 92.5 g; protein, 0.6 g; fat, 0.1 g; carbohydrates, 5.7 g; fiber, 0.3 g; ash, 0.8 g; calcium, 8 mg; phosphorus, 12 mg; vitamin A, 600 Int. units; thiamine, 0.04 mg; riboflavin, 0.04 mg; niacin, 1.5 mg; ascorbic acid, 25 mg. Information from J. Morton.

Naranjilla fruits look something like tomatoes, but they have yellow skin and green flesh. Their greenish juice provides one of the culinary delights of Colombia, Ecuador, and Venezuela. (W.H. Hodge)

Naranjilla is a "heavy feeder" and responds well to fertilization. Pruning old woody stems at the end of a fruiting cycle results in vigorous regrowth and prevents fruit size from diminishing.

Andean farmers mainly grow naranjilla on rainy slopes, where, as long as temperatures remain moderate, fruits are produced year-round. To prevent fungal and bacterial root infections, well-drained soils are imperative. A common practice is to plant naranjilla in openings in the forest or to interplant it with banana, tamarillo (see page 307), or achira (see page 27). The taller plants help protect the naranjilla's brittle branches from wind damage.[6]

HARVESTING AND HANDLING

The fruits are closely borne in the axils of the leaves on stems and branches. They are easy to pull off by hand and are normally picked when about half ripe—that is, when they have started to color. (They subsequently ripen normally.) The stiff hairs can irritate the skin, so the fruits are handled with gloves until ripe enough for the fuzz to be wiped off with a towel.

The picked fruits have a shelf life of up to two weeks without

[6] It has been suggested that on a commercial scale, the naranjilla could be interplanted between babaco (see page 257), mandarin orange, passionfruit (on wire lattices), or other crop with the naranjilla's climatic requirements. Suggestion from D. Endt.

refrigeration, making naranjilla an ideal truck crop. Cold storage lengthens the storage period considerably.

Yields are high. Individual plants may produce up to 10 kg of fruit a year, and on a per-hectare basis may yield 27 tons of fruit or 47,000 liters of juice.

If not handled correctly, much of the flavor can be lost in canning and the juice can turn muddy looking. Proper cooling, storage, and the use of antioxidants are necessary.

LIMITATIONS

Naranjilla's major problems have already been discussed. They are its climatic restrictions and susceptibility to pests. More information is given below.

Adaptability   This plant has received little attention from horticultural researchers; its environmental adaptations, therefore, are not well known, but seem to be narrow. It is possible that it needs a long growing season, as well as high humidity. It cannot tolerate frost. Heat and dryness can also cause crop failure.

Moreover, there are possible pollination problems. Naranjilla appears to be a short-day plant; pollen abortion occurs when days are long.[7] Pollinators may be absent in locations outside its native range. The effects of shade and altitude are also uncertain. The plant is said to perform poorly under 1,200 m elevation in the Andes.

Pests and Diseases   As noted, the plant is extremely susceptible to root-knot nematodes (*Meloidogyne* species). Plantings often fail to reach full fruiting size as a result. The naranjilla also suffers from insect pests; in particular, a wide variety of coleopterans (beetles and weevils) chew the leaves.

The plants can succumb to diseases such as bacterial wilt and fungal infections. Root and stem rots can be particularly severe. Viruses, too, can be troublesome.

RESEARCH NEEDS

Germplasm Collection   Replicate germplasm collections should be established in Ecuador, Colombia, Peru, Costa Rica, and other countries. A germplasm collection of 185 samples is already being evaluated at the Instituto Colombiano Agropecuario (ICA) in Colombia.[8]

---

[7] Information from J. Soria.
[8] Information from L.E. Lopez J.

Nematodes   The nematode problem is the major one to be addressed. As a starter, nematologists should determine the varieties of the offending pests. Also, the relation between nematode resistance and temperature should be checked. (Recent research has shown that nematode-resistant tomatoes become susceptible as temperature rises.)

Although nematode-killing chemicals can be used to treat the soil, these tend to be toxic and expensive. An alternative approach could be biological control.[9] Screening for strains resistant to nematodes (and viruses) seems promising as well, although development of horticulturally viable types could take years.

Alternative approaches include the following:

• Hybridizing naranjilla with closely related, nematode-resistant species. Hybrids with *Solanum hirtum* and *S. macranthum*, for example, have good nematode resistance. Backcrossing these to naranjilla has produced a range of plants that have shown resistance and have borne fairly good fruit.[10]

• Grafting naranjilla on related plants with nematode-resistant rootstock. When cleft grafted on species such as *S. macranthum* and *S. mammosum*, naranjilla plants have survived for about three years and fruited successfully. In tropical Africa, naranjilla has done well when grafted to its nematode-resistant local relative, *S. torum*.[11]

• Improving plant vigor by better management.

• Producing the crop in beds of sterilized soil.

• Growing a cover crop or rotation crop of plants, such as velvet bean or *Indigofera* species, that help eliminate nematode infestations.

• Inducing somaclonal variation in regenerated plants as a way of unmasking inherent nematode resistance that is now hidden.

• Educating farmers about nematodes and the means of keeping sites free of infestation. (This is important, whatever other approach may be used.)

SPECIES INFORMATION

Botanical Name   *Solanum quitoense* Lamarck
Family   Solanaceae (nightshade family)

---

[9] For example, the bacterium *Pasteuria penetrans* has proved as effective as current nematocides on other crops in Britain. Spores can be produced in vivo on nematodes in roots, without the necessity of sterile production systems. Information from B.R. Kerry.
[10] Information from L.E. Lopez J.
[11] Both hybridizing and grafting also bestow resistance to root and collar rot. However, further evaluation is needed to ensure that alkaloids in the fruits do not rise to hazardous levels. Information from J. Soria.

Synonyms   *Solanum hirsutissimum; Solanum angulatum*
Common Names
  Quechua: lulo, lulu puscolulu
  Spanish: naranjilla, naranjillo; naranjilla de Quito, naranjita, lulo,
  lullo, toronja, tomate chileno (Peru)
  English: Quito orange, naranjilla, lulo
  French: naranjille, morelle de Quito, orange de Quito
  German: Lulo-Frucht

Origin.   Naranjilla's origin is unknown. Its wild progenitor may
yet be discovered—probably in Colombia. It is thought that the plant
was domesticated within the last few hundred years, because there is
no evidence that it was cultivated in pre-Columbian times. The first
records of naranjilla cultivation are from the mid–1600s in Ecuador
and Colombia. Traditionally, areas of major cultivation have been the
valleys of Pastaza and Yunguillas in Ecuador and the mountain areas
of Cauca and Nariño in Colombia.

Description.   The plant is a perennial, herbaceous shrub, 1–1.5 m
high, with stout, spreading, brittle stems. Its dark-green, purple- or
white-veined leaves are often more than 30 cm long and, like the
stems, are densely pubescent. The pale-lilac flowers are covered with
a thick "felt" of light-purple hairs.
  The spherical, yellow-orange fruit is 3–8 cm in diameter. It has a
leathery skin, densely covered with fine, brittle, easily removed, white
to brown hairs. Internally, its structure resembles a tomato. The
acidulous, yellow-green flesh contains a greenish pulp with numerous
seeds and green-colored juice.

Horticultural Varieties.   On the whole, the species is unusually
uniform for a cultivated plant. However, two geographically separated
varieties are recognized. Variety *quitoense* is the common, spineless
form found in southern Colombia and Ecuador. Variety *septentrionale*
has spines, is hardier, and grows mainly at altitudes of 1,000–1,900 m
in central Colombia and Costa Rica.
  In the last few years, an apparently new variety has appeared in
Quito markets. Although the fruits are smaller than normal, they are
rapidly becoming the dominant commercial type.[12] Investigation of
this may help open a new era in naranjilla use. Is this a new hybrid?
Is it being grown because of greater nematode resistance? Or is the
fact that its fruits have few hairs the driving force behind its production?

---

[12] Information from J. Soria.

Environmental Requirements

Daylength.   As noted, the plant seems to require short days for pollination. However, this is not certain because satisfactory fruit set has been noted in south Florida at any time of year.[13]

Rainfall.   Naranjilla requires considerable moisture. It is commercially grown in the Andean areas where annual rainfall is 1,500–3,750 mm. The lower moisture limits are uncertain, but even moderately dry conditions check its growth.

Altitude.   This is not a restriction. Samples have been collected near 2,000 m elevation in Ecuador, and the plant grows near sea level in New Zealand and California.

Low Temperature.   Below 10°C the plant's growth is severely checked. It is frost sensitive.

High Temperature.   Above 30°C the plant grows poorly. It does not set fruit in areas with high night temperature—a possible reason why it has failed in some lowland tropical and subtropical areas.

Soil Type.   In Ecuador, the naranjilla grows best on fertile, well-drained slopes. It requires soils that hold moisture but that drain well enough to avoid waterlogging. It seems particularly sensitive to salt.[14]

Related Species.   Naranjilla has several relatives that produce desirable fruits and deserve more attention. They could be useful in their own right as well as perhaps for genetically improving naranjilla.[15]

Solanum pectinatum This plant, although undomesticated, produces high-quality fruits, almost as good as those of the naranjilla. Compared with the naranjilla, the fruits are slightly smaller, but their hairs rub off more readily, and they taste somewhat sweeter. This is a lowland species, widely distributed from Peru to southern Mexico, and from sea level to 1500 m. Throughout the region, people gather the wild fruits for making juice. It, too, deserves wider appreciation; a little horticultural investigation might produce a new crop as popular as the naranjilla, but suitable for cultivation in areas too warm for naranjilla.

Solanum vestissimum Native to Colombia and Venezuela, this is another wild species with pleasantly flavored fruits. It grows at higher altitudes than naranjilla. Known as "lulo de la tierra fria" (naranjilla of the cold lands), toronja, or tumo, it is a small tree that bears fruit about the size of duck eggs. The chief objection is that the fruit's hairs are quite bristly and the juice is difficult to extract. For all that, however, it has an excellent flavor and deserves research attention.

---

[13] Information from J. Morton.
[14] Information from L. Davidson.
[15] Information in this section is mainly from C. Heiser.

Inga edulis

Pacay (Ice-Cream Beans)

All legumes have their seeds encased in pods, and a few of these pods are widely eaten as vegetables—green beans and snow peas, for example. A few lesser known legumes produce pods containing a sweet, mealy pulp and are eaten as fruits. Carob, tamarind, and honey locust are the best known.[1] However, trees of the genus *Inga* also produce sweet pods. They deserve much greater recognition and could become widely known and enjoyed.

For people in Central and South America, inga pods have long been favorite snacks. The pods are mostly narrow, straight, and in some species as long as a person's forearm. They are easily cracked open to expose the white, sugar-rich pulp—reminiscent of cotton candy—surrounding the seeds. In English they have been called "ice-cream beans" because this white pulp has a sweet flavor and smooth texture.

Most *Inga* species occur in the lowland tropics of the Americas, but several occur in the upland Andes. The main Andean one is *Inga feuillei* (pronounced "few-*i*-lee"). It is widely grown in highland valleys as well as in coastal lowlands of Peru and Ecuador where it is often employed as a shade tree or street tree. Its pods, prized as snacks, are found in markets and on street vendors' carts and are often consumed by children.

Most commonly called "pacay" and "guama,"[2] this particular inga has long been popular. Its pods are depicted in ancient ceramics. The Incas had pacay pods carried to their mountain capital of Cuzco. Pedro Pizarro reports that the Inca emperor Atahualpa sent to Francisco Pizarro a basketful of guamas as a gift.

It is surprising that this plant (as well as other *Inga* species[3]) is not

[1] See National Research Council, 1979.
[2] Pacay (pacae) is a Quechua word used in Spanish specifically for the upland Andes species. "Guama" or "guaba" are names widely used for any *Inga* species whose pods have sweet pulp.
[3] Most *Inga* species are tropical lowland species unknown in the Andes. *Inga edulis* is considered the best species for shade in coffee plantations in Colombia; *I. vera* is common in Central America. (*I. vera* and related species are covered in the companion report, *Firewood Crops Volume I*. National Research Council. 1980. National Academy Press, Washington, D.C.)

more widely known. *Inga* species are dependable, they produce in abundance, and they provide sustenance in bad times. They are a source of snacks for their owners and cash for the enterprising.[4] They grow rapidly, are tolerant of diverse soils, and are resistant to diseases and fire. They are easy to establish, spread their shade quickly, and provide fruit for years.

Inga trees produce abundant root nodules, which fix nitrogen, and benefit the land by raising fertility levels. They can produce food without occupying the farmland used for food crops, because they can grow on sites neglected by agriculture.

Pacay is cultivated specifically for its fruits; the fruits of other *Inga* species in the Andes are only a by-product of trees whose main purpose is to shade plantations of coffee and cacao. Other common species in the Andes are *I. edulis*, *I. vera*, *I. adenophylla*, and *I. densiflora*. The fruits of *I. densiflora* are sold in markets and fruit stalls, especially in Colombia. So far, however, the others are more often used for shade, not food.

Pacay and other inga trees have important futures. They are multipurpose trees and are potentially valuable additions to gardens, orchards, fields, hedgerows, or wayside wastelands throughout most warm parts of the world. They also have outstanding prospects as urban trees for much of the tropics.[5]

PROSPECTS

The Andes. Although pacay pods are already widely consumed in rural areas of the Andes, there are still many possibilities for developing inga fruits as cash crops. One potential for expanding markets exists in cities, particularly among newly urbanized campesinos. As with other fruit crops, pacay lends itself to new entrepreneurial ventures, small-business enterprise, and economic development among the poorest levels of society.

Other Developing Areas. This delectable snack that comes in its own natural wrapper should become much better known in coming decades. Inga trees can promote self-reliance, and they have potential not just in Latin America but throughout the tropics. Some species are already grown in the West Indies, Hawaii, and East Africa, but

---

[4] In Mexico, coffee-plantation workers can double their annual salary by selling the pods from the inga trees used to shade the coffee plants. In Central America, the seeds are cooked and eaten as a vegetable. In Mexico, at least, the seeds are roasted and sold outside theaters to moviegoers. Information from J. Roskoski.

[5] If the pods are not eaten, however, during years of bumper crops of pods, the sidewalk stroller may have to wade through pods half-a-foot deep. This may make this species less than "outstanding." Information from J. Roskoski.

Pacay is one of the most unusual fruits on earth. Inside these huge pods is a sweet, frothy pulp that is a special favorite of children. (H. Popenoe)

outside Latin America the edible pods at present are hardly exploited. The development of markets for the pods could contribute additional income to local farmers.

Fruit trees such as ingas are underutilized in reforestation efforts. Give a peasant a pine tree and it's likely to be neglected, but give him a fruit tree and he'll protect it with his life, especially if it is productive.

Inga has major potential for acid soils in the lowland humid tropics, especially in Southeast Asia and Africa. As a nitrogen-fixing, multi-purpose tree that can be used to produce fruit, fuel, and green manure, it should find ready acceptance by farmers. The wide range of agricultural systems within which it can be integrated include hedgerow intercropping,[6] live fences, shade trees for plantation crops or animals, and fuelwood plantations.

Industrialized Regions. Most ingas are frost-sensitive tropical trees with little or no potential for cultivation in Europe, North America, Japan, or most of Australasia. Their pods do not ship well, so the possibility for ice-cream beans to be on dinner tables is slight.

[6] This involves growing trees in rows and crops in the alleys formed between the tree rows. The trees are pruned periodically to provide mulch and nutrients for the crops while minimizing competition for nutrients and water. Ingas seem ideally suited for such purposes. Information from P.A. Sánchez.

The pacay tree is a nitrogen-fixing legume that is promising in forestry, both in plantations and village backyards. It provides the benefits of shade, erosion control, and wood products—all in addition to food. (W.H. Hodge)

USES

Most ice-cream beans are eaten fresh. They are merely split open and the pulp dipped out with the fingers. However, ice-cream beans can also be handled as a processed product. They are washed and split and the contents removed and strained to separate the pulp from the seeds.

As noted, inga trees have many other uses, including those discussed below:

• Forage. Cows and other livestock eat the foliage. In Mexico, farmers cut and carry the leaves to feed their livestock.

• Beautification. The trees can be planted along urban streets and in home orchards. Inga trees are also frequently grown as single trees near homes. Their pods add to their value as shade and shelter.

• Lumber. The wood is moderately heavy (that of *I. vera* has a specific gravity of 0.57; the others are probably similar). It can be used for furniture, boxes, crates, light construction, and general carpentry. However, it is highly susceptible to drywood termites and when in contact with the ground decays readily.

• Fuel. Inga wood makes excellent fuel and is utilized for charcoal throughout the West Indies. Certain species pollard and coppice well.[7]

• Shade. *Inga* species grow rapidly and are common shade trees used on coffee and cacao plantations in Central America and the West Indies, where their cultivation and characteristics are well established.

• Soil Improvement and Intercropping. One of the Andean species, *I. edulis*, is being used in trials of alley cropping in Chile.

Like most legumes, inga trees fix nitrogen and improve the soil around them. The leaves have extrafloral nectaries, which may support beneficial insects, such as parasitoid wasps, that are natural enemies of crop pests.

NUTRITION

Although pleasing, the pulp is not particularly nutritious. It contains about 1 percent protein and 15 percent carbohydrate—mainly sugars.[8]

SILVICULTURE

In the forest, ingas grow naturally from seeds. In silviculture, they are usually planted from seedlings grown in nurseries.

All species studied so far are self-incompatible—they need at least two genetically different individuals for fruit set. Vegetative propagation would lead to little or no fruit unless cuttings are taken from more than one individual. Ingas usually grow very quickly, with a trunk diameter sometimes increasing more than 2.5 cm a year. A seedling normally begins to provide sufficient canopy to cover plantation crops within three years. Once established, the trees need little attention.

---

[7] The trait is exceptionally valuable. In pollarding, the tree is pruned to 2-3 m (that is, just above the reach of grazing livestock). The tree then puts out a dense flush of new growth. In coppicing, the tree is cut near its base, and its stump regenerates new shoots. Both management systems allow repeated and frequent harvest of wood without the cost and effort of replanting seedlings each time.

[8] Pacay seeds are high in protein, but whether they are safe to eat is not certain. In Mexico, boiled inga seeds were once a common vegetable, but their use has declined, probably because of the advent of new vegetables in the markets. Information from J. Roskoski.

In their native habitats, ingas thrive on many soil types, even limestone. Root nodules are formed by slow-growing rhizobia.[9]

Pacay, the Andean species, is a tree of moist areas, but others seem to have some drought tolerance.

LIMITATIONS

As noted, inga trees are restricted to tropical or subtropical climates. Tropical species produce pods nearly continuously, but in the subtropics fruit production is seasonal and heavy. Spoilage is sometimes a problem. Because of the sugars in the pulp, the pods ferment in about 3–5 days. With cool storage, however, they will keep three weeks.

Inga trees are generally healthy. If stressed, however, they can be affected by many common fungi, as well as a mosaic virus and "witches broom." Fruit predators are numerous. (They include porcupines, which love the immature pods.) Also, the pods are sometimes destroyed by larvae of various insects. Seed predators include wasps, beetles, and organisms that ingest and digest soft seeds. The foliage is susceptible to *Psylla* species—insects that defoliate some other legume trees.

There is little likelihood that inga pods could become a major commercial fruit crop in their own right; they are likely to remain as an additional and occasional source of cash to small landowners who have planted the trees for other purposes.

As noted, fruit set requires that more than one genetically different tree be in the same vicinity. This is why solitary ornamental trees normally fail to set fruit.

One of inga's biggest limitations is seed viability. Seed storage is very poor. The seeds often start germinating inside the pods. If seed were easier to handle, *Inga* species would possibly be as widely planted as trees such as leucaena and gliricidia. This is a limitation to dispersal or movement of inga germplasm to new areas in Africa or Asia, which modern communications might overcome.

RESEARCH NEEDS

Topics for future study (both for pacay and for ingas in general) include the following:

- Pest control, pruning, and maintenance.
- Agronomic research to increase and standardize fruit production.

[9] Measurements of nitrogen fixation by *I. inicuil* in Mexico showed that the amount of nitrogen fixed was about 50 kg per hectare per year. Information from J. Roskoski.

• Provenance selection for adaptability and fast growth, and for uniform fruit quality: large size and good taste, for example.

• Study of traditional methods to gain insights into production.

• Methods that prolong viability of inga seeds.

• The use of the leaves for fodder, and their nutritional qualities.

• Development of methods to properly and economically harvest, transport, and store the pods.

• Toxicological analyses of seeds to determine edibility.

• Use in intercropping. (Although trials have begun, the overall potential isn't well documented yet.)

• Methods of pollination.

• Effect on soil fertility (amounts of nitrogen fixed).

• Food science and processing of the pods.

SPECIES INFORMATION

Botanical Name As noted, several species occur in the Andes. The main ones, however, are *Inga feuillei* de Candolle (used since pre-Columbian times) and its widespread relative, *I. edulis* von Martius.

Family Leguminosae (Mimosoideae)

Common Names

Quechua: pa'qay, paccai (Cuzco)

Aymara: pa'qaya

Spanish: pacay, pacae, pacay de Perú, guama, guamo

English: ice-cream beans, food inga

French: pois sucre

Portuguese: ingá cipó, rabo de mico

Origin. It seems likely that pacay (*Inga feuillei*) originated on the eastern slope of the Andes, and, like other fruit crops of that area, was introduced to coastal Peru. That must have happened long ago, because the use of these fruits in the mountains and on the coast is ancient. Inga pods are portrayed on pre-Columbian pottery, and their pods and seeds have been found in tombs dating back to about 1000 B.C.

Description. *Inga* species are usually small trees, normally less than 15 m, although some of them can reach up to 40 m. They can be either evergreen or deciduous. The simply pinnate, dark-green leaves have large, oval leaflets in pairs without a terminal one. Many species have a green wing (rachis) between each pair of leaves. There is a nectary (small pit containing nectar) between each pair of leaves. Ants frequent these nectaries and probably play an important role in protecting the trees from insect pests such as aphids.

The fragrant flowers are arranged in crowded heads, spikes, or panicles at the stem tips or axils. Because they are rich in nectar, they attract bees, hummingbirds, and a variety of beetles.

Fruit begins to mature a few months after pollination, and production may be nearly continuous (twice a year in *I. feuillei* in the Andes). The pods may be flat, twisted, or cylindrical. They are often flattened or four-sided, the margins frequently overhanging. They grow up to 70 cm long and 1–3 cm in diameter, usually with the seeds buried in the white, sweet pulp.

Horticultural Varieties.   In pacay, vegetative selections (apparently made in ancient times) exist,[10] but no true horticultural varieties are known. As noted, a single selection is of little use in fruit production because of self-incompatibility. Because cuttings establish roots poorly, vegetative propagation has not yet been developed.

Environmental Requirements

Daylength.   Pacay is apparently daylength neutral, at least within the subtropics, which in any case are its outer limits.

Rainfall.   Pacay requires a subtropical climate with plenty of moisture. At Yurimaguas (Peruvian Amazon), both *I. feuillei* and *I. edulis* are very productive (wood, leaf, and fruit) at 1,500–2,700 mm annual rainfall.[11]

Altitude.   Up to 1,500–1,800 m in Peru's inter-Andean valleys.

Low Temperature.   Most species are damaged by low temperatures and killed by extended freezing weather.[12]

High Temperature.   Pacay thrives at 25°C; and, with sufficient moisture, seems capable of withstanding short periods above 33°C.

Soil Type.   These trees are apparently widely adaptable. They withstand soils from pH 4.0 to 8.0. They tolerate high aluminum saturation (70–90 percent) in acid soils. *I. edulis* and *I. feuillei* nodulate profusely even at pH 4.5 and are also heavily mycorrhizal (vesicular arbuscular mycorrhizae). These mycorrhizae play an important role in enabling ingas to take up phosphorus even though phosphorus is in very short supply in acid soils.

Some are reported to exhibit tolerance to waterlogging and are

---

[10] For example, the *I. feuillei* from Lima is different from *I. feuillei* from Cuzco, which seems to be the Andean variety. Information from J. León.

[11] Information from P.A. Sánchez.

[12] However, at least one (*I. affinis*) has a hardiness rating to −4°C.

common in riverine thickets and in wooded swamps, sometimes even below the high-water mark. They exhibit better growth rates on clayey or loamy soils.

Related Species. *Inga* is a large, widespread genus of about 350–400 species, and its taxonomic distinctions are uncertain at present. Most are shrubs and trees of tropical and subtropical America. The majority are similar to *I. feuillei*, differing mainly in the fruit form and other botanical details. Most have the characteristic sweet pulp. A few examples are mentioned below.

Inga vera This is the best known of all *Inga* species. It is commonly used as a plantation shade tree throughout Central America, and its pods are widely eaten.

Inga edulis As noted, research at Yurimaguas, Peru, indicates that this is a highly promising species for alley cropping on acid soils in the lowland humid tropics.

Inga paterno One of the best fruits of Costa Rica, it is planted widely for shade and fruit.

Inga spectabilis The huge pods (about 70 cm long) are sold in the markets of Costa Rica.

Inga marginata An attractive Costa Rican ornamental, this species has large spikes of fragrant flowers that bloom more than once a year. For this reason, it is much appreciated by beekeepers.

Inga brenesii, Inga punctata, Inga densiflora, Inga oerstediana All are used as shade trees in coffee and cacao plantations, and all have edible fruit.

Inga mortoniana This is a good ornamental with attractive reddish new leaves and tasty fruit.

The most popular inga fruits in Mexico (especially in Veracruz, where there are many coffee plantations) are *I. jinicuil* (or *inicuil*), the most well known, followed by *I. sapendoides* and *I. paterno*.

There are numerous species in the Amazon that are edible.

galupa

Passionfruits

An estimated 3,000 different fruits occur in the tropics, but apart from banana, pineapple, papaya, and mango, they seldom appear regularly on the world's tables. However, given the current enthusiasm for the passionfruit (*Passiflora* species), that group of fruits may soon have another addition.

Because of its strong, pleasing aroma and flavor, passionfruit juice is becoming an increasingly popular ingredient in drinks. Future production could be sizable because the universal demand for fruit-based soft drinks is rising. (In the United States, for instance, fruit-based drinks are expected to be the next "colas" that will soon become a major part of the soft-drink industry.) Already, passionfruit cultivation is spreading, and a number of large corporations are investing in it heavily.[1]

So far, most commercial passionfruit projects are based on a single species: the purple or yellow common passionfruit[2] (*Passiflora edulis*), which is native to Brazil and nearby tropical areas. However, at least 40 *Passiflora* species produce fruits, of which 11 are cultivated on at least a small scale.[3] Several that occur in the Andes are highlighted in this chapter.[4] Research to select strains with high yields and good fruit quality could transform these into significant commercial resources.

All these plants are vines that produce round or ellipsoid fruits about the size of goose eggs. They are usually propagated by seed and are grown on trellises or fences. Their horticultural requirements are like those of the common passionfruit. All require moderate moisture, but vary in their tolerance to cooler temperatures. Depending on the

[1] One measure of the enthusiasm for the passionfruit is the "craze" sweeping Puerto Rico. Starting from nothing in 1976, Puerto Rico now produces more than 3,000 tons of passionfruit a year. Two dozen juices and drinks are being sold, and the island's annual sales are estimated at $10 million, and rising. Hawaiian Punch is one U.S. drink that has included passionfruit juice for decades.
[2] This is known in Spanish as maracuja or granadilla. However, both names are applied, in different parts of the tropics, to several species of passionfruit.
[3] Many Quito backyards, for example, contain one or two passionfruit plants hung on trellises for their shade and their fruit.
[4] Other promising species found in lower altitude parts of the Andes include *P. serrulata*, *P. laurifolia*, and *P. quadrangularis*. These are basically tropical lowland species, and for this reason are not highlighted in this report.

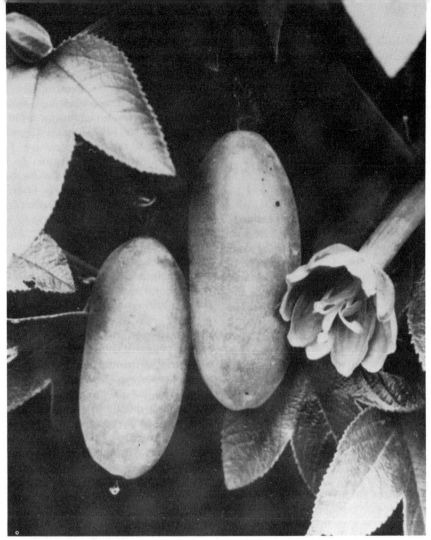

Curuba. (A. Ospina)

species, pollination is by bees, wasps, hummingbirds, or man. All can be eaten out of hand, or the juice extracted for use in fruit salads, desserts, or cold beverages.

SPECIES

Curuba.[5] This species (*Passiflora mollissima*), often called the banana passionfruit in English, is found in the wild at midelevations

[5] In its Latin American homeland, it is known as curuba, curuba de Castilla, or curuba sabanera blanco (Colombia); tacso, tagso, tauso (Ecuador); parcha (Venezuela); tumbo or curuba (Bolivia); tacso, tumbo, tumbo del norte, trompos, or tintin (Peru). Information from J. Morton.

in Andean valleys from Venezuela and eastern Colombia to Bolivia and Peru. It is native to that area and seems to have been domesticated shortly before the Spanish Conquest. Today, it is cultivated in home gardens and commercial orchards, and the highly prized fruits are regularly available in local markets. Colombia has some outstanding varieties; it has begun exporting the fruits, and has established a national committee to study the biology and agronomy of this species.[6]

Curuba juice is considered the finest of all passionfruit juices, and a wine is made from it. The fruits are also used in jams, jellies, and gelatin desserts. In addition, the pulp is strained (to remove the seeds), blended with milk and sugar, and served as a drink called *"sorbete de curuba."* It is also made into ice cream. Combined with alcoholic liquors (aguardiente) and sugar, it is served as a cocktail.

The plant seems suited to colder conditions than the common passionfruit. In the Andes, it prospers at elevations up to 3,400 m (in Cuzco, for instance) and briefly tolerates temperatures of –5°C. Under cultivation, it is high yielding. When densely spaced, well weeded, and fertilized, annual harvests in Colombia are said to reach 300 fruits per vine, amounting to some 500,000 fruits per hectare weighing about 30,000 kg. The hard-shelled, golden-yellow fruits[7] are up to 15 cm long and weigh 50–150 g.

Although little known outside the Andes, the plant has escaped from cultivation and become naturalized in Hawaii, New Zealand, Australia, and Papua New Guinea. In Hawaii it is permitted only on the island of Hawaii, where it grows wild in the forests and its vigorous vines strangle trees.

Sweet Granadilla.   This passionfruit (*Passiflora ligularis*) is native and common from western South America to Central Mexico. Its white translucent pulp is almost liquid, acidulous, and sweet smelling. The rind is strong, so that the fruit transports well without injury. Indeed, Colombia is now exporting this fruit to Europe.[8]

Since the plant grows at moderate elevations, it seems sufficiently cold resistant to withstand light frost, although probably not extended periods of temperatures below about –1°C. In Ecuador, it is cultivated mainly between 2,200 and 2,700 m, but in Bolivia and Colombia its cultivation has been extended to as low as 800 m and as high as 3,000 m.

The sweet granadilla was introduced into Hawaii probably in the late 1800s and was naturalized there by 1929. It flourishes in many

[6] Information from L. de Escobar,
[7] A form called "curuba quiteña" in Colombia has dark-green fruit (even when fully ripe), with orange pulp.
[8] It is widely cultivated in the Urrao area of Colombia (Department of Antioquia).

moist, woody areas, and its fruits are sold in local markets. It is also being grown in a small way in New Zealand.

This plant sets fruits less abundantly than the common passionfruit, but can produce two crops a year. Because of its resistance to root and collar rot, it is also a useful rootstock for other passionfruit species. The fruits are more than acceptable for eating fresh and for drinks. However, some people find it too sweet and flat, which is why lime juice is often added.

Colombian Passionfruit. This little-known species (*Passiflora antioquiensis*) is a pretty and exotic-looking plant—distinguished from other cultivated passionfruits by its flowers, which are bright red—and very ornamental. Also known as the banana passionfruit, its fruits are similar in appearance to those of the curuba but are sharper pointed and are juicier and of better eating quality. A native of Colombia, it is grown at higher altitudes in the northern Andes, but still at a somewhat lower level than that where curuba is grown.

The sweet fruits are 4–5 cm long, greenish yellow outside and pale yellow inside. They are eaten fresh, in beverages, and in desserts. When eaten out of hand they have a delicious sweet flavor, although they lose much of their flavor when cooked. Their thick skin can make it difficult to extract the pulp.

So far, this plant is not commercially cultivated. Its need to grow in the shade, the long peduncle on the fruits, and the long time the fruit take to develop are obstacles to its production on a large scale.

Curubejo. Also known as granadilla de Quijos, this vine (*Passiflora popenovii*) has fragrant red, white, and blue flowers. It is quite rare and is mostly found growing wild. However, at about 1,300 m elevation on the eastern slopes of the Ecuadorian Andes, it is cultivated on a small scale. Its fruits are found in markets around Baños in central Ecuador. The plant is also cultivated in El Tambo Municipio in the Cauca Department of Colombia where it is regularly propagated by cuttings.[9]

The sweet juice is highly regarded and is enjoyed for its rich aroma and taste. Less acidic than the common passionfruits, this species is particularly good for eating out of hand. The rind is thick but tender, and the outer color is green even on ripe fruit.

Given the commercial success of the common passionfruit, it seems possible that this rare fruit—so far unimproved, yet more flavorful—will do even better. However, it reportedly sets fruits only once a year

[9] Information from L. de Escobar.

The yellow-skinned curuba is common in the markets of the Andes. (T. Johns)

(around Easter in the Andes), which may be a limitation in some situations.

Galupa.[10] This plant (*Passiflora pinnatistipula*) is grown for its edible fruits in small, scattered areas all the way from Chile to Colombia. It is produced in climates of average temperature between 16°C and 22°C, and is easily and quickly propagated by seed. Its origin seems to be the Andes of Peru and Chile between 2,500 and 3,000 m altitude.

The fruits are eaten raw or used to prepare beverages. They are shaped like common passionfruits, but are smaller, being 4–5 cm in diameter. The brittle rind turns purple at maturity and encloses numerous seeds surrounded by a yellow pulp that is fragrant and has a sweet-acidulous flavor.

Although the fruit is agreeable, its yields are said to be low, and it is seldom found in the markets. Given research, that might change. In its wild form, the plant grows as a tangle. Because of its quick and rampant growth, galupa may find a place as a pioneer and specialty cash crop.

Chulupa. The chulupa (*Passiflora maliformis*)[11] is a little-known passionfruit that could have a big future. Its fruits have excellent flavor

---

[10] Common names for this species include galupa, bejuco (which is also used for other passionfruits), tintin, puru-puru, and chulupa.
[11] In the Andes this fruit is sometimes known as granadilla de piedra or granadilla de hueso, for the hard rind, and in the West Indies it is known as sweetcup or conch apple.

Sweet granadilla. (N. Vietmeyer)

and are commonly marketed in Bogotá. When PROEXPO, a government agency that promotes Colombia's exports, sent collections of different passionfruits to food fairs in Europe, the chulupa received high marks for flavor.

This plant occurs naturally at midelevations (1,200–1,500 m) in the Andes of Colombia. It is a more tropical species than the others highlighted in this chapter, and is already well known in the West Indies.

Although still a wild plant, the chulupa can probably be cultivated. In trials at La Mesa, near Bogotá, the plant thrived and yielded well under the standard methods used to grow the common passionfruit.[12] The woody vine, sometimes climbing to 10 m or more, often drapes trees, walls, and small buildings. It is noted for its resistance to pests and diseases that affect its relatives.

The fruits, borne continuously and prolifically, are apple-shaped. Fully ripe, they are green or yellow, with numerous white dots. The rind varies from flexible and leathery to hard and brittle.

The pulp is normally pale orange-yellow, juicy, sweet or subacid, and pleasingly aromatic. It contains many small flat, black seeds. The juice is excellent for cold drinks. Jamaicans, for instance, serve it with wine and sugar.

[12] Information from L. de Escobar.

At present this fruit has one drawback: the often tough rind can frustrate consumers. Research to select and propagate easy-opening varieties would do much to advance this species. Also, hybridization—for example, with its close relative *P. serrulata*, which has a papery pericarp—may provide a new fruit with a thinner rind.

Rosy Passionfruit. In Bogotá, another well-known passionfruit is *Passiflora cumbalensis*.[13] This banana-shaped "curuba bogotana" is much like the true curuba (*Passiflora mollissima*, see above), but it has a bright, attractive, red skin.[14] Its aromatic, mildly biting orange flesh is used to flavor drinks, ice cream, yogurt, or other products. Some judge the fruit to be inferior to selected strains of *P. mollissima*, but it still is of sufficient quality to place the fruits in local markets and supermarkets. Selecting improved varieties could likely boost its acceptance even more.

The rosy passionfruit is fairly widespread in the northern Andes[15] but is most extensively cultivated in the central and eastern mountains of Colombia, especially in the Department of Cundinamarca. It grows naturally from 1,800 to more than 3,000 m throughout Colombia and into northern Ecuador. It seems to hybridize with other varieties as well as with *P. mixta*.

Hybridizations have also been effected between the rosy passionfruit and the curuba itself, and this line of research seems particularly promising. In fact, hybridization is a promising road to valuable new fruits among many of the different passionfruit species. It might, for instance, lead to much larger fruits and maybe even to seedless ones.

Related Species.[16] Exploratory research might also include other wild Andean passionfruits, some of which are mentioned below.

Passiflora schlimiana This species grows from 2,000–3,500 m in the Sierra Nevada of Colombia. The juice, said to taste like blackberry juice, is used to prepare sherbets and jams. The fruits are yellow and measure up to 12 cm long. Although now hardly known, this is an unusual species much deserving of research.

---

[13] It was until recently considered a separate species, *Passiflora goudotiana*. Now it is considered *P. cumbalensis* var. *goudotiana*. Escobar, 1987.

[14] This fruit, which we have termed the "rosy passionfruit," currently has no common name in English.

[15] Seven varieties of *P. cumbalensis* can be found from northern Colombia to south-central Peru in cool to cold areas between about 1,800 and 4,100 m. Other common names for the species include taxo, tauso, and gulian in Ecuador, and puru puru in Peru. The variety found in the Bogotá area is considered the best of the species.

[16] Information from M. Tapia and L. de Escobar.

Passiflora ampullacea   Although this species is very little known, some botanists who have eaten the fruits and seen the plant's vigorous growth consider it to be a promising future crop. The fruits are much larger than those of the curuba and the rind is thicker, which protects the fruit on the way to market. The common name in central Ecuador, to which the species is endemic, is gulin. This species also hybridizes easily with *P. mollissima.*[17]

Passiflora tripartita   This plant (which may not be a distinct species but a variety of the curuba) grows all over Colombia and has fruits as good as those of curuba.

Passiflora mixta   (the curuba de Indio of Colombia).

Passiflora ambigua   (Colombia, a fruit much like the curubejo).

Passiflora mandonii   (Bolivia) a close relative of the galupa, but with longer fruit.

PROSPECTS

Andean Region.   With the rising international interest in passion-fruits, the Andean native species are a resource of large potential. They seem ripe for research and could be increasingly used locally as well as becoming valuable exports. Horticultural and marketing analyses, trials, and comparisons should be undertaken quickly. It seems probable that many of the Andean species initially will be low yielding, although they are likely to show outstanding response to modern methods and research.

There is little factual information available on the Andean passion-fruits. This must be generated by comparative tests in the areas where these species are available. Research institutes in Bolivia, Peru, Ecuador, Colombia, and Venezuela could provide information that may be useful as the basis for industrial development. The future of these fruits will depend upon horticultural development. The production of pulp and concentrate has extremely good prospects if commercial-scale production can be established and maintained.

Passionfruits seem particularly suited to countries with inexpensive labor because the fruits have to be picked by hand and must be grown on supports such as trellises.

---

[17] See Escobar, 1981. It has been used in experimental crosses at the MAG (Ministerio de Agricultura y Ganderría) Agricultural Station, Guaslan, near Riobamba, Ecuador.

Other Developing Areas. Countries such as Brazil, Sri Lanka, and Kenya that have commercial passionfruit production should explore the potential of the other species, including those from the Andes. Among such lesser known species are some with bigger and juicier fruits as well as different flavors from today's common passionfruit.

These vines are possibly useful in agroforestry systems with trees used as supports. The combination might be good for both tropical highlands and lowlands throughout Africa, Asia, and Latin America.

Many tropical and subtropical countries with land suitable for cultivating passionfruits are not now exploiting them. The knowledge to grow the plants is available and trials should be undertaken, but caution should be exercised because the economics of their production is uncertain, given their generally primitive horticultural state and because some species are so prolific they might become weeds.

Industrialized Regions. New Zealand, Australia, South Africa, and a few other warm-temperate regions produce the common passionfruit for commerce as well as for backyard planting. Species from the Andes could strengthen such enterprises by possibly adding a few degrees of extra cold tolerance,[18] by providing sweeter or different tasting fruits, as well as by perhaps increasing the fruit size. In drinks and other processed products, a great future is likely to be found.

Hybrids between the different passionfruit species may provide important new fruits, perhaps larger, maybe seedless, and probably more robust in their growth.

---

[18] Hybrids with the North American maypop (*P. incarnata*) are another source for cold resistance. Information from R. Knight.

Pepino

The pepino dulce[1] (*Solanum muricatum*) is a common fruit in the markets of Colombia, Ecuador, Peru, Bolivia, and Chile. It comes in a variety of shapes, sizes, colors, and qualities. Many are exotically colored in bright yellow set off with jagged purple streaks. Most are about as big as goose eggs; some are bigger. Inside, they are somewhat like honeydew melons: watery and pleasantly flavored, but normally not overly sweet.[2]

Despite the fact that South Americans enjoy this fruit, there seems to be a curious lack of awareness for its commercial possibilities elsewhere. Although pepinos are related to, and grown like, tomatoes, they nevertheless remain a little-known crop, and their various forms are currently unexplored and underexploited.

This plant's obscurity may not last much longer. In Chile, New Zealand, and California, the pepino (pronounced peh-*pee*-noh) is beginning to be produced under the most modern and scientifically controlled conditions. As a result, international markets are opening up. For example, the fruit has recently been successfully introduced to up-scale markets in Europe, Japan, and the United States.

In Japan, consumers have an insatiable appetite for pepinos, and in recent years they have bought them at prices among the highest paid for any fruit in the world. Pepinos are offered as desserts, as gifts, and as showpieces. Often they are individually wrapped, boxed, and tied with ribbons. Some trendy stores display pepinos whether they sell or not.

Its success in Japan is perhaps an indication of its future: the pepino is attractive, it has a good shelf life, it is tasty, and its shape and compact size are ideal for marketing.

---

[1] In Spanish, "pepino dulce" means "sweet cucumber." Regrettably, the shortened name "pepino" is becoming the common name for this fruit in English, for in Spanish "pepino" refers only to the cucumber. This fruit, however, is botanically related to tomatoes and is nothing like a cucumber.

[2] Cieza de León, the Spanish chronicler of the Incas, related that "in truth, a man needs to eat many before he loses his taste for them."

PROSPECTS

The Andes. Pepino is an ideal home garden plant; it grows readily from cuttings and is cheap to produce, and increased demand could greatly benefit home producers. Given attention by horticulturists, a colorful array of pepino types—both traditional and newly bred—could bring increased appeal to consumers from Colombia to Argentina.

The transition to more extensive production has already begun. In the coastal valleys of Peru, there are some large fields of pepinos (usually rotated with potatoes, corn, and other crops). Lima is provided with the fruits year-round, and a small export trade has begun. In Ecuador, too, a few fields are grown under advanced agricultural conditions. In Chile, more than 400 hectares of pepinos are planted in the Longotoma Valley, and increasing quantities are being exported, notably to Europe. Formation of cooperatives to develop markets, coordinate transport, and control quality could lead to greater local and export earnings.

There are parts of the Andes that are unaware of this crop. In Colombia, for instance, it is hardly known in most of the highland departments, although in San Agustín (Valle) and Manizales (Caldas), there are large farms (*fincas*) that specialize in pepinos.

Other Developing Areas. In addition to its wide cultivation in South America, the plant has been introduced to Central America, Morocco, Spain, Israel, and the highlands of Kenya. Relatively unknown in other nations but worth trying in all warm-temperate areas, this seems to be a crop with a big future fast approaching. Commercial pepino production has been suggested for southern Brazil, Uruguay, Paraguay, the highlands of Haiti, Puerto Rico, Guatemala, and Mexico—as well as for the cooler areas of Africa and Asia (particularly China).

Industrialized Regions. This crop has potential for production in many parts of Europe, North Africa, the eastern Mediterranean, North America, Australasia, South Africa, and Japan, although in some areas it may have to be grown under glass or plastic to produce the sweet, unblemished fruits demanded by the top-paying markets.

As noted, pepino is already an established crop in New Zealand. In the United States, it is grown on a small scale in Hawaii and California, where several hundred hectares are now under commercial cultivation. This seems to be the beginning of a promising new addition to the horticultural resources of much of the temperate zones.

The pepino has been called "a decadent fruit for the '90s." It is sweet, succulent, and melts in the mouth. (Frieda's Finest)

USES

The pepino is so versatile that it can be a component of any part of a meal: refreshment, appetizer, entree, or dessert. South Americans and Japanese eat it almost exclusively as a fresh dessert. It is highly suited to culinary experimentation. For instance, New Zealanders have served it with soups, seafood, sauces, prosciutto, meats, fish, fruit salads, and desserts. The fruits can also be frozen, jellied, dried, canned, or bottled.

Pepinos are often peeled because the skin of some varieties has a disagreeable flavor. It pulls off easily, however. The number of seeds depends on the cultivar, but even when present, the seeds are soft, tiny, and edible, and because they occur in a cluster at the center of the fruit they are easily removed.

NUTRITION

As a source of vitamin C, the pepino is as good as many citrus fruits, containing about 35 mg per 100 g. It also supplies a fair amount of vitamin A. Otherwise, it is 92 percent water and only 7 percent carbohydrates.

The fruits are normally subacid. Levels of 10–12 Brix (sugar concentration) are common.[3]

AGRONOMY

All pepino cultivars are propagated vegetatively. Cuttings establish roots so easily that mist sprays or growth hormones are usually unnecessary. Tissue culture is also possible.[4]

By and large, pepino is grown like its relatives, tomato and eggplant. With its natural upright habit of growth and fruiting, it may be cultivated as a free-standing bush or as a pruned crop on trellises. (Supports can be used to keep the weight of the fruit from pulling the plant to the ground.) The plant grows quickly and can flower and set fruit 4–6 months after planting. It is a perennial but is usually cultivated as an annual.

Undemanding in its basic requirements, the plant has wide adaptability to altitude, latitude, and soils. When young, it is intolerant of weeds, but it later smothers any low-growing competition. Established bushes show some tolerance to drought stress, quickly recovering vegetative growth, although their yield may be depressed. In dry regions, irrigation is normally used.

The plants are parthenocarpic, which means they need no pollination to set fruit. However, self-pollination or cross-pollination greatly encourages fruiting.

HARVESTING AND HANDLING

Pepinos are harvested when fruits have a pale yellow or cream background color (at least in the popular cultivars El Camino and Suma). Fruits left on the plant until overripe often have poor flavor. Harvesting must be done carefully because the fruits bruise easily and finger markings show up. With current varieties, the fruits on a single bush mature at different times, and several pickings are necessary throughout the warm season. Yields of 40–60 tons per hectare are not

---

[3] Any dessert-quality fruit should be sweet, with Brix levels above 8—preferably 12 or even more. Information from S. Dawes.

[4] Pepinos are easily propagated by seed, but usually the seedlings are inferior to their parents. Seedlings, however, normally differ widely from each other, which allows breeders to search for superior new strains.

Auckland, New Zealand. Over the last 20 years, New Zealand horticulturists have taken up the pepino as a commercial crop and have developed it, probably to a greater extent than in any other country. Their varieties derive from clonal material, introduced from Chile (following heat treatment to remove viruses). Today, pepinos are being grown on many hectares, much of it under glass, and the fruits are shipped to North America, Japan, and Europe. In fact, since 1984, pepinos have been one of New Zealand's most lucrative fruit exports. (*New Zealand Herald*)

uncommon, and even more may be possible under greenhouse conditions.

The fruits are susceptible to chilling injury and are stored at 10–12°C. At this temperature they may keep in good condition for 4–6 weeks. (Sea freighting may be possible from many countries.) A fruit taken out of storage has a shelf life of several weeks at room temperature.

LIMITATIONS

The pepino is a little-studied crop, with sparse factual data or commercial field experience behind it. Particular areas of uncertainty include the following.

Fruit Quality Few sweet varieties also have good horticultural and marketing qualities; the skin, although edible, is often tough and bitter; and improperly ripened fruit have a bad aftertaste.

Lack of Adaptability The best fruit candidates are insufficiently hardy for cultivation in many cool areas and are susceptible to nematodes. High temperatures retard their growth and reduce the quality of their fruits, and drought readily kills the bushes because of their shallow roots.

Horticulture Cultural conditions and plant nutrition can greatly affect fruit color, sweetness, taste, and overall quality in ways that are not yet fully understood.

Fruit Set Poor fruit set is often a problem. The causes seem to include over-fertilizing, which fosters vegetative growth rather than flowering, and high temperatures, which cause the flowers to abort.[5]

Pests and Diseases The plant's susceptibility to pests and diseases in regions of intensive agriculture is scarcely known. Although attacks have rarely been of economic importance, more intensive cultivation of larger areas may intensify disease and pest problems. Aphids, spider mites, and whitefly already have been serious problems in California and New Zealand. Nematodes and root rot have also been concerns. In addition, the plants have shown susceptibility to viruses.

RESEARCH NEEDS

Fruit Quality Research is needed to better understand the causes of the insipid flavor of many pepinos. If the flavor can be sharpened and strengthened, the crop's future will be more secure. Approaches might include analysis of the effects on flavor of different varieties, stages of picking, postharvest handling, fertilization, and perhaps the use of salt.[6]

Cultivation Cropping systems have not been investigated in depth, and most commercial growers rely on tomato technology. Future agronomic research should include analysis of different cultivation practices, stress tolerance, plant nutrition and irrigation, light and temperature requirements, pollination, and methods for training and supporting the plants (such as trellising).

Plant Physiology The physiological problems relating to fruit set need to be better understood. Also, a convenient method for judging ripeness, other than using fruit color, would be extremely valuable.

Genetic Development Because pepino reproduces easily by seed, it can be improved readily through selection of sexual variants from cross-pollination. The mixed genetic composition (heterozygosity)

[5] Hermann, 1988.
[6] Tomatoes grown with saline irrigation have become a premium export of Israel because of their tangy taste.

allows considerable range in character selection. Added to this, vegetative propagation is simple, which means that any mutant type can be perpetuated without difficulty and clonal lines established.[7]

Market Development The creation of a new crop requires the development of marketing as well as horticulture. Because pepinos are new to consumers outside the Andes, markets are unstable. Furthermore, there is a lack of basic marketing knowledge, consumer acceptability is unknown, and ultimate market demand is uncertain. Promotion and market development could do much to assure the steady advancement this crop deserves.

Species Information

Botanical Name *Solanum muricatum* Aiton
Family Solanaceae (nightshade family)
Botanical Synonyms *Solanum variegatum* Ruíz and Pavón; *Solanum guatemalense* Hort., and others
Common Names
Quechua: cachun, xachun
Aymara: 'kachan, kachuma
Spanish: pepino, pepino dulce, pepino blanco, pepino morado, pepino redondo, pepino de fruta, pepino de agua, mataserrano, peramelon (Canary Islands)
English: pepino, Peruvian pepino, pear melon, melon pear, melon shrub, tree melon, sweet cucumber, mellowfruit, "kachano" (an Aymara derivative that has been suggested to avoid confusion with melons or cucumbers)

Origin. The place and time of the pepino's domestication are unknown, but the plant is native to the temperate Andean highlands. It is known only in cultivation or as an escaped plant. It is an ancient crop, and is frequently represented on pre-Columbian Peruvian pottery.

Description. This highly variable species is a sprawling, perennial herb that reaches about 1 m in height, with a woody base and fibrous roots. Several stems may arise from the base, and they may establish roots where they contact soil.

The leaves may be simple or compound; when compound, the number of leaflets may vary from 3 to 7. The white to pale-purple to

---

[7] Recent studies show that even immature seed is viable if germinated in nutrient agar. Information from J.R. Martineau.

bright-blue flowers occur in clusters. As noted, fruits can be produced without pollination (such parthenocarpic fruits are seedless), but fruit set is much greater when self- or cross-pollination occurs. Pollen is not usually abundant. As the stigma is longer than the anthers, pollination is unlikely to occur unless pollen is transferred by an insect or human hand.

The fruit varies from globose to pointed oval. When ripe, the skin background color may be creamy to yellow-orange. Purple, gray, or green striping or blush colorations give the fruit distinctive appearance. The flesh may be greenish, yellow, salmon, or nearly clear.

Horticultural Varieties.   Pepinos appear in markets throughout the Andes, but although there are many distinct strains, few have been stabilized into named cultivars.

In Chile, however, there are named varieties. All produce similar purple-striped, egg-shaped fruits. These are only slightly sweet, with a Brix rating generally less than 8. The purple stripes mask the bruise marks so common on the golden, unstriped pepinos. Chile is a major exporter, and its varieties are now also grown in California and New Zealand.

In New Zealand, the most common cultivated varieties are El Camino and Suma. El Camino has medium to large egg-shaped fruit with regular purple stripes. For reasons that possibly have to do with mineral nutrients given to the plant, it sometimes produces off-flavored fruits (these are identifiable by their brownish green color). Suma is a vigorous cultivar producing heavy crops of medium to large globose fruits, with regular purple stripes and attractive appearance. Their flavor is mild and sweet.

In California, New Yorker is the most widely grown cultivar. Since 1984, however, Miski Prolific has become equally popular. Its flesh is deep-salmon color, and its skin creamy white with light-purple stripes. There are a few seeds in each fruit.

Environmental Requirements.   Although this plant is native to equatorial latitudes, it is typically grown on sites that are cool. Thus, it is found in upland valleys, in coastal areas cooled by fog, and parts of Chile where the summers are not hot.

Daylength.   Since the pepino fruits well at many latitudes, it appears to be photoperiod-insensitive.

Rainfall.   1,000 mm minimum, well distributed over several months. As noted, the pepino has little drought resistance, and in Chile and Peru irrigation is often used.

Altitude. The plant seems unaffected by altitude. It grows from sea level in Chile, New Zealand, and California to 3,300 m in Colombia.

Low Temperature. Once established, the plant experiences frost damage at temperatures below −3°C. Seedlings are even more sensitive. Cool, wet weather during the harvest season results in skin cracking.

High Temperature. The plant performs best at 18–20°C. With adequate moisture, it can tolerate intermittent temperatures above 30°C. However, fruit production then declines, particularly if both day and night temperatures are high.

Soil Type. The plant thrives in moderately moist soils with good drainage. Soils should be above pH 6.0 to avoid disorders such as manganese toxicity and iron deficiency. If soil is too fertile, there can be problems of fruit set and fruit quality.

Related Species. *Solanum caripense* (tzimbalo) is a possible wild ancestor, which crosses readily with pepino and bears edible fruit. It is a sprawling plant, more open and smaller than the pepino, that is fairly widespread in the Andes between 800 and 3,800 m elevation. Its fruits are elongate and slightly smaller than ping-pong balls. There is, however, little flesh to eat, for they are mostly juice and seeds. Some are rather intensely flavored, sweet, and occasionally leave a bitter aftertaste. The plant's advantages are early fruiting, abundant yields, and fairly tough-skinned fruit.

Solanum tabanoense is a rare plant found between 2,800 and 3,500 m in southern Colombia, Ecuador, and Peru. The fruit has an appreciable amount of flesh and is similar to the pepino in size and taste.

Tamarillo (Tree Tomato)

The tamarillo or tree tomato[1] (*Cyphomandra betacea*) is native to the Andes, and has been cultivated on mountainsides since long before Europeans arrived. Today, it is grown in gardens from Chile to Venezuela, and its fruits are among the most popular of the region. On the Colombian and Ecuadorian uplands, for instance, it is found in every city, including Bogotá and Quito.

Although distributed throughout much of the world's subtropics, this little tree is usually produced only on a small scale in home gardens. Few people have ever considered it for large-scale orchard production, and it has hardly been developed commercially as yet.

However, development of this crop is under way in one country, New Zealand, where tamarillos have been popular for more than 50 years. They got a particular boost during World War II, when bananas and oranges could not be imported. During the last half-century, New Zealand horticulturists have made selections, developed improved fruit types, and created a commercial industry. Indeed, the fruits have become widely popular—children take them to school in their lunch boxes, and until the kiwifruit "boom" of the 1960s, more New Zealand acreage was devoted to tamarillos than to kiwifruit. New Zealand is already demonstrating this fruit's international potential. It is airfreighting tamarillos to North America, Japan, and Europe, and the trade is said to be expanding.

Tamarillos are egg shaped, and have attractive, glossy, purplish-red or golden skins. Inside, they look somewhat like a tomato, but their succulent flesh has a piquancy quite unlike that of its more famous cousin.

This fruit seems to have a bright future; it is flavorful, pretty, and has flexible uses and a long production season. Moreover, its "tropical" tang could be an asset in the increasing international markets for exotic

---

[1] In English, the plant is traditionally called "tree tomato." The name "tamarillo" was devised in New Zealand in 1967. it is becoming the standard designation for the fruit in international commerce, and is even becoming common in the Andes, where the name "tomate de árbol" (tree tomato) has been traditional.

The tamarillo looks much like an oval-shaped tomato, but it grows on a small tree and has a sharp, tangy, unique flavor. (N. Vietmeyer)

fruit juices as well as for processed products such as jams, chutneys, sauces, and flavorings for ice creams. For warm-temperate areas, an additional advantage is that tamarillos fetch premium prices because they ripen late in the growing season when few other fresh fruits are available.

All in all, this unusual fruit should not be unusual much longer. At present, through lack of horticultural and scientific attention, its true potential has scarcely been touched.

PROSPECTS

The Andes. The tamarillo, already well known in the Andes, is little promoted, and its potential is far greater than is recognized at present. The trees, by and large, receive little management, and by export standards, the fruits in Andean markets are far from premium quality. Selection and propagation of elite cultivars,[2] better management of the plants, and better handling of the fruits would enhance this crop's prospects.

[2] The red-fleshed strain is already being widely cultivated in Colombia because farmers receive premium prices for it. Information from L.E. Lopez J.

With sufficient effort, Andean countries could be at the forefront of an expanding international commercial industry. Colombia has already begun exporting a juice concentrate as well as fresh fruits. If the plant is given research and promotion, this could spark a new Andean industry. Cooperatives of small growers may be particularly adapted to providing the fruits.

Other Developing Areas. The tamarillo has much promise for frost-free, subtropical, and warm-temperate areas. A few plants in the home garden can add good nutrition and a novel taste to the family diet. It is particularly good for home use because it is easily propagated, and when grown in warm regions it provides fruits year-round. Also, the trees are very high yielding. Undoubtedly, it has an important future in cool highlands of Third World countries, but true frost-free lands that are not too hot for this plant are uncommon.

Industrialized Regions. New Zealand's selection of attractive cultivars and the development of shipping and storage techniques has resulted in a relatively obscure fruit entering international trade within the past decade. This experience is demonstrating that there is a real future for the tamarillo. Research is needed, however, to improve production, particularly in such matters as fruit type and palatability, virus control in the orchards, and yield variability (there is a specific need to create uniform fruit set).

The rather acid taste of the fruits arriving in international markets is hindering greater acceptance of the tamarillo as a fruit to be eaten raw. Most likely, sweeter fruit will soon become available from selected cultivars or from improved handling, and this will certainly increase its popularity. As they enter the markets and are promoted, tamarillos could become a common commodity in the produce markets of North America, Europe, Japan, and other regions.

The richly colored juice of tamarillos (especially of the deep red types) seems to have much potential for blending with grapefruit and other juices whose consumer appeal may be increased by the added color.

USES

Ripe tamarillos have fine eating qualities and can be used in many ways. They are usually cut in half, and the flesh scooped out. As with common tomatoes, the seeds are soft and edible. The skin is easily removed—it peels off when dipped briefly in hot water. The fruits are especially good on desserts such as cakes and ice cream, in fruit salads, or (like tomatoes) in sandwiches and green salads. The whole fruit (including skin) can be liquidized and drunk. In South America, such juices are frequently blended with milk, ice, and sugar to make tasty drinks.[3]

Tamarillos are also cooked and eaten in stews, soups, baked goods, relishes, and sweet and savory sauces. In New Zealand, diced tamarillos, with onion, breadcrumbs, butter, and appropriate seasonings, are becoming popular as a stuffing for roast lamb—the national dish. Being high in pectin, tamarillos make good jellies, jams, preserves, or chutneys, but they oxidize and discolor unless treated.

The fruit freeze well, either whole (peeled) or pureed, and can be stored this way almost indefinitely.

NUTRITION

Tamarillos are excellent sources of provitamin A (carotene–150 International Units per 100 g), vitamin B_6, vitamin C (25 mg per 100 g), vitamin E, and iron. They are low in carbohydrates; an average fruit contains less than 40 calories.

AGRONOMY

The plant is usually grown from seed. Seedlings develop a straight, erect trunk about 1.5 m tall before branching. New Zealand practice is to prune seedlings early to encourage multiple stems, which makes the fruit easier to reach and reduces chances of heavily laden trees toppling over. Normally, the trees begin bearing within 18 months of planting out. Peak production is reached usually in 3–4 years.

The plant is self-compatible; it can set fruit without cross-pollination. However, the fragrant flowers attract many bees, the trees produce flowers abundantly, and pollination seems to improve fruit set.

Because of a shallow, spreading root system, deep cultivation is not possible close to the trees. Also, perhaps as a result of the shallow roots, the plant cannot tolerate prolonged drought. Mulches help conserve moisture and suppress weeds, but during dry periods, ample water must be available.

Although adaptable and easy to grow, the tree seems to be short-lived; the life of a commercial plantation is usually no more than eight years.

Pruning can be used to make the harvest coincide with periods of peak demand, such as in the off-season.

---

[3] For example, a thick, rich drink made of the fruit and served with chopped fruit on top is popular in Cali, Colombia, and sells under the New Zealand name "tamarillo."

Bay of Plenty, New Zealand. Tamarillo harvest. (P. Sale)

HARVESTING AND HANDLING

Fruits occur in clusters. They can be formed year-round, but in seasonal countries (such as New Zealand) there is a distinct harvesting season. Because the fruits do not mature simultaneously (unless the tree has been pruned), several pickings are usually necessary. The fruits pull off readily, carrying a short stem still attached.

The trees bear prolifically and yield generous harvests. A single tree may produce 20 kg of fruit or more each year; commercial yields from mature orchards in New Zealand can reach 15–17 tons per hectare.

In handling tamarillos, the most serious problem is fungal rot (glomorella) on the stem, which can quickly extend into the fruit. Copper sulfate (bordeaux mixture) controls this adequately. The New Zealand Department of Scientific and Industrial Research has developed a cold-water dipping process that is also effective. Storage of 6–10 weeks is then possible.

LIMITATIONS

To develop tamarillo into a practical, widespread crop will not be quick or easy. Even selected New Zealand cultivars usually produce a variable product, insufficiently reliable to assure importers that shape, color, or sugar:acid ratio will remain consistent lot by lot and year by year. Nor are future yield levels predictable.

Large leaves and brittle wood make the trees prone to wind damage. Even light winds easily break off fruit-laden branches. In most locations, permanent windbreaks should be established at least two years before the plants are set out. Even then, staking may be needed to keep branches from breaking under the weight of fruit.

Small, hard, irregular, "stones," containing large amounts of sodium and calcium, occasionally appear in the fruits. They usually occur in the outmost layers of the fruit and do not present much of a problem in fresh fruits because those layers are not eaten. In canning, however, they are a concern because the whole fruit is used. Probably the best solution is a breeding program to select for types without these concretions.

The tree is generally regarded as fairly pest resistant. However, some pests of concern are given below.

Viruses In most places, viruses are the most significant diseases. They reduce the plant's vigor and can leave unattractive splotches on the fruit. Newly emerged seedlings are, as a rule, virus free, but unless precautions are taken they soon become infected. (The vectors are mainly aphids.)[4]

Nematodes Root-knot nematodes damage plants, particularly in sandy soils.

Insects The trees are attacked by aphids and fruit flies, and whitefly is sometimes a serious pest.[5] In the Andes, the tree-tomato worm (which also infests the tomato and the eggplant) feeds on the fruits, sometimes causing heavy losses.

Fungi The principal fungal disease is powdery mildew, which can cause serious defoliation.

Dieback A dieback of unknown origin is at times lethal to the flowers, fruits, twigs, and new shoots.

RESEARCH NEEDS

Germplasm Collection More types need to be collected and information on their economic traits developed. Identifying and selecting

---

[4] Paradoxically, tree tomatoes are notable for their apparent resistance to tobacco mosaic—a virus that affects tobacco, potatoes, tomatoes, and other solanaceous crops. The plant is perhaps a source of virus-resistant genes for those crops.

[5] In warmer areas, biological control with the wasplike insects *Encarsia formosa* and *E. pergandiella* appears promising. Information from P. Sale.

superior germplasm—especially that found in the Andes—could assist breeding programs and speed the tamarillo industry toward success.

Breeding   To support the orderly genetic development of this crop, more information is needed on the plant's reproduction habits and genetics. Long-term breeding aims should include virus-resistant trees, low-growing trees, and deeper-rooted trees. To increase yields, improved fruit set is needed. This is probably not genetic, but the reason for flower and fruit abortion is not established.

Other breeding goals should be:

- The development of pure seed lines to enable the growing of desirable types from seed;
- Raising sugar:acid ratios;
- Developing fast-maturing and cold-hardy types to increase the area of potential use in temperate climates; and
- Creating dwarf cultivars.

Plantation Management   Improvements are needed in orchard practice and management. Among specific research needs are the following:

- Controlling viruses;
- Increasing fruit set;
- Developing fertilizer requirements;
- Learning the ideal pH level;
- Applying trickle and overhead irrigation;
- Developing growing systems;
- Understanding the plant's nutrition; and
- Breaking seed dormancy.[6]

Biological control of the tree-tomato worm should be sought for use in the Andes. (Perhaps *Bacillus thuringensis* would be useful.)

Sweetness   There seems to be a good possibility that the sweetness of the deep-red fruits can be raised. Already, sweet types are known; however, these are mostly smaller or less eye-catching. Nevertheless, low-acid red specimens have been reported.[7] These deserve greater attention because the combination of the spectacular red color with a sweet taste would cause this appealing fruit to take off.

Stones   The cause of the "stones" should be determined, and methods developed to eliminate or circumvent this nuisance.

---

[6] Information in this list is mostly from J. Laurenson and represents current goals of New Zealand tamarillo research.
[7] Information from S. Spangler. One New Zealand variety, Oratia Red, is sweeter than normal, but even sweeter types seem possible.

Postharvest Handling   Work should continue on storage and handling to increase shelf life and to enable tamarillos to be shipped by sea—a development expected to greatly improve the export situation for most countries. Important advances have already been made.

Hybrids   Several closely related species (see below) produce good fruit, and there is the potential of developing hybrid tamarillos, perhaps with seedless fruits and different flavors. There seem to be substantial genetic barriers to interspecific hybrids, however.[8] Nonetheless, cell fusion and gene splicing, which seem particularly easy in Solanaceae, might prove practical.

Tissue Culture   Tissue-culture techniques have been developed for the tamarillo. Although their main use will be in multiplying improved material for planting, they are possibly useful in genetic selection and in propagating of difficult hybrid crosses and haploids. Colchicine treatment to produce fertile plants from haploids should be attempted.

Work is currently under way to select a virus-resistant strain with the aid of tissue culture. In the meantime, however, virus resistance should be sought in wild species.

SPECIES INFORMATION

Botanical Name   *Cyphomandra betacea* (Cavanilles) Sendtner
Family   Solanaceae (nightshade family)
Common Names

Spanish: tomate de árbol, tomate extranjero, lima tomate, tomate de palo, tomate francés

Portuguese: tomate de érvore, tomate francês

English: tree tomato, tamarillo

Dutch: struiktomaat, térong blanda

German: Baumtomate

Italian: pomodoro arboreo

Origin.   The tamarillo is unknown in the wild state, and the area of its origin is at present unknown. It is perhaps native to southern Bolivia (for example, the Department of Tarija) and northwestern Argentina (the provinces of Jujuy and Tucumán).

Description.   The plant is a fast-growing herbaceous shrub that reaches a height of 1–5 m (rarely 7.5 m). It generally forms a single upright trunk with spreading lateral branches. The leaves are large, shiny, hairy, prominently veined, and pungent smelling.

---

[8] Information from L. Bohs.

Flowers and fruits hang from the lateral branches. The pinkish flowers are normally self-pollinating, but can require an insect pollinator; unpollinated flowers drop prematurely.

The fruits are egg shaped, pointed at both ends, 4–10 cm long and 3–5 cm wide, smooth, thin-skinned, and long-stalked. The skin color may be yellow or orange to deep red or almost purple, sometimes with dark, longitudinal stripes. The flesh inside is yellowish, orange, deep red, or purple. It has a firm texture and numerous flat seeds. The most flavorful and juicy flesh lies toward the center of the fruit, becoming more bland toward the skin.

Horticultural Varieties. Although there is much variety in the fruits and many local preferences based on color, there are apparently at present few named cultivars. Growers normally select their own trees for seed selection.

In New Zealand, where the most extensive selection has taken place, two strains are cultivated: red and yellow. Oratia Red (or Oratia Round) was the first recognized cultivar. The red strain has a stronger, more acid flavor and is more widely grown. Yellow fruits have a milder flavor and are preferred for canning.

A dark-red strain (called "black") was selected in New Zealand in about 1920 as a variation from the yellow and red types. It was propagated and reselected thereafter.

Environmental Requirements

Daylength.   Unknown; probably daylength-insensitive.

Rainfall.   Cannot tolerate prolonged drought, nor waterlogged soils or standing water.

Altitude.   Unrestricted. Grows at 1,100–2,300 m at the equator in the Andes; near sea level in New Zealand and other countries.

Low Temperature.   This species is injured by frost. Short periods below −2°C kill all but the largest stems and branches.

High Temperature.   In tropical lowlands, tamarillos grow poorly and seldom set fruit. (Fruit set seems to be affected strongly by night temperature.) The plant seems to do best in climates where daytime temperatures range between 16 and 22°C during the growing season.

Soil Type.   Fertile, light, well-drained soil seems best.

Related Species.   The genus *Cyphomandra*, native to South and Central America and the West Indies, contains about 40 species. Many that are grown for their fruits remain to be investigated. Several are said to produce fruits as good as those of the cultivated tamarillo. These may hold potential as economic crops in their own right, or as germplasm sources for improving the tamarillo.

Fruit-bearing species deserving special studies include the following:

Cyphomandra casana (C. cajanumensis) Found growing wild on the edge of rain forests in the highlands of Ecuador, especially in the Loja Province. Like the tamarillo, the casana grows rapidly to a small tree, 2 m tall. Its large, furry, deep-green leaves make it an interesting ornamental, but it also produces heavy crops of mild-flavored fruits in about 18 months. The fruits are spindle shaped and golden yellow when ripe. They are sweet, juicy, and said to be rather like a blending of peach and tomato in flavor.

The casana seems to need cool growing conditions. Unlike the tamarillo, the casana fruit is a poor shipper. A breeding program for selection of firmer, larger, and better colored fruit could yield a fruit of commercial value. It deserves special attention from horticulturists and scientists as a source of genetic material for such qualities as nematode resistance, root rot resistance, fragrance, flavor, color, and yield.

Cyphomandra fragrans Compared with the tamarillo, this tree has greater tolerance to powdery mildew, a smaller and more robust stature, basal fruit abscission (the fruits break away clean without the stalk as found on tamarillos), and a greater degree of cold hardiness. Its fruits resemble small tamarillos, but have a thick and leathery orange skin. Like most solanaceous fruit, it is somewhat acid.

Cyphomandra hartwegi This species has an extensive natural range, but is not yet commercially cultivated. Apparently, it has not even been tested as a potential crop. It has a yellow berry about the size of a pigeon's egg.

PART VI
Nuts

The cultivation of nut trees is one of the most neglected aspects of tropical agriculture and forestry.[1] Few plants are more versatile and desirable. In addition to providing edible, nourishing fruits, they beautify cities, wastelands, and drab landscapes. Nut trees could improve the environment by preventing soil erosion in vulnerable areas. Once planted, they usually require little labor or expense and will provide food for decades, if not centuries. Should famine ever occur, nuts provide a local reserve of concentrated food. Some are also potentially valuable export commodities.

With rising international interest in massive plantations of long-lived trees to lower global warming trends, nut trees could become much more widely planted in the future. They absorb carbon dioxide, as well as provide shade, erosion control, and, above all, a premium food product.

The following two chapters describe promising nut crops of the Andes: the Quito palm and the Andean walnut.

[1] For more details, see F. Rosengarten, Jr. 1984. *The Book of Edible Nuts*. Walker and Company, New York.

Quito Palm

The Quito palm (*Parajubaea cocoides*) is a graceful, elegant tree native to the Andes. For a palm, it grows at remarkably high elevations, occurring at altitudes between 2,000 and 3,000 m along the middle-level uplands of Ecuador and southern Colombia.

Because of its beauty, this species is cultivated as an ornamental in the principal Ecuadorian cities as well as in Pasto, Colombia. It is best known, however, as the palm of Quito, a mountain city at 2,800 m above sea level. Along the road leading from the airport into Quito, handsome plantings may be seen, and the tree is common in squares (for instance, the Parque Bolívar, Plaza de Independencia, and the Catholic University), as well as in private yards.

But this plant is much more than ornamental. It bears long clusters of 30–50 edible fruits that look like little coconuts,[1] with three eyes and hard, thick shells. All these "minicoconuts"—which are smaller than a golfball—mature at about the same time, and fall off when ripe. They are then broken open and eaten raw. The kernel is the size of a macadamia nut. Its fleshy mesocarp is sweet and contains usable oil. These nuts are so popular—especially with children—that you can hardly find one unless you look early in the morning.

Despite its popularity, the Quito palm has scarcely been grown outside Ecuador. There are, however, scattered trees in San Francisco, California; in northern New Zealand; and in Sydney, Australia, where some very old specimens are to be found in the Royal Botanical Gardens.

Beautifully shaped, like a dainty coconut palm, this plant makes a spectacular ornamental. Its trunk is slender, sometimes curved like the coconut's, and can top 8 m or more. Above this is a graceful, spreading circle of fronds.

Like other palms, this is not an easy plant to propagate. As a rule, palms are slow to germinate and very slow to grow.[2] The Quito palm,

---

[1] They are locally called "coquitos" (little coconuts), but that name is also applied to different nuts in Chile and other countries.
[2] Quito palm seeds must be thoroughly dried before germination is attempted. They are available from the abundant production of the palms in Quito and can be gathered along the streets.

The streets, parks, plazas, and patios of Quito are dominated by the Quito palm. But these graceful trees are more than just ornamental. The "minicoconuts" that fall from them are much appreciated, especially by the city's children. (W.H. Hodge)

however, is relatively quick growing, taking only 3–4 years to produce its first seeds—lightning fast for a palm.

Little is known about the Quito palm's requirements. It seems to thrive on ample water, but also can be cultivated in dry areas. Its Ecuadorian habitat has a short (1 or 2 month) dry season. Apparently, its deep roots grow straight down, so they usually reach layers with subsoil containing moisture year-round. It is a sun-loving species, but (at least for a palm) shows a high resistance to cold. In its habitat— 3,000 m up in the Andes—night temperatures are in the range of 5– 10°C. Indeed, it is probable that the plant requires cool nights: in areas where night temperatures are consistently over 13–16°C, it seems to lose its vigor and health.

PROSPECTS

Andean Region.   These palms should make handsome ornamentals in cool and dry areas in many parts of the Andes. Moreover, the nuts could be an increasingly useful product for home consumption, for sale in markets, and perhaps to feed guinea pigs (they dehusk the nuts with their sharp teeth).

The plant also has increased potential as an ornamental. Although naturally occurring stands are not known, they may yet be found. The Andes should be searched for germplasm because the specimens growing in Quito and other cities seem to be very uniform—they even seem to be all the same age.[3]

Other Developing Areas.   Because this is a cool-weather palm, it is unsuited to most of the normal palm-growing regions. However, it seems unsurpassed for cool, mild areas such as the lower Himalayas, the uplands of eastern Africa, and the mountains of New Guinea. These regions have the right habitat for what is one of the most beautiful and spectacular palm species known. It is not likely to become a major economic crop, but the nuts would be enjoyed and would provide some nourishment, especially to children.

Industrialized Regions.   The Quito palm should make a handsome ornamental in cool and dry areas of warm-temperate regions. It is promising for use in large public areas such as parks. It is now becoming established as a new landscape feature in northern and central California. Coconuts cannot grow in such areas. All in all, it is a unique and beautiful species that answers the home gardener's need for a long-lived, elegant, vigorous, and tough ornamental tree. The small, edible nuts are a bonus. One caution must be noted, however. This plant may prove to be quite restricted in its adaptability and may, for example, strictly require a cool summer climate.

---

[3] The Quito palm's little-known relative, the janchicoco (*Parajubaea torallyi*), occurs in the ravines of sandstone mountains in central and southern Bolivia (mostly to the east of the city of Sucre in Chuquisaca) where no rain falls for 10 months of the year. It is considered to be an endangered species. However, it, too, is potentially valuable—perhaps even more valuable than the Quito palm. Its fleshy, sweet nuts are eaten, and they contain usable oil. The plant is also used for hearts of palm; the fiber in its frond is woven into ropes, baskets, saddles, and mats; its fruits serve as animal feed; its leaf midrib is burned for fuel; and its leaf stalk is used for construction. It is also cultivated as an ornamental.

Monique Endt.

Walnuts

Although most walnuts have their origins in North America and Asia, a handful of species are found in the Andes. One of these (*Juglans neotropica*) is so prized for its nuts, its fine and beautiful wood, and other products,[1] that it is grown in nearly every highland town in western Venezuela, Colombia, Ecuador, and northern Peru.

The "Andean walnuts"[2] that come from these trees are black shelled and larger than commercial walnuts elsewhere, although the size is somewhat misleading because their shells are unusually thick. The kernels have a fine flavor and are often used in pastries and confections. Women in the Ecuadorian town of Ibarra prepare a famous sweetmeat, the nogada de Ibarra, out of sugar, milk, and these walnuts.

Despite its value, this is a species in difficulty. Its hard, attractive wood is so highly prized for carving, cabinetmaking, and general woodwork, that demand for it has resulted in most of the sizable trees being felled. Throughout the Andean region, large specimens are now scarce, and commercial plantings are not being established. Because of the need for cooking fuel, many of these valuable trees are even being sacrificed for firewood.

Yet there are indications that this species could make an excellent plantation and village crop for the Andes and elsewhere. Seed nuts collected in Ecuador in 1977 have been planted in New Zealand and have grown rapidly. In the Auckland region, they have reached as much as 1.5 m growth per year during the first few years. That is comparable to the growth rate of *Pinus radiata*, New Zealand's fastest growing plantation timber. After 10 years, trees raised from these seeds were more than 10 m high and were bearing their third annual crop of nuts.[3]

The Andean walnut differs from better-known walnut species in at least two ways: the tree is almost evergreen (it grows virtually year-

[1] These notably include dyes and a decoction of the leaves that is considered a valuable tonic.
[2] Common names are tocte, nogal, nogal silvestre, cedro grande, and cedro negro. The name "tocte" is mostly applied just to the fruit; and the name "nogal" (walnut), just to the tree.
[3] Information from D. Endt.

323

round) and it has no chilling requirement, probably because its native habitat straddles the equator where there are no true summers or winters.

Although it is native to an equatorial region, the Andean walnut occurs at an average altitude of 2,500 m, where the climate is temperate. Temperatures vary between −3°C and 25°C (such extremes can occur even in the same day). The cold tolerance is unknown, but frosts occur regularly where the Andean walnut grows. It seems resistant to pests and diseases. The trees are often found growing along stream banks and the boundaries of fields, where they regenerate freely.

PROSPECTS

Andean Region.   This is potentially an extremely valuable species for the whole Andean region. Germplasm collections should be made throughout. Types should be selected for large nuts, thin shells, fast growth, and other qualities.

At the same time, other indigenous walnuts should be collected. These are probably less promising as nut crops, but they, too, are potentially valuable timber species. These include:

● Argentine walnut[4] (*Juglans australis*). Argentina and southern Bolivia. The nut is small and its shell is very thick, making the meat difficult to extract. However, the wood is prized for its fine qualities and is sought after for making guitars.
● Bolivian walnut[5] (*J. boliviana*). Mountains of northern Bolivia as well as southern and central Peru. Similar to *J. neotropica*, this species has grown well in Costa Rica.[6]
● Venezuelan walnut[7] (*J. venezuelensis*). Coastal mountains of northern Venezuela. The trees once occurred frequently in the mountains near Caracas but are now extremely rare, although they still exist between Junquito and the Colonia Tovar cloud forest.[8]

No cross-pollination is required for nut production in the Andean walnut, but apparently hybrids can be made if they are desired. For example, hybrids between *J. neotropica* and the common ("English")

---

[4] Spanish names: nogal cayure, cayuri, nogal cimarrón, nogal criollo, nogal silvestre, nogal de monte.
[5] Spanish names: nogal de la tierra, nogal negro, and nogal blanco.
[6] Information from W.E. Manning.
[7] Local names include nogal, nogal de Caracas, cedro negro, nogal plance, and laurel.
[8] Information from J. Steyermark via W.E. Manning.

Turrialba, Costa Rica. Forty-year-old Andean walnut. This very large tree is about 27 m tall and 0.66 m in diameter. (M. Mora)

walnut have been reported.[9] The exploration of various hybrids might well lead to valuable new types.

Other Developing Regions.   Andean walnuts deserve to be included in forestry and agroforestry trials in areas of upland Central America and Brazil, the hill regions of India, Pakistan, and Nepal, and other seemingly suitable subtropical and tropical highland regions.

Edible nuts are rare in many parts of the tropics, and their commercial production is practically nonexistent. However, the Andean walnut may provide a new, multipurpose tree crop for the middle-elevation areas of the tropics and for the subtropics.

Industrialized Regions.   Any walnut species that can reach 10 m tall in 10 years in a temperate area such as New Zealand deserves testing as a plantation crop. Worldwide demand for walnut timber outpaces current supplies, and outside the Andes this crop's potential is probably greater for timber than for nuts. Silvicultural trials should be initiated with this and other *Juglans* species in a search for the combination of fast growth and high-quality timber. Perhaps, also, robust hybrids can be created.

For example, hybridizing the tropical species with some of the best timber and nut selections from the temperate walnut (*J. nigra*) might produce trees for the tropics that produce both high-grade furniture wood and fine-quality nuts.

[9] Popenoe, 1924.

Selected Readings

In this appendix we have attempted to reference the most recent or readily available literature on little-known Andean food plants. Unfortunately, many of the publications were produced in small quantity and can be extremely difficult to find. Some theses and dissertations are available for a fee from University Microfilms International, 300 North Zeeb Road, Ann Arbor, Michigan 48106, USA. In many cases an author can be found listed as a research contact in Appendix C. Direct correspondence is often the best way to locate information on the availability of these publications and other relevant information.

This appendix follows the same organization as the text. Publications discussing several species are listed under a general section heading, such as "Fruits."

GENERAL

Nutritional information on native Andean crops is often given in *Archivos Latinoamericanos de Nutrición*. Further information is available from Instituto de Nutrición de Céntro América y Panamá (INCAP), Apartado Postal 1188, Guatemala City, Guatemala. National governments often produce consumer information as well.

Antuñez de Mayolo R., S.E. 1981. *La Nutrición en el Antiguo Perú*. Oficina Numismática, Banco Central de Reserva del Perú, Lima. 189 pp.

Beyersdorf, M. and O. Blanco. 1984. *Diccinario Quechua-Español Términos Agrícolas (de Cuzco Colla)*. Proyecto Investigación de los Sistemas Agrícolas Andinos (PISA). Instituto Interamericano de Cooperación para la Agricultura (IICA), Lima. 83 pp.

Brücher, H. 1977. *Tropische Nutzpflanzen: Ursprung, Evolution und Domestikation* (Commercial Tropical Plants: Origin, Evolution and Domestication). Springer Verlag, Berlin.

Brücher, H. In press. *Useful Plants of Neotropical Origin*. English translation from German.

Cárdenas, M. 1969. *Manual de Plantas Económicas de Bolivia*. Imprenta Methodista "ICTHUS," Cochabamba, Bolivia. 421 pp.

Cardich, A. 1987. Native agriculture in the highlands of the Peruvian Andes. *National Geographic Research* 3(1):22–39.

Centro Internacional de Recursos Fitogenéticos (IBPGR). 1983. *El Germoplasma Vegetal en los Países Andinos*. IBPGR Secretariat, c/o Food and Agriculture Organization of the United Nations (FAO), Rome.

Cook, O.F. 1925. Peru as a center of domestication. *Journal of Heredity* 16:33–46; 95–110.

Friere, W. 1965. *Tabla de Composición de los Alimentos Ecuatorianos*. Instituto Nacional de Nutrición, Quito.

Fries, A.M. and M.E. Tapia. 1986. *Los Cultivos Andinos en el Perú*. Programa Nacional de Sistemas Andinas de Producción Agropecuaria, Instituto Nacional de Investigación y Promoción Agropecuaria (INIPA), Lima. 36 pp. (For ordering information, contact M.E. Tapia; see Research Contacts.)

Gade, D.W. 1975. *Plants, Man, and the Land in the Vilcanota Valley of Peru.* W. Junk B.V. Publishers, The Hague. 240 pp. (Extensive bibliography.)

Holle, M. 1986. La conservación ex-situ de la variabilidad genética de los cultivos andinos (1958–1986). Pages 53–64 in *Anales V Congreso Internacional de Sistemas Agropecuarios Andinos.* (See below.)

Instituto Nacional de Nutrición. 1983. *Tabla de Composición de Alimentos para Uso Práctico* (revised). Publication No. 42. Instituto Nacional de Nutrición, Caracas.

International Board for Plant Genetic Resources (IBPGR). 1982. *Plant Genetic Resources in the Andean Region.* IBPGR, Lima. IBPGR Secretariat, c/o FAO, Rome. 82 pp.

IBPGR. *Newsletter for the Americas.* IBPGR, c/o CIAT, Apartado 6713, Cali, Colombia.

Latcham, R.E. 1936. *La Agricultura Precolombiana en Chile y los Países Vecinos.* Ediciones de la Universidad de Chile, Santiago.

León, J. 1964. *Plantas Alimenticias Andinas.* Boletín Técnico No. 6. IICA, Agrícolas Zona Andina, Lima. 112 pp.

León, J. 1968. *Fundamentos Botánicos de los Cultivos Tropicales.* Textos y Materiales de Enseñanza No. 18. IICA, San José, Costa Rica.

Leung, W.-T.W. and M. Flores. 1961. *Food Composition Tables for Use in Latin America.* U.S. Department of Health, Education, and Welfare, Bethesda, Maryland.

Mateo, N. and M.E. Tapia. 1988. High mountain environment and farming systems in Latin America. Status of valuable native Andean crops in the high mountain agriculture of Ecuador. In *Proceedings of Workshop on Mountain Agriculture and Crop Genetic Resources.* International Centre for Integrated Mountain Development (ICEMOD). India Book House, New Delhi.

Montaldo B., P. 1988. El area de distribución y actual de los cultivos andinos en Chile. Pages 332–336 in *Memorias: VI Congreso Internacional de Cultivos Andinos.* (See below.)

Nieto, C., J. Rea, R. Castillo, and E. Peralta. 1984. *Guia para el Manejo y Preservación de los Recursos Fitogenéticos.* Publicación Miscelánea No. 47. Estación Experimental Santa Catalina, Instituto Nacional de Investigaciones Agropecuarias (INIAP), Quito.

Parodi, L.R. 1935. Relaciones de la agricultura prehispanica con la agricultura argentina actual. *Anales de la Academia Nacional de Agronomía y Veterinaria* (Buenos Aires) 1:115–167.

Patiño, V.M. 1964. *Plantas Cultivadas y Animales Domésticos en América Equiñoccial, vol. 2: Plantas Alimentícias.* Imprinta Departamental, Cali.

Pérez-Arbeláez, E. 1978. *Plantas Útiles de Colombia.* Litografía Arco, Bogotá.

Piedra B., M.B. and J.T. Esquinas-Alcazar. 1983. *Situación Actual del Germoplasma Vegetal de los Países Andinos.* AGPG/IBPGR/83/50. IBPGR Secretariat, c/o FAO, Rome.

Querol, D. 1988. *Recursos Genéticos, Nuestro Tesoro Olvidado: Aproximación Técnica y Socioeconómica.* 218 pp. (For ordering information, contact D. Querol, Av. Javier Prado (Este) 461, San Isidro, Lima, Peru; or G. Nabhan, see Research Contacts.)

Rea, J. 1985. *Recursos Fitogenéticos Agrícolas de Bolivia. Base para establecer el sistema.* Oficina Regional para América Latina, Comite Internacional de Recursos Fitogenéticos (IBPGR), La Paz. IBPGR Secretariat, c/o FAO, Rome.

Rengifo, G. 1983. *Bibliografía sobre Agricultura Andina.* PISA. IICA, Lima.

Sánchez-Monge y Parellada, E. 1980. *Diccionario de Plantas Agrícolas.* Servicio de Publicaciones Agrarias, Ministerio de Agricultura, Madrid. 467 pp.

Sauer, C.O. 1952. Cultivated plants of South and Central America. Pages 487–543 in *Handbook of South American Indians vol. 6: Agricultural Origins and Dispersals.* Bureau of American Ethnology Bulletin 143, New York.

Soukup, J. 1970. *Vocabulario de los Nombres Vulgares de la Flora Peruana y Catálogo de los Géneros.* Colegio Salesiano, Lima. 380 pp.

Tapia, M.E. 1982. *El Medio, Los Cultivos y Los Sistemas Agrícolas en los Andes del Sur del Perú.* PISA (Apartado 110697), Lima. 80 pp.

Tola C., J., C. Nieto, E. Peralta, and R. Castillo. 1988. Status of valuable native Andean crops in the high mountain agriculture of Ecuador. In *Proceedings of Workshop on Mountain Agriculture and Crop Genetic Resources.* ICEMOD. India Book House, New Delhi.

Other general works include an excellent series of technical reports entitled *Cultivos Andinos* produced by the Programa de Investigación en Cultivos Andinos (PICA), Universidad Nacional de San Cristóbal de Huamanga, Ayacucho, Peru.

In addition, there have been six international congresses on Andean crops. The published proceedings for these, listed below, contain a wealth of information, particularly on tubers and grains, as well as on indigenous agriculture and related topics.

1. 1978. *Primer Congreso Internacional sobre Cultivos Andinos, 1977.* M.E. Tapia and M.V. Teran, eds. Universidad Nacional de San Cristóbal de Huamanga. Serie Informe de Cursos, Conferencias, y Reuniones No. 117. IICA, OEA, Ayacucho, Peru.

2. 1980. *Segundo Congreso Internacional sobre Cultivos Andinos.* Facultad de Ingenería Agron-

ómica, Escuela Superior Politécnica de Chimborazo, IICA, Riobamba, Ecuador. 323 pp.

3. 1982. *Tercer Congreso Internacional de Cultivos Andinos*. Instituto Boliviana de Tecnología Agropecuaria (IBTA), Centro Internacional de Investigaciones para el Desarrollo (CIID), La Paz, Bolivia. 544 pp.

4. 1984. *Cuarto Congreso Internacional de Cultivos Andinos: Memorias*. Instituto Colombiano Agropecuario (ICA), CIID, Pasto, Colombia. 484 pp.

5. 1987. *V Congreso Internacional de Sistemas Agropecuarios Andinos: Anales*. Universidad Nacional del Altiplano. CORPUNO, CINIPA, CIID, Puno, Peru. 514 pp.

6. 1988. *VI Congreso Internacional de Cultivos Andinos: Memorias*. INIAP, LATINRECO, FUNDAGRO, CIID, Quito, Ecuador. 558 pp.

ROOTS AND TUBERS

Arbizu Avellaneda, C. and E. Robles García. 1986. *Catálogo de los Recursos Genéticos de Raíces y Tubérculos Andinos*. PICA, Universidad Nacional de San Cristóbal de Huamanga, Ayacucho, Peru.

Kay, D.E. 1973. *Root Crops*. TPI Crop and Product Digest, No. 2. The Tropical Products Institute (56/62 Gray's Inn Road, London WC1X 8LU).

Montaldo, A. 1977. *Cultivo de Raíces y Tubérculos Tropicales*. IICA, San José. 284 pp.

PICA. 1987. *Cultivos Andinos: Sistemas Agrícolas Andinos: Tecnología de Producción de Alimentos en Condiciones de Alto Fresco Climático III(1)*. PICA, Universidad Nacional de San Cristóbal de Huamanga, Ayacucho, Peru.

Razumov, V.I. and R.S. Limar. 1974. Effect of night temperature on tuber formation and flowering of certain plant species under short day conditions. *Botanichrskii Zhurnal* 59(11):1648–1657.

Rea, J. and D. Morales. 1980. *Catálogo de Tubérculos Andinas*. Programa de Cultivos Andinos, IBTA, Ministerio de Asuntos Campesinos y Agropecuaria, La Paz. 47 pp.

Tapia, M.E. 1981. Los tubérculos andinos. Pages 45–61 in *Avances en los Investigaciones sobre Tubérculos Alimenticios de los Andes*. PISA. IICA, Lima. 114 pp.

Zimmerer, K. 1988. *Seeds of Peasant Subsistence: Agrarian Structure, Crop Economics, and Quechua Agriculture in Reference to the Loss of Biodiversity in the Southern Peruvian Andes*. Ph.D. dissertation, Department of Geography, University of California, Berkeley. (Extensive bibliography.)

Achira

Chaparro, R.B. and H. Cortés. 1979. *La Achira. Cultivo. Industrialización. Utilidad forrajera*. Orientación Agropecuaria No. 131. ICA, Bogotá.

Chung, H.L. and J.C. Ripperton. 1924. *Edible Canna in Hawaii*. Hawaii Agricultural Experimental Station Bulletin No. 54. Honolulu, Hawaii. 16 pp.

Gade, D.W. 1966. Achira, the edible canna, its cultivation and use in the Peruvian Andes. *Economic Botany* 20(4):407–415.

Imai, K. and T. Ichihashi. 1986. Studies on dry matter production of edible canna (*Canna edulis* Ker.) I. Gas exchange characteristics of leaves in relation to light regimes. *Japanese Journal of Crop Science* 55(3):360–366.

Khoshoo, T.N. and I. Guha. n.d. *Origin and Evolution of Cultivated Cannas*. Vikas Publishing House Pvt. Ltd., New Delhi.

Le Dividich, J. 1977. Feeding value of *Canna edulis* roots for pigs. *The Journal of Agriculture of the University of Puerto Rico* 61(3):267–274.

Mukerjee, L. and T.N. Khoshoo. 1971. Genetic-evolutionary studies on cultivated cannas V: Intraspecific polyploidy in starch yielding *Canna edulis*. *Genetica Iberica* 23:35–42.

Oka, M., O. Boonseng, and W. Watananonta. 1987. Characteristics of dry matter and tuber production of Queensland arrowroot (*Canna edulis* Ker.). *Japanese Journal of Tropical Agriculture* 32(3):173–178.

Tu, L. and H.-D. Tscheuschner. 1981. Untersuchung von wichtigen Eigenschaften der Dong-Rieng-Stärke (Testing of important characteristics of achira starch). *Lebensmittelindustrie* 28:515–516.

Ufer, M. 1972. *Canna edulis* Ker., a neglected root crop. *Tropical Root Crop Newsletter* (Rome) 5:32–34.

Ugent, D., S. Pozorski, and T. Pozorski. 1984. New evidence for ancient cultivation of *Canna edulis* in Peru. *Economic Botany* 38(4):417–432.

Ahipa

Brücher, H. 1985. Der Reiche Gen-Fundus Vergfssener Tropischer Nutzpflanzen (The rich gene-resources of forgotten tropical commercial plants). *Geowissenschaften in Unserer Zeit* 3(1):8–14.

Clausen, R.T. 1944. A botanical study of the yam beans (*Pachyrrhizus*). *Cornell University Memoir 264*. Ithaca, New York. 38 pp.
Herrera, F.L. 1942. *Plantas Tropicales Cultivadas por los Antiguos Peruanos* 2(2):179–195.
Sørensen, M. 1988. A taxonomic revision of the genus *Pachyrhizus* (Fabaceae—Phaseoleae). *Nordic Journal of Botany* 8(2):167–192. (Extensive bibliography.)
Yacovleff, E. 1933. La jíquima, raíz comestible extinguida en el Perú. *Revista del Museo Nacional* (Lima) 2(1):51–56.

Arracacha

Arracacha has been featured in symposiums held in Brazil in 1984 and 1987. Articles based on the first symposium appear in *Informe Agropecuario* (Belo Horizonte, Brasil) 10(120)(December 1984). (For information on availability, contact A.C.W. Zanin; see Research Contacts.)
Barrantes del A., F. and E. Mendoza G. 1987. Nuevas enfermedades en cultivo de mashua (*Tropaeolum tuberosum*), olluco (*Ullucus tuberosus*), arracacha (*Arracacia xanthorrhiza*), y camote (*Ipomoea batatas*) en Ayacucho. Pages 36–51 in *Cultivos Andinos: Sistemas Agrícolas Andinos: Tecnología de Producción de Alimentos en Condiciones de Alto Fresco Climático III(1)*. (See above under Roots and Tubers.)
Canahua Z., A. 1978. Cultivo de la arracacha en Puno. Pages 268–271 in *Primer Congreso Internacional sobre Cultivos Andinos, 1977*. (See above under General heading.)
Castillo T., R. 1984. La zanahoria blanca: otro importante alimento andino. *Desde el Surco* (Ecuador) 42:39–42.
Empresa de Assistência Técnica e Extensão Rural do Estado me Minas Gerais (EMBRATER MG). 1982. *Sistema de Produção para a Cultura da Mandioquinha-salsa*. Série Sistema de Produção No. 009. EMBRATER MG, Empresa Brasileira de Pesquisa Agropecuária (EMPRAPA). 36 pp.
Faillace, P.G. 1972. *El Apio Andino*. División de Proyectos, Corporación de los Andes, Mérida, Venezuela. 95 pp.
Franco P., S. and J. Rodríguez C. 1988. Evaluación del Germoplasma de Arracacha o Racaha (*Arracacia xanthorriza*) en el Valle de Cajamarca. Mimeo. 6 pp. (For information on availability, contact S. Franco P.; see Research Contacts.)
Higuita M., F. 1968. El cultivo de la arracacha en la sabana de Bogotá. *Agricultura Tropical* (Bogotá) 24(3):139–146.
Hodge, W.H. 1954. The edible arracacha—a little-known food plant of the Andes. *Economic Botany* 8:195–221.
León, J. and J. Rea. 1967. Clasificación de clones de arracacha (*Arracacia xanthorrhiza*). *Informe Anual*. IICA, San José.
Mathias, M. and L. Constance. 1976. Umbelliferae. *Flora of Ecuador* 5(145):43–49.
Rea, J. 1984. *Arracacia xanthorrhiza* en los países andinos de Sud América. Pages 387–396 in *Cuarto Congreso Internacional de Cultivos Andinos: Memorias*. (See above under General heading.)
Roth, Q. 1977. *Arracacia xanthorrhiza*: el tubérculo comestible del apio criollo y su estructura interna. *Acta Botanica Venezuelica* 12:147–170.
Sánchez, I. and G. Plascencia. 1981. *Informe Técnico sobre Conservación de Clones de Arracacha*. Universidad Técnica de Cajamarca, Peru. 4 pp.

Maca

Bonnier, E. 1986. Utilisation du sol à l'époque préhispanique: le cas archéologique du Shaka-Palcamayo (Andes centrales). *Cahiers des Sciences Humaines* 22(1):97–113.
Chacon R., G. 1961. *Estudio fitoquímico de Lepidium Meyenii Walp*. Tesis de Bachillerato. Universidad de San Marcos, Lima.
IBPGR, 1982. (See above under General heading.)
Johns, T. 1981. The añu and the maca. *Journal of Ethnobiology* 1(2):208–212.
King, S.R. 1988. *Economic Botany of the Andean Tuber Crop Complex: Lepidium meyenii, Oxalis tuberosa, Tropaeolum tuberosum, and Ullucus tuberosus*. Ph.D. dissertation, Graduate Faculty in Biology, City University of New York. (Extensive bibliography.)
León, J. 1964. The "maca" (*Lepidium meyenii*), a little-known food plant of Peru. *Economic Botany* 18(2):122–127.
Matos M., R. n.d. La maca: una planta peruana en extincion. *Cielo Abierto* n.v.:3–9. (Photocopies available from N. Vietmeyer, National Academy of Sciences, 2101 Constitution Avenue, Washington, DC 20418, USA.)

Mashua

Barrantes and Mendoza, 1987. (See above under Arracacha.)

Bateman, J. 1961. Una prueba exploratoria de la alimentación usando *Tropaeolum tuberosum*. *Turrialba* 11:98–100.

Beckett, K. 1979. Growing perennial tropaeolums. *The Garden* (*Journal of the Royal Horticultural Society*) 104:147–151.

Cortés, H.B. 1984. Avances en la investigación en tres tubérculos andinos: oca (*Oxalis tuberosa*), olluco (*Ullucus tuberosus*), maswa, isaño o añu (*Tropaeolum tuberosum*). Pages 62–83 in *Avances en los Investigaciones sobre Tubérculos Alimenticios de los Andes*. (See above under Roots and Tubers.)

Delgado, N.C. 1978. Características morfológicas de la planta de mashua (*Tropaeolum tuberosum*) asociadas al rendimiento bajo condiciones de Allpachaka (3.600 m.s.n.m.). Pages 250–254 in *Primer Congreso Internacional sobre Cultivos Andinos, 1977*. (See above under General heading.)

Fernandez, J. 1973. Sobre la dispersión meridional de *Tropaeolum tuberosum* R. et P. *Boletín de la Sociedad Argentina de Botánica* 15:106–112.

Gibbs, P.E., D. Marshall, and D. Brunton. 1978. Studies on the cytology of *Oxalis tuberosa* and *Tropaeolum tuberosum*. *Notes from the Royal Botanic Garden Edinburgh* 37:215–220.

Hodge, W.H. 1957. Three native tubers of the high Andes. *Economic Botany* 5(2):185–201.

Johns, 1981. (See above under Maca.)

Johns, T. and G.H.N. Towers. 1981. Isothiocyanates and thioureas in enzyme hydrolysates of *Tropaeolum tuberosum*. *Phytochemistry* 20:2687–2689.

Johns, T., W.D. Kitts, F. Newsome, and G.H.N. Towers. 1982. Anti-reproductive and other medicinal effects of *Tropaeolum tuberosum*. *Journal of Ethnopharmacology* 5:149–161.

Kalliola, R., P. Jokela, L. Pietilä, A. Rousi, J. Salo, and M. Yli-Rekola. In press. Influencia del largo del día en al crecimiento y formación de tubérculos de ulluco (*Ullucus tuberosus*, Basellaceae), oca (*Oxalis tuberosa*, Oxalidaceae) y añu (*Tropaeolum tuberosum*, Tropaeolaceae). (For information on availability, contact A. Rousi; see Research Contacts.)

King, 1988. (See above under Maca.)

King, S.R. and S.M. Gershoff. 1987. Nutritional evaluation of three underexploited Andean tubers: *Oxalis tuberosa* (Oxalidaceae), *Ullucus tuberosus* (Basellaceae), and *Tropaeolum tuberosum* (Tropaeolaceae). *Economic Botany* 41(4):503–511.

Rea, J. 1983. Germoplasma Boliviano y calidad bromatológica de *Tropaeolum tuberosum*. In *Tercer Congreso Internacional de Cultivos Andinos*. (See above under General heading.)

Rea, J. 1984. Germoplasma Boliviano y calidad alimenticia del Isaño (*T. tuberosum*). Pages 7–13 in *Cuarto Congreso Internacional de Cultivos Andinos: Memorias*. (See above under General heading.)

Robles G., E. 1981. *Origen y Evolución de la Oca, Ulluco, y Mashua*. Universidad Nacional Agraria, La Molina, Lima. 26 pp.

Mauka

De Montenegro, L. and S. Franco. 1988. Evaluación de nutrientes en tres variedades de *Mirabilis expansa* "chago." Pages 268–273 in *VI Congreso Internacional de Cultivos Andinos: Memorias*. (See above under General heading.)

Franco, S. and J. Rodríguez. 1988. Multiplicación por esquejes del chago, miso o mauka (*Mirabilis expansa*). Pages 265–267 in *VI Congreso Internacional de Cultivos Andinos: Memorias*. (See above under General heading.)

Rea, J. 1968. *El Miso en Ecuador. Su Calidad Bromatológica*. Informe Técnico. Programa de Cultivos Alimenticios Andinos, IICA, San José.

Rea, J. 1982a. El miso: *Mirabilis expansa*, una contribución de la agricultura pre-Inca de Ecuador y Bolivia. *Desde el Surco* (Quito) n.v.:23–26.

Rea, J. 1982b. Informe de colección de cultivos andinos en Ecuador. IPBGR 82/142. Convenio IBPGR-INIAP, Quito. 35 pp.

Rea, J. and J. León. 1965. La mauka (*Mirabilis expansa* Ruiz y Pavon), un aporte de la agricultura andina prehispanica de Bolivia. *Anales Científicos de la Universidad Agraria* (La Molina, Peru) 3(1):38–41.

Seminario, J. 1988a. Cajamarca: primer centro de variabilidad genética de chago. *Pulso Norteño* (Chiclayo, Peru) 29/30:80.

Seminario, J. 1988b. El chago o mauka *Mirabilis expansa* R y P en Cajamarca. Pages 257–264 in *VI Congreso Internacional de Cultivos Andinos: Memorias*. (See above under General heading.)

Oca

Blanco, O. 1977. Investigación en el mejoramiento de tubérculos menores. In *Primer Congreso Internacional sobre Cultivos Andinos, 1977*. (See above under General heading.)

Consejo Internacional de Recursos Fitogenéticos (IBPGR). 1982. *Descriptores de Oca*. IBPGR Secretariat, c/o FAO, Rome. 23 pp.

Cortés, H.B. 1977. Avances en la investigación de la oca. In *Primer Congreso Internacional sobre Cultivos Andinos, 1977*. (See above under General heading.)
Cortés, 1984. (See above under Mashua.)
de Azkue, D. and A. Martínez. 1987. *Chromosome Numbers of the* Oxalis tuberosa *Alliance*. 9 pp. (Available from the authors; see Research Contacts).
Kalliola et al., 1988. (See above under Mashua.)
King, 1988. (See above under Maca.)
King and Gershoff, 1987. (See above under Mashua.)
King, S.R. and H.H.C. Bastien. In press. Oxalis tuberosa Mol. (Oxalidaceae) in Mexico: the distribution, diversity, cultivation, utilization, nutritional value, and introduction of an Andean tuber crop in Meso-America. *Advances in Economic Botany*.
Orbegoso, A.G. 1960. Estudio sobre la oca (*Oxalis tuberosa* Mol.) con especial referencía a su structura y variabilidad. *Agronomía* (Lima) 27(1):28–38.
Stegemann, H., S. Majino, and P. Schmiediche. 1988. Biochemical differentiation of clones of *Oxalis tuberosa* by their tuber proteins and the properties of these proteins. *Economic Botany* 42(1):37–44.
White, J.W. 1975. *Notes on the Biology of* Oxalis tuberosa *and* Tropaeolum tuberosum. B.A. honors thesis. Harvard University, Cambridge, Massachusetts.

Potatoes

Brush, S.B., H.J. Carney, and Z. Huamán. 1981. Dynamics of Andean potato agriculture. *Economic Botany* 35:70–88.
Correll, D.S. 1962. *The Potato and Its Wild Relatives*. Texas Research Foundation, Renner, Texas. 606 pp.
Hanneman, R.E., Jr. and J.B. Bamberg. 1986. *Inventory of tuber-bearing* Solanum *species*. Wisconsin Agricultural Experiment Station Bulletin 533. 216 pp. (Available from J.B. Bamberg, Potato Introduction Station, Sturgeon Bay, Wisconsin 54235, USA.)
Hawkes, J.G. In press. *Potatoes of Bolivia*. Oxford University Press, Oxford. (Extensive bibliography.)
Hoekstra, R. and L. Seidewitz. 1987. *Evaluation Data on Tuber-bearing* Solanum *Species* (2nd edition). Institut für Pflanzenbau und Pflanzenzüchtung der FAL (Federal Research Centre of Agriculture), Braunschweig, West Germany, and Stichting voor Plantenveredeling, Wageningen, Netherlands.
Hoopes, R.W. and R.L. Plaisted. 1987. Potato. Pages 385–436 in W.R. Fehr, ed., *Principles of Cultivar Development, vol. 2, Crop Species*. Macmillan, New York.
Huamán, Z. 1975. *The Origin and Nature of* Solanum ajanhuiri *Juz. et Buk., a South American Cultivated Diploid Potato*. Ph.D. dissertation, University of Birmingham, Birmingham, UK. 193 pp.
Huamán, Z. 1983. The breeding potential of native Andean potato cultivars. Pages 96–97 in W.J. Hooker, ed., *Proceedings of the International Congress on Research for the Potato in the Year 2000*. International Potato Center (CIP), Lima.
Huamán, Z. 1987. *Inventory of Andean Potato Cultivars with Resistance to Some Pests and Diseases and Other Desirable Traits*. CIP, Lima. 22 pp.
Huamán, Z., J.G. Hawkes, and P.R. Rowe. 1980. *Solanum ajanhuiri*: an important diploid potato cultivated in the Andean Altiplano. *Economic Botany* 34:335–343.
Jackson, M.T., J.G. Hawkes, and P.R. Rowe. 1977. The nature of *Solanum X chaucha* Juz. et Buk., a triploid cultivated potato of the South American Andes. *Euphytica* 26:775–783.
Johns, T. and S.L. Keen. 1986. Ongoing evolution of the potato on the Altiplano of western Bolivia. *Economic Botany* 40(4):409–424.
Muñoz, F.J. and R.L. Plaisted. 1981. Yield and combining abilities in andigena potatoes after six cycles of recurrent phenotypic selection for adaptation to long day conditions. *American Potato Journal* 58:469–479.
Ochoa, C.M. In press. *The Potatoes of South America: Bolivia (vol. 1)*. Cambridge University Press, New York. (Extensive bibliography.)
Ross, H. 1986. Potato breeding—problems and perspectives. *Fortschritte der Pflanzenzüchtung (13)*. Verlag Paul Parey, Berlin. 132 pp.
Schmiediche, P.E., J.G. Hawkes, and C.M. Ochoa. 1980. Breeding of the cultivated potato species *Solanum X juzepczukii* Buk. and *Solanum X curtilobum* Juz. et Buk. *Euphytica* 29:685–704.
Zimmerer, 1988. (See above under General heading.)

Ulluco

Arbizu, C. and J. Valladolid. 1982. Lista de descriptores para el ulluco. Pages 335–344 in *Tercer Congreso Internacional de Cultivos Andinos*. (See above under General heading.)

Barrantes and Mendoza, 1987. (See above under Arracacha.)
Benavides, A. 1967. Variabilidad clonal en ulluco (*Ullucus tuberosus* Loz.). *Fitotécnia Latino-americana* 4(2):91–98.
Brunt, A.A., S. Phillips, R.A.C. Jones, and R.H. Kenten. 1982a. Viruses detected in *Ullucus tuberosus* (Basellaceae) from Peru and Bolivia. *Annals of Applied Biology* 101:65–71.
Brunt, A.A., R.J. Barton, S. Phillips, and R.A.C. Jones. 1982b. Ullucus virus C, a newly recognized comovirus infesting *Ullucus tuberosus* (Basellaceae). *Annals of Applied Biology* 101:73–78.
Cáceda, F. and J. Rossel. 1985. *Entomología de los Cultivos Andinos: Estudio en Comunidades Campesinas*. Editorial Universitaria 85, Universidad Nacional Agraria, La Molina, Lima.
Cortés, 1984. (See above under Mashua.)
Hodge, 1957. (See above under Mashua.)
Kalliola et al., 1988. (See above under Mashua.)
King, 1988. (See above under Maca.)
King and Gershoff, 1987. (See above under Mashua.)
Montaldo, A. 1977. In *Cultivo de Raíces y Tubérculos Tropicales*. (See above under Roots and Tubers.)
Rousi, A., P. Jokela, R. Kalliola, L. Pietilä, J. Salo, and M. Yli-Rekola. 1989. *Morphological Variation Among Clones of ulluco* (Ullucus tuberosus, *Basellaceae) Collected in Southern Peru*. *Economic Botany* 43(1):58–72.
Rousi, A., M. Yli-Rekola, P. Jokela, R. Kalliola, L. Pietilä, and J. Salo. 1988. The fruit of *Ullucus* (Basellaceae), an old enigma. *Taxon* 37:71–75.
Sperling, C.R. 1987. *Systematics of the Basellaceae*. Ph.D. dissertation, Department of Organic and Evolutionary Biology, Harvard University, Cambridge. (Extensive bibliography.)
Stone, O.M. 1982. The elimination of four viruses from *Ullucus tuberosus* by meristem-tip culture and chemotherapy. *Annals of Applied Biology* 101:79–83.
Vallenas R., M. 1984. Ollucos silvestres en Puno—Peru. Pages 340–345 in *Cuarto Congreso Internacional de Cultivos Andinos: Memorias*. (See above under General heading.)
Zimmerer, 1988. (See above under General heading.)

Yacon

Calvino, M. 1940. A new plant *Polymnia edulis* for forage or alcohol. *Industria Sacchariferra Italiana* (Genova) 33:95–98.
Endt, A. and R. Endt. 1986. *Growing Yacon*. Pamphlet. (Available from authors; see Research Contacts.)
Kierstan, M. 1980. Production of fructose syrups from inulin. *Process Biochemisty* 15(4):2,4,32.
Montaldo, A. 1967. Bibliografía de raíces y tubérculos tropicales. *Revista Facultad Agrícola Universidad Central de Venezuela* 13:489–490.

GRAINS

Risi C., J. and N.W. Galwey. 1984. The *Chenopodium* grains of the Andes: Inca crops for modern agriculture. *Advances in Applied Biology* 10:145–216. (Extensive bibliography.)
Rodriguez, I. 1986. *En Su Mesa: Cultivos Andinos*. Proyecto Fitomejoramiento y Producción de Semilla de Cultivos Andinos para el Desarrollo Rural (FIPS), Cuzco. 71 pp. (Nutritional information and recipes.)
Tapia, M., H. Gandarillas, S. Alandia, A. Cardozo, A. Mujica, R. Ortiz, V. Otazu, J. Rea, B. Salas, E. Zanabria. 1979. *La Quinua y la Kañiwa: Cultivos Andinos*. Serie Libros y Materiales Educativos No. 40. IICA, Turrialba, Costa Rica. (Extensive bibliography.)

Kaniwa

Briceño P., O. and F. Canales M. 1976. La cañihua (*Chenopodium pallidicaule* Aellen) como sucedáneo del maíz en raciones para pollo, parrilleros. *Anales Científicos* (Lima) 14:151–163.
Carmen, M.L. 1984. Acclimatization of quinoa (*Chenopodium quinoa* Willd.) and canihua (*Chenopodium pallidicaule* Aellen) to Finland. *Annales Agricultural Fennial* 23:135–144.
Gade, D.W. 1970. Ethnobotany of cañihua (*Chenopodium pallidicaule*), rustic seed crop of the Altiplano. *Economic Botany* 24(1):55–61
Risi and Galwey, 1984. (See above under Grains.)
Tapia et al., 1979. (See above under Grains.)
Trinidad S., A., F. Gómez L., and G. Suarez R., eds. 1986. *El Amaranto: Amaranthus spp. (Alegria): Su Cultivo y Aprovechamiento*. Memoria, Primer Seminario Nacional del Amaranto. (For ordering information, contact A. Sanchez M.; see Research Contacts.)

Kiwicha

For a general account of the promise of *Amaranthus* species, see National Research Council, 1984, *Amaranth: Modern Prospects for an Ancient Crop*, National Academy Press, Washington, DC. Ordering information can be found at the end of this book.

Two newsletters dealing broadly with all the *Amaranthus* species are currently published. These are: *Amaranth Today*, published by Rodale Press, Inc., 33 East Minor Street, Emmaus, Pennsylvania 18098, USA; and *Amaranth Newsletter*, published by Editorial Office, Archivos Latinoamericanos de Nutrición, Instituto de Nutrición de Céntro América y Panamá (INCAP), Apartado Postal 1188, Guatemala City.

In addition, the following volume is a guide to the planting and management of amaranths: Weber, L.E., E.S. Hubbard, L.A. Nelson, D.H. Putnam, and J.W. Lehmann. 1988. *Amaranth Grain Production Guide*. Rodale Research Center and the American Amaranth Institute. 24 pp. (Published annually; copies and amaranth bibliography available from Rodale Research Center, RD 1, Box 323, Kutztown, Pennsylvania 19530, USA.)

Bressani, R., J.M. Gonzales, J. Zuniga, M. Breuner, and L.G. Elías. 1987a. Yield, selected chemical composition and nutritive value of 14 selections of amaranth grain representing four species. *Journal of the Science of Food and Agriculture* 38:347–356.

Bressani, R., J.M. Gonzales, L.G. Elías, and M. Melgar. 1987b. Effect of fertilizer application on the yield, protein and fat content, and protein quality of raw and cooked grain of three amaranth species. *Qualitas Plantarum/Plant Foods for Human Nutrition* 37:59–67.

Goldberg, A.D. and G. Covas. 1988. *Actas de las Primeras Jornadas Nacionales sobre Amarantos*. Facultad de Agronomía, Universidad Nacional de La Pampa, Santa Rosa, La Pampa, Argentina.

Imeri, A., R. Flores, L.G. Elías, and R. Bressani. 1987a. Efecto del procesamiento y de la suplementación con aminoacidos sobre la calidad proteinica del amaranto (*Amaranthus caudatus*). *Archivos Latinoamericanos de Nutrición* 37(1):160–173.

Imeri, A., J.M. Gonzales, R. Flores, L.G. Eliás, and R. Bressani. 1987b. Variabilidad genética, y correlaciones entre rendimiento, tamano del grano, composición química y calidad de la proteina de 25 variedades de amaranto (*Amaranthus caudatus*). *Archivos Latinoamericanos de Nutrición* 37(1):132–146.

Kulakow, P.A. 1987. Genetics of grain amaranths II. The inheritance of determinance, panicle orientation, dwarfism, and embryo color in *Amaranthus caudatus*. *The Journal of Heredity* 78:293–297.

Morales, E., J. Lembcke, and G.G. Graham. 1988. Nutritional value for young children of grain amaranth and maize-amaranth mixtures: effects of processing. *Journal of Nutrition* 118(1):78–85.

Nieto, C. 1986. Analysis of growth and response to photoperiod in six *Amaranthus* species. Master's thesis, Universidad de Costa Rica. CATIE-Turrialba. 101 pp.

Pedersen, B., L.S. Kalinowski, and B.O. Eggum. 1987a. The nutritional value of amaranth grain (*Amaranthus caudatus*): 1. Protein and minerals of raw and processed grain. *Qualitas Plantarum/Plant Foods for Human Nutrition* 36(4):309–324.

Pedersen, B., L. Hallgren, I. Hansen, and B.O. Eggum. 1987b. The nutritional value of amaranth grain (*Amaranthus caudatus*): 2. As a supplement to cereals. *Qualitas Plantarum/Plant Foods for Human Nutrition* 36(4):325–334.

Robles, G., E. and E. Nuñez A. 1987. Comparativo del rendimiento de 13 accesiones de achita (*Amaranthus caudatus* L.) en las localidades: Centro Experimental de Wayllapampa (2,450 m.s.n.m.) y Canaan (2,720 m.s.n.m.) Pages 1–9 in *Cultivos Andinos: Sistemas Agrícolas Andinos: Tecnología de Producción de Alimentos en Condiciones de Alto Fresco Climático III(1)*. (See above under Roots and Tubers.)

Sumar K., L. 1985. *El Pequeño Gigante*. Oficina de Area, United Nations Children's Fund (UNICEF)-Perú, Lima. 24 pp.

Sumar K., L., O. Blanco G., and J. Pacheco N. 1986a. *Descriptores Para Amaranthus (con orientación para* Amaranthus caudatus *L.)*. Tercera Edición, corregida y aumentada. Programa de Investigación Amaranthus Reporte 86–3. Programa Nacional de la Kiwicha, Universidad Nacional del Cusco. 20 pp.

Sumar K., L., J. Pacheco N., and F. Florés L. 1986b. *Correlaciones Simples y Multiples del Rendimiento en Grano de la "Kiwicha"* (Amaranthus caudatus), *con Algunas Variables Bio-Metricas*. Programa de Investigación Amaranthus Reporte 86–2. Programa Nacional de la Kiwicha, Universidad Nacional del Cusco. 54 pp.

Vaidya, K.R. and S.K.J. Jain. 1987. Response to mass selection for plant height and grain yield in amaranth (*Amaranthus* spp.). *Plant Breeding* 98:61–64.

Quinoa

Canahua M., A. and J. Rea. 1980. Quinuas resistentes a heladas, avances a la investigación. Pages 143–149 in *Segundo Congreso Internacional sobre Cultivos Andinos*. (See above under General.)

Carlsson, R. 1983. Leaf protein concentrate from plant sources in temperate countries. Pages 52–80 in L. Telek and H.D. Graham, eds., *Leaf Protein Concentrates*. AVI Publishing Co., Westport, Connecticut, USA.

Castillo, R. 1987. *A Study of the Long-Term Storage Behaviors of* Chenopodium quinoa *Seeds*. Master's thesis, Plant Biology Department, University of Birmingham, U.K.

Consejo Internacional de Recursos Fitogenéticos (IBPGR). 1981. *Descriptores de Quinua*. IBPGR Secretariat, c/o FAO, Rome. 18 pp.

Cusack, D. 1984. Quinua: grain of the Incas. *The Ecologist* 14(1):21–31.

Galwey, N.W. and J. Risi. 1984. *Suggestions for the Cultivation of Quinoa in Britain*. Mimeo. 7 pp. (Copies available from authors; see Research Contacts.)

Gandarillas S.C., H. 1968. *Razas de Quinua*. Boletín No. 34. Instituto Boliviano de Cultivos Andinos, División de Investigaciones Agrícolas, Ministerio de Agricultura, La Paz.

Gandarillas S.C., H. 1974. *Genética y Origen de la Quinua*. Boletín No. 9. Instituto Nacional de Trigo, Ministerio de Asuntos Campesinos y Agropecuarios, La Paz.

Gandarillas S.C., H. 1984. *Obtención Experimental de* Chenopodium quinoa *Willd.*. IBTA, Ministerio de Asuntos Campesinos y Agropecuarios, La Paz. 21 pp.

Hernández, S. 1987. *Incorporation of High-Protein Endemic Cultigens into the Formal Economic Sector of Ecuador*. Ph.D. dissertation, Department of Urban Planning, Michigan State University, Ann Arbor, Michigan, USA.

Lescano, R., J.L. 1981. *Cultivo de la Quinua*. Universidad Nacional Técnica del Altiplano, Centro de Investigaciones en Cultivos Andinos, Puno, Peru.

López de Romaña, G., G. Graham, M. Rojas, and W.C. McLean, Jr. 1981. Digestibilidad y calidad proteínica de la quinua: estudio comparativo en niños entre semilla y harina de quinua. *Archivos Latinamericanos de Nutrición* 31:485–497.

Risi and Galwey, 1984. (See above under Grains.)

Tapia et al., 1979. (See above under Grains.)

Torres, H.A. and I. Minaya. 1980. *Escarificadora de quinua. Diseño y Construcción*. Publicaciones Miscelaneas No. 243. IICA, Lima.

Wilson, H.D. 1981. Genetic variation among South American populations of tetraploid *Chenopodium* sect. *Chenopodium* subsect. *Cellulata. Systematic Botany* 6(4):380–398.

Wilson, H.D. 1988. Quinua biosystematics I: domesticated populations. *Economic Botany* 42(4):461–477. (Extensive bibliography.)

Wood, R. 1989. *Quinoa the Super Grain*. Japan Publications (Kodansha International), New York. 200 pp. (Extensive bibliography.)

Zanabria, E. and A. Mujica S. 1977. Plagas de la quinua. Pages 129–142 in *Curso de Quinua*. Fondo Simón Bolívar, Ministerio de Alimentación. IICA, Universidad Nacional Técnica del Altiplano, Puno, Peru.

LEGUMES

For a general account of the promise of legumes, see National Research Council. 1979. *Tropical Legumes: Resources for the Future*. National Academy of Sciences, Washington. (Ordering information can be found at the end of this book.)

Basul

Peralta V., I.C. 1983. *Integración de la Leguminosa Basul* (Erythrina edulis) *en la Alimentación Humana*. Mimeo. 19 pp. (Copies available from author; see Research Contacts.)

Pérez, C., D. de Martínez, and E. Díaz. 1979. Evaluación de la calidad de la proteina de la *Erythrina edulis* (Balu). *Archivos Latinoamericanos de Nutrición* 29(2):193–207.

Romero-Castañeda, R. 1961. Pages 96–97 in *Frutas silvestres de Colombia* (vol. 1). Editora "San Juan Eudes," Bogotá. 342 pp.

Nuñas (Popping Beans)

Brücher, H. 1988. The wild ancestor of *Phaseolus vulgaris* in Middle America. Pages 185–214 in P. Gepts, ed., *Genetic Resources of* Phaseolus *Beans*. Kluwer Academic Publishers, Boston.

Gepts, P.L., T.C. Osborn, K. Rashka, and F.A. Bliss. 1986. Phaseolin protein variability in wild forms and landraces of the common bean (*Phaseolus vulgaris*): evidence for multiple centers of domestication. *Economic Botany* 40(4):451–468.

Kaplan, L., and L.N. Kaplan. 1988. *Phaseolus* in archaeology. Pages 125–142 in P. Gepts, ed., *Genetic Resources of* Phaseolus *Beans*. Kluwer Academic Publishers, Boston.

Zimmerer, K. 1985. *Agricultural Inheritances: Peasant Management of Common Bean* (Phaseolus vulgaris) *Variation in Northern Peru*. Master's thesis, Department of Geography, University of California, Berkeley. (Extensive bibliography.)

Zimmerer, K. 1987. La nuña. Pages 233–234 in *V Congreso Internacional de Sistemas Agropecuarios Andinos: Anales*. (See above under General heading.)

Tarwi

The International Lupin Association (ILA) publishes a newsletter and a series of proceedings of its biennial conferences. Many issues contain literature on tarwi ("Andean lupin"). These are available from the ILA, PO Box 3048, Córdoba, Spain.

Antuñez de Mayolo, S. 1982. Tarwi in ancient Peru. Pages 1–11 in R. Gross and E.S. Bunting, eds., *Agricultural and Nutritional Aspects of Lupines*. Schriftenreihe der Deutsche Gesellschaft für Technische Zusammenarbeit (GTZ) No. 125, (Dag-Hammarskjöld-weg 1, Postfach 5180, D-6236) Eschborn, West Germany.

Brücher, H. 1968. Die genetischen reserven Südamerikas für die kulturpflanzenzüchtung. *Theoretical and Applied Genetics* 38:9–22.

Frey, F. and E. Yabar. 1983. *Enfermedades y Plagas de Lupinos en el Perú*. Schriftenreihe der GTZ No. 142, (Dag-Hammarskjöld-weg 1, Postfach 5180, D-6236) Eschborn, West Germany.

Gross, R. 1982. *El Cultivo y la Utilización del Tarwi* (Lupinus mutabilis *Sweet*). Producción y Protección Vegetal No. 36. Estudio FAO. FAO, Rome. 236 pp.

Gross, U., R.G. Galindo, and H. Schoeneberger. 1983. The development and acceptability of lupine (*Lupinus mutabilis*) products. *Qualitas Plantarum/Plant Foods for Human Nutrition* 32:155–164.

International Board for Plant Genetic Resources (IBPGR). 1981. *Lupin Descriptors*. IBPGR Secretariat, c/o FAO, Rome. 68 pp.

Ortiz, C., R. Gross, and E. von Baer. 1975. Protein quality of *Lupinus mutabilis* compared to *Lupinus albus, Lupinus luteus* and soybeans. *Zeitschrift für Ernährungswissenschaft* 14:230–234.

Pakendorf, K.W., D.J. Van Schalkwyk, and F.J. Coetzer. 1973. Mineral element accumulation in Lupinus II. *Zeitschrift Acker- und Pflanzenbau* 138:46–62.

Rodriguez, I., 1986. (See above under Grains.)

Römer, P. and W. Jahn-Deesbach. 1986. Developments in breeding of *Lupinus mutabilis*. Pages 31–39 in *Proceedings of the Fourth International Lupin Conference*. ILA, Córdoba.

Römer, P. and W. Jahn-Deesbach. In press. Developments in *Lupinus mutabilis* breeding. *Proceedings of the Fifth International Lupin Conference*. ILA, Córdoba.

Van Jaarsveld, A.B. and P.S. Knox-Davies. 1974. Resistance of lupins to *Phomopsis leptostromiformis*. *Phytophylactica* 6:55–60.

Villareal, J.A. and F. Angstburger. 1986. El cultivo de tarhui (*Lupinus mutabilis*), fijación y aporte de nitrogeno al suelo y su efecto residual en cebada (*Hordeum vulgaris*). Pages 774–785 in *V Congreso Internacional de Sistemas Agropecuarios Andinos: Anales*. (See above under General heading.)

von Baer, E. and D. von Baer. In press. *Lupinus mutabilis*: cultivation and breeding. *Proceedings of the Fifth International Lupin Conference*. ILA, Córdoba.

VEGETABLES

Peppers

Andrews, J. 1984. *Peppers: The Domesticated Capsicums*. University of Texas Press, Austin, Texas.

Eshbaugh, W.H. 1982. Variation and evolution in *Capsicum eximium* (Solanaceae). *Baileya* 21(4):193.

Eshbaugh, W.H., S.I. Guttman, and M.J. McLeod. 1983. The origin and evolution of domesticated *Capsicum* species. *Journal of Ethnobiology* 3(1):49–54.

Heiser, C.B. 1985. How many kinds of peppers are there? Pages 142–154 in Heiser, 1985a. (See below under Fruits.)

Martin, F.W., J. Santiago, and A.A. Cook. 1979. *The Peppers, Capsicum Species*. Vegetables for the Hot, Humid Tropics, Part 7. U.S. Department of Agriculture. 18 p. (Science and Education Administration, PO Box 53326, New Orleans, Louisiana 70153, USA.)

Squashes and Their Relatives

Barrantes, F. 1988. Cucurbitáceas nativas de Ayacucho (2750 msnm) estudios básicos en colecciones locales. Pages 294–299 in *VI Congreso Internacional de Cultivos Andinos: Memorias*. (See above under General heading.)

Esquinas-Alcazar, J.T. and P.J. Gulick. 1983. *Genetic Resources of Cucurbitaceae*. AGPG:IBPGR/82/48. IBPGR Secretariat, c/o FAO, Rome. 99 pp.

Jeffrey, C. 1980. A review of the Cucurbitaceae. *Botanical Journal of the Linnean Society* 81:233.

Morton, J.F. 1975. The sturdy Seminole pumpkin provides much food with little effort. *Proceedings of the Florida State Horticultural Society* 88:137–142. (Reprints available from the author; see Research Contacts.)

Whitaker, T.W. 1980. The potential economic importance of some unexploited species of Cucurbitaceae and the use of cucurbits as experimental material. Paper presented at Conference on

Biology and Chemistry of the Cucurbitaceae, August 3–6, 1980, Cornell University, Ithaca, New York. (For information on availability, contact the author; see Research Contacts.)

FRUITS

Books that deal in a general way with fruits or that include information on several of the species in this report include the following.

Dawes, S.N. and G.J. Pringle. 1983. Fifteen subtropical fruits from South and Central America. Pages 123–139 in G.S. Wratt and H.C. Smith, eds., *Plant Breeding in New Zealand*. Butterworths, Wellington, New Zealand.
Heiser, C.B. 1969. *Nightshades: The Paradoxical Plants*. W.H. Freeman and Company, San Francisco. 200 pp.
Heiser, C.B. 1985. *Of Plants and People*. University of Oklahoma Press, Norman, Oklahoma.
Martin, F.W., C.W. Campbell, and R.M. Ruberté. 1987. *Perennial Edible Fruits of the Tropics: An Inventory*. Agriculture Handbook No. 642. U.S. Department of Agriculture. U.S. Government Printing Office, Washington, DC. 252 pp.
Morton, J.F. 1987. *Fruits of Warm Climates*. Published by the author. (Available from Creative Resource Systems, Inc., Box 890, Winterville, North Carolina 28590, USA.) 505 pp.
Patiño, V.M. 1963. *Plantas Cultivadas y Animales Domésticos en América Equinoccial, Vol. 1. Frutales*. Imprinta Departamental, Cali. 403 pp.
Popenoe, W. 1924. *Economic Fruit-Bearing Plants of Ecuador*. Contributions from the United States National Herbarium vol. 24, No. 5, Washington, DC.
Romero-Castañeda, R. 1961. *Frutas Silvestres de Colombia* (vol. 1). Editora "San Juan Eudes," Bogotá. 342 pp.
Romero-Castañeda, R. 1969. *Frutas Silvestres de Colombia* (vol. 2). Editora "San Juan Eudes," Bogotá. 384 pp.
Romero-Castañeda, R. 1985. *Frutas Silvestres del Choco*. Instituto Colombiano de Cultura Hispánica, Bogotá. 122 pp.
Sarmiento G., E. 1986. *Frutas en Colombia*. Ediciones Cultural Colombiana Ltda., Bogotá. 173 pp.
Schneider, E. 1986. *Uncommon Fruits and Vegetables: A Commonsense Guide*. Harper & Row, New York. 546 pp.
Schultes, R.E. and R. Romero-Castañeda. 1962. Edible fruits of *Solanum* in Colombia. *Botanical Museum Leaflets* 19:235–286.

A quarterly newsletter, *Solanaceae*, which often contains information on goldenberry, naranjilla, pepino, tamarillo, and peppers, is available from J. Riley; see Research Contacts.

Andean Berries

At present, there is little information readily available on the horticulture of either ugni or the Andean blueberry.

Bautista, D. 1977. Observaciones sobre el cultivo de la mora (*Rubus glaucus* Benth.) en los Andes venezolanos. *Agronomía Tropical* 27(2):253–260.
Darrow, G.M. 1952. *Rubus glaucus*, the Andes blackberry of Central America and northern South America. *Ceiba* 3:97–101.
Darrow, G.M. 1955. The giant Colombian blackberry of Ecuador. *Fruit Varieties and Horticultural Digest* 10:21–22.
Darrow, G.M. 1967. The cultivated raspberry and blackberry in North America—breeding and improvement. *American Horticulture Magazine* 46:203–218.
De la Cadeña, J.I. and A. Orellana. 1985. *Cultivo de la Mora, vol. 1*. Unidad de Capacitación Fruticultura, Instituto Nacional de Capacitación Campesina, Quito. 116 pp. (Available from USAID, Washington, DC 20523)
De la Cadeña, J.I. and A. Orellana. 1985. *Cultivo de la Mora, vol. 2*. Unidad de Capacitación Fruticultura, Instituto Nacional de Capacitación Campesina, Quito. 52 pp. (Available from USAID, Washington, DC 20523)
Jennings, D.L. 1978. The blackberries of South America—an unexplored reservoir of germplasm. *Fruit Varieties Journal* 32(3):61–63.
Lavín A., A. and C. Muñoz S. 1988. Propagación de la murtilla (*Ugni molinae* Turcz.) mediante estacas apicales semi leñosas [Propagation of "murtilla" (*Ugni molinae* Turcz.) using semi-hardwood apical cuttings]. *Agricultura Tecnica* (Chile) 48(1):58–59.
Morton, J.F. 1984. The Andes berry (*Rubus glaucus* Benth.) in Haiti. *Proceedings of the American Society of Horticultural Science, Tropical Region* 28:87–89.
Osorio B., I.A.J. 1975. El cultivo de la mora (zarzamora). *ICE-Información* 11(11):7–20.

Popenoe, W. 1920. La mora gigante de Colombia. *Revista del Ministerio de Agricultura y Comercia* (Colombia) 6:503–507.
Popenoe, W. 1921a. The Andes berry. *Journal of Heredity* 12:387–393.
Popenoe, W. 1921b. The Colombian berry. *Journal of Heredity* 11:195–202.
Popenoe, 1924. (See above under Fruits.)
Rodríguez Z., E. and J. Duarte B. n.d. Mora de castilla. Pages 399–420 in D. Ríos-Castaño and R. Salazar C., eds., *Frutales, vol. 2.* ICA, Bogotá.
Servicio Nacional de Aprendisahe (SENA). n.d. *El Cultivo de la Mora.* División de Communicaciones, Regional Boyacá, SENA, Boyacá, Colombia. 30 pp.
Sherman, W.B. and R.H. Sharpe. 1971. Breeding *Rubus* for warm climates. *Horticultural Science* 6:147–149.
Williams, C.F., B.W. Smith, and G.M. Darrow. 1949. A Pan-American blackberry hybrid. Hybrids between the Andean blackberry and American varieties. *Journal of Heredity* 40:261–265.

Capuli Cherry

Acosta-Solis, M. 1973. Origen y geografía del capulí. *Boletín de Informaciones Científicas Nacionales* (Casa de la Cultura Ecuatoriana) 14(105–106):28–32.
Castillo T., R. 1988. *Capulí.* Mimeo. 7 pp. (Copies available from author; see Research Contacts.)
Centauro, S.A. 1985. Capulí. Pages 21–23 in *Apuntes Sobre Algunas Especies Forestales Natives de la Sierra Peruana.* Edited by J. Pretell C., D. Ocaña V., R. Jon L., and E. Barahona C. Proyecto FAO/Holanda/INFOR (GCP/PER/027/NET), Lima, Peru.
McVaugh, R. 1951. A revision of the North American black cherry (*Prunus serotina* Ehrh. and relatives). *Brittonia* 7:279–315.
McVaugh, R. 1952. Suggested phylogeny of *Prunus serotina* and other wide-ranging phylads in North America. *Brittonia* 7:317–346.
Raimondo, A. 1981. Il capulin, una pianta da frutto acclimatata in Sicilia (The capulin, a plant with fruit acclimatized in Sicily). *Estratto da Frutticoltura* (Gruppo Giornalistico Edagricole) 18(5):47–49.

Cherimoya

Banks, A.G. and G.M. Sanewski. 1988. *Growing Custard Apples.* Queensland Department of Primary Industries (QDPI). (For ordering information, contact QDPI Information Centre, GOP Box 46, Brisbane 4001, Australia.)
Bender, G.S. and N.C. Ellstrand. 1986. *Cherimoya Culture in California.* University of California Cooperative Extension Publication #CP-465(150):7–86.
Cañizares Z., J. 1966. *Las Frutas Annonaceas.* Ediciones Fruticuba, Havana, Cuba. 63 pp.
Carlos L., F. 1982. Polinización artificial en chirimoyos—técnica que aumenta rendimientos y calidad. *El Campesino* January-February 1982.
Dawes, S.N. 1985. Cherimoya—subtropical fruit with potential. *Growing Today,* August 1985.
Dawes and Pringle, 1983. (See above under Fruits.)
Endt, R. 1984. A commercial future for the cherimoya? *Southern Horticulture* September, 49–50.
Gardiazábal, F. and G. Rosenberg. 1986. *Cultivo del Chirimoyo.* Escuela de Agronomía, Universidad Católica de Valparaíso, Chile. (For information on availability, contact F. Gardiazábal; see Research Contacts.)
George, A.P. and R.J. Nissen. 1987. Propagation of *Annona* species: a review. *Scientia Horticulturae* 33:75–85.
Geurts, F. 1981. *Annonaceous Fruits (Annona spp.): Aspects Related to Germ Plasm Conservation: A Preliminary Study.* AGP:IBPGR/81/11, January. Mimeo. Royal Tropical Institute, Amsterdam, Netherlands.
Guirado Sánchez, E. *Chirimoyo. Comparación de Metodos de Polinización Artificial. Análisis de Componentes del Fruto. Efecto del Acido Giberélico en el Desarrollo del Fruto y de sus Componentes.* Escuela Universitaria de Ingenieria Técnica Agricola (E.U.I.T.A.) "La Rabida," Seville. 116 pp. (Most recent of a series of reports done at the Estación Experimental "La Mayora" of the Universidad de Málaga. For information on topics and availability, contact J.M. Farré; see Research Contacts.)
Hopping, M.E. 1982. Pollination and fruit set of cherimoya. *The Orchardist of New Zealand* 55(2):56–60.
New Zealand Tree Crops Association. 1985. *Proceedings of a Seminar on Cherimoyas Held at Carrington Technological Institute.* New Zealand Tree Crops Association (Auckland Branch), New Zealand.
Schroeder, C.A. 1951. Fruit morphology and anatomy of the cherimoya. *Botanical Gazette* 112:436–446.
Schroeder, C.A. 1971. Pollination of cherimoya. *California Avocado Society Yearbook* 42:119–122.

Schroeder, C.A. 1988. The pollination problem in cherimoya. *Indoor Citrus and Rare Fruit Society Newsletter* 26(Summer):12–13.
Thomson, P.H. 1970. The cherimoya in California. *Californian Rare Fruit Growers Yearbook* 2:20–34.

Goldenberry

Cailes, R.L. 1952. The cultivation of cape gooseberry. *Journal of Agriculture of Western Australia* 1:363–365.
Chia, C.L., M.S. Nishina, and D.O. Evans. 1987. *Poha*. Commodity Fact Sheet Poha-3(A), Hawaii Cooperative Extension Service, Hawaii Institute of Tropical Agriculture and Human Resources. University of Hawaii at Manoa, Honolulu, Hawaii. 2 pp.
Council of Scientific and Industrial Research (CSIR). 1969. Physalis. Pages 38–40 in *The Wealth of India: Raw Materials, vol. 8*. Publications & Information Directorate, CSIR, New Delhi.
Davis, B.R. and V.B. Whitehead. 1974. *Cape Gooseberry Cultivation*. Information Bulletin Number 210. Fruit and Fruit Technology Research Institute, Stellenbosch, South Africa. Mimeo. 4 pp.
Dremann, C.C. 1987. *Ground Cherries, Husk Tomatoes, and Tomatillos*. 22 pp. (Available for a nominal charge from the author; see Research Contacts.)
Jaw-Kai Wang. 1966. *Equipment for Husking Poha Berries*. Technical Bulletin No. 60. Hawaii Agricultural Experiment Station, Honolulu. 18 pp.
Klinac, D.J. 1986. Cape gooseberry (*Physalis peruviana*) production systems. *New Zealand Journal of Experimental Agriculture* 14:425–30. (Reprints available from the author; see Research Contacts.)
Klinac, D.J. and F.H. Wood. 1986. Cape gooseberry (*Physalis peruviana*). *The Orchardist of New Zealand* 60(4):103.
Legge, A.P. 1974. Notes on the history, cultivation, and uses of *Physalis peruviana*. *Journal of the Royal Horticultural Society* 99:310–314.
Mazumdar, B.C. 1979. Cape-gooseberry, the jam fruit of India. *World Crops* 31(January/February):19,23.
Morton, J.F. and O.S. Russell. 1954. The cape gooseberry and the Mexican husk tomato. *Proceedings of the Florida State Horticultural Society* 67:261–266. (Reprints available from J.F. Morton; see Research Contacts.)

Highland Papayas

Badillo, W.M. 1967. Acerca de la naturaleza híbrida de *Carica pentagona*, *C. chrysopetala* and *C. fructifragrans*, frutales del Ecuador y Colombia. *Revista Facultad Agronomía Venezuelano* 4(2):92–103.
Badillo, W.M. 1971. *Monografía de la Familia Caricaceae*. La Asociación de Profesores, Universidad Central de Venezuela, Facultad de Agronomía, Maracay, Venezuela.
Camacho, S. and V. Rodríguez. 1982. El cultivo comercial del babaco (*Carica pentagona* Heilb.). *Proceedings of the American Society for Horticultural Science (Tropical Region)* 26:17–20.
Cossio, F., ed. 1987. *Atti del Iº* Convegno Nazionale sul Babaco. Proceedings of a conference held March 28, 1987, in Ferrara, Italy. (Information available from F. Cossio; see Research Contacts.)
Cossio, F. 1988. *Il Babaco*. Edagricole (Via Emilia Levante, 31), Bologna, Italy.
Dawes and Pringle, 1983. (See above under Fruits.)
Endt, A. 1981. *One Hundred One Ways of Using Pawpaws*. Landsendt Tropical Fruits (108 Parker Road, Oratia) Auckland, New Zealand.
Fabara G., J., N. Bermeo C., and C. Barberán B. 1985. *Manual del Cultivo del Babaco*. Universidad Técnica de Ambato/Consejo Nacional de Ciencia y Tecnología, Ambato, Ecuador.
Heilborn, O. 1921. Taxonomical and cytological studies on cultivated Ecuadorian species of *Carica*. *Arkiv för Botanik* 17(12):1–17.
Litz, R.E. 1986. Papaya (*Carica papaya* L.) Pp. 220–232 in Y.P.S. Bajaj, ed., *Biotechnology in Agriculture and Forestry vol. 1: Trees I*. Springer-Verlag Berlin Heidelberg, West Germany.
Mata, J. and N. Rivas. 1981. Especies del género *Carica* como fuente de extración de papaína. Pages 51–54 in *Informe de Investigación 1980*. Instituto de Genética, Universidad Central de Venezuela, Caracas.
Mekako, H.U. and H.Y. Nakasone. 1977. Sex inheritance in some *Carica* species. *Journal of the American Society of Horticultural Science* 102(1):42–45.
Popenoe, 1924. (See above under Fruits.)
Silva, E., V. Fuentes, and L.A. Lizana. 1980. Azucares y ácidos constitutivos y su evolución durante la maduración de papayas (*Carica candarmarcensis* Hook.). *Proceedings of the American Society for Horticultural Science (Tropical Region)* 24:97–102.
Vega de Rojas, R. 1988. *Somatic Embryogenesis of Babaco Carica pentagona (Heilborn) Badillo*. Master's thesis, Faculty of Plant Science, University of Delaware, Newark.

Lucuma

Calzada B., J., V. Bautista C., J. Bermudez R., and M. Moran R. 1972. *Cultivo del Lúcumo*. Boletín Técnica 2. Universidad Nacional Agraria, La Molina, Lima. 44 pp. (For information on availability, contact the authors; see Research Contacts.)

Dawes and Pringle, 1983. (See above under Fruits.)

Ebel, C. 1935. Los lúcumos en Chile. *Revisa Chilena de Historia Natural* 34:183–203.

Martin, F.W. and S.E. Malo. 1978. *The Canistel and Its Relatives*. Cultivation of Neglected Tropical Fruits With Promise: Part 5. Science and Education Administration, U.S. Department of Agriculture, Washington, DC. 12 pp.

Saavedra, E. 1975. Efecto de la temperature y reguladores de cremiento sobre la germinación y crecimiento de la plantula del lúcumo (*Lucuma bifera* Mol.). *Investigaciones Agricultura* (Chile) 1(3):147–152.

Saavedra, E. 1975. Injertación del lúcumo. *Investigaciones Agricultura* (Chile) 1(3):191–193.

Naranjilla (Lulo)

Campbell, C.W. 1965. The naranjilla. *Bulletin of the Rare Fruit Council of South Florida* 2:3–5.

García R., E.H. and M.A. García D. 1985. *Colleción y Establecimiento de un Banco de Germoplasma de Lulo*, Solanum quitoense, *y Especies Relacionadas en el Suroeste Colombiano*. Tesis Ingeniero Agrónomo, Facultad de Ciencias Agropecuarias, Universidad Nacional de Colombia, Palmira, Colombia. 100 pp.

Giraldo, J.A. and C. Bolívar. 1977. *Enfermidades del Lulo en Tres Zonas Produciones de Colombia*. Tesis Ingeniero Agrónomo, Facultad de Ciencias Agropecuarias, Universidad Nacional de Colombia, Palmira, Colombia. 103 pp.

Heiser, C.B. 1972. The relationships of the naranjilla, *Solanum quitoense. Biotropica* 4(2):77–84.

Heiser, C.B. 1985a. Little oranges of Quito. Pages 60–81 in Heiser, 1985a. (See above under Fruits.)

Heiser, C.B. 1985b. Ethnobotany of the naranjilla (*Solanum quitoense*). *Economic Botany* 39:4–11.

Instituto Colombiano Agropecuario (ICA). 1973. *Las Plagas Del Lulo y Su Control*. Boletín Técnico No. 25. Programa Nacional de Entomología, ICA, Ministerio de Agricultura, Bogotá.

Instituto Nacional de Investigaciones Agropecuarias (INIAP). 1982. *Memorias de la Primera Conferencia Internacional de Naranjilla*. INIAP, Quito. 177 pp.

Jaramillo Vásquez, J. 1984. *El Cultivo del Lulo*. Mimeo. Programa Nacional de Hortalizas, ICA, Palmira. 11 pp. (For information on availability, contact the author; see Research Contacts.)

Lobo A., M., E. Girard O., J. Jaramillo V., and G. Jaramillo S. n.d. El cultivo del lulo o naranjilla. Pages 10–21 in *Contribución de los Programas de Hortalizas y Frutales*. ICA, Bogotá. (Information on availability from J. Jaramillo V.; see Research Contacts.)

Pacheco, R. and J. Jiménez. 1968. *El cultivo de la naranjilla en el Ecuador*. Ministerio de Agricultura y Ganadería, Quito.

Reyes, C.E. 1987. *Descripción de la Información Existente sobre el Lulo y/o Naranjilla*, Solanum quitoense *Lamarck y de las Prácticas Realizadas por los Agricultores en Diferentes Zonas de Colombia*. Tesis Ingeniero Agrónomo, Facultad de Ciencias Agropecuarias, Universidad Nacional de Colombia, Palmira, Colombia. 280 pp.

Romero-Castañeda, R. 1961. El lulo: una fruta de importancia económica. *Agricultura Tropical* 17(4):214–18.

Wolff, L.D. n.d. *El Cultivo del Lulo o Naranjilla* (Solanum quitoense *Lam.*). Facultad de Agronomía, Universidad Nacional de Colombia, Medellín, Colombia. 97 pp.

Pacay (Ice-Cream Beans)

There is much literature on the genus *Inga*. However, little is available concerning the specific cultivation and use of pacay as food. General *Inga* articles include the following:

Cavalcante, P.B. 1976. *Frutas Comestíveis da Amazônia* (2nd edition). Falangola, Belém, Brazil.

Holdridge, L.R. and L.J. Poveda, Jr. 1975. *Arboles de Costa Rica vol. 1*. Centro Cientifico Tropical, San José, Costa Rica.

Koptur, S. 1983a. The ecology of some Costa Rican *Inga* species. Pages 259–261 in D.H. Janzen, ed., *Costa Rican Natural History*. University of Chicago Press, Chicago.

Koptur, S. 1983b. Flowering phenology and floral ecology of *Inga. Systematic Botany* 8(4):354–368.

Koptur, S. 1984. Pollinator limitation of fruit set-breeding systems of *Inga. Evolution* 38(5):1130–1143.

León, J. 1966. Central American and West Indian species of *Inga* (Leguminosae). *Annals of the Missouri Botanical Garden* 53:265–359.

Marin M., F. 1983. Informe preliminar sobre el género *Inga* en el sur oriente peruano, I parte. *Revista Universitaria* 70(132).

National Research Council. 1980. *Firewood Crops: Shrub and Tree Species for Energy Production.* National Academy Press, Washington, D.C.

Passionfruits

Escobar, L.K.A. de. 1980. *Interrelationships of the edible forms of* Passiflora *centering around* Passiflora mollisima *(HBK) Bailey Subgenus Tacsonia.* Ph.D. dissertation, Department of Botany, University of Texas, Austin.

Escobar, L.K.A. de. 1981. Experimentos preliminares en la hibridación de especies comestibles de *Passiflora. Actualidades Biológicas* 10(38):103–111.

Escobar, L.K.A. de. 1985. Taxonomía, biología reproductiva y mejoramiento de la curuba. Pages 200–206 in *II Seminario Recursos Vegetales Promisorios.* Faculdad de Ciencia Agropecuarias, Universidad Nacional de Colombia, Palmira, Colombia.

Escobar, L.K.A. de. 1986. Hibridación de la curuba. In *Resúmenes del Seminario Recursos Vegetales Promisorios.* Facultad de Agronomía, Universidad Nacional de Colombia, Medellín. 2 pp.

Escobar, L.K.A. de. 1987. A taxonomic revision of the varieties of *Passiflora cumbalensis* (Passifloraceae). *Systematic Botany* 12(2):238–250.

Escobar, L.K.A. de. 1988. Passifloraceae *Passiflora,* subgeneros: *Tacsonia, Rathea, Manicata y Distephana.* Monograph No. 10 in P. Pinto and G. Lozanos, eds., *Flora de Colombia.* Universidad Nacional de Colombia, Bogotá.

Fouqué, A. 1972. Espèces fruitières d'Amérique tropicale: genre *Passiflora. Fruits* 27(5):368–382.

Knight, R.J., Jr. 1969. Edible-fruited passionvines in Florida: the history and possibilities. *Proceedings of the American Society for Horticultural Science (Tropical Region)* 13:265–274.

Martin, F.W. and H.Y. Nakasone. 1970. The edible species of *Passiflora. Economic Botany* 24(3):333–343.

Popenoe, 1924. (See above under Fruits.)

Programa Interciencias de Recursos Biológicos (PIRB). 1986. *Recursos Biológicos Nuevos.* Boletín Informativo No. 03. 12 pp. (This issue is devoted mostly to the curuba. Copies are available from Editores Boletín PIRB, Apartado Aéreo 51580, Bogotá 2.)

Schoeniger, G. 1988. *La Curuba; Técnicas para el Mejoramiento de su Cultivo.* Fondo Colombiano de Investigaciones Científicas y Proyectos Especiales "Francisco José de Caldas" (COLCIEN-CIAS) and Editora Guadalupe Ltda. (Apartado Aéreo 29765), Bogotá.

Schultes, R.E. 1986. Algunos apuntes botánicos sobre *Passiflora curuba.* Pages 253–255 in *Mesa Redonda de la Red Latino-Americana de la Industria de Frutas Tropicales, II.* Federación Nacional de Cafeteros y Oficina Regional de la Foro para América Latina y el Caribe, Bogotá.

Whittaker, D.E. 1974. Passion fruit: agronomy, processing and marketing. *Tropical Science* 14:59–77.

Pepino

Burge, G.K. 1982. Pepinos: fruit set. *New Zealand Commercial Grower* 37(8):33.

Corporación de Fomento de la Producción. 1982. *Cultivo del Pepino Dulce: Antecedentes, Agronómicos, Económicos.* AA 82/34. Corporación de Fomento de la Producción, Gerencia de Desarrollo, Faculdad de Agronomía de la Pontificia Universidad Católica de Chile, Santiago.

Dawes, S.N. 1984. Pepino breeding and selection. *The Orchardist of New Zealand* 58(5):172.

Dawes and Pringle, 1983. (See above under Fruits.)

Dennis, D.J., G.K. Burge, and R. Lill. 1985. *Pepino Cultural Techniques: An Introduction.* Aglink Horticultural Produce and Practice No. 208. 2 pp. (Pamphlet available from Ministry of Agriculture and Fisheries Information Services, Private Bag, Wellington, New Zealand.)

Heiser, C.B. 1964. Origin and variability of the pepino (*Solanum muricatum*): a preliminary report. *Baileya* 12:151–158.

Heiser, C.B. 1985a. Green "tomatoes" and purple "cucumbers." Pages 128–141 in Heiser, 1985a. (See above under Fruits.)

Hermann, M. 1988. *Beiträge zur Okologie der Frucht- und Ertragsbildung von* Solanum muricatum *Ait.* (Contributions to the ecology of fruit- and yield-increase of *S. muricatum* Ait.). Doctor of Agricultural Science dissertation, Technical University of Berlin, West Germany. 156 pp. (Extensive bibliography.)

Lizana, L.A. and B. Levano. 1977. Caracterisación y comportamiento del pepino dulce, *Solanum muricatum. Proceedings of the American Society for Horticultural Science (Tropical Region)* 21:11–15.

Morley-Bunker, M.J.S. n.d. *Pepino Growing in Canterbury.* Pepino Project, Horticulture Department, Lincoln College, Christchurch, New Zealand. 19 pp. (Copies available from author; see Research Contacts.)

Morley-Bunker, M.J.S. 1983. A new commercial crop, the pepino (*Solanum muricatum* Ait) and suggestions for further development. *Royal New Zealand Institute of Horticulture Annual Journal* 11:8–19.

Seidel, H. 1974. Erfahrungen mit dem Anbau von *Solanum muricatum* in Südspanien (Experiences with cropping of *Solanum muricatum* in Southern Spain). *Tropenlanwirt* 75(5):24–30.

Tamarillo (Tree Tomato)

Bilton, J. 1986. *Tamarillo Cookbook*. 64 pp. (Available from Irvine Holt, P.O. Box 28019, Auckland 5, New Zealand.)

Bohs, L. 1986. *The Biology and Taxonomy of* Cyphomandra *(Solanaceae)*. Ph.D. dissertation, Harvard University, Cambridge, Massachusetts.

Bohs, L. 1989. The ethnobotany of the *Cyphomandra*. *Economic Botany* 43(2).

Carnevali, A. 1976. Il tamarillo: una planta da trutto coltivabile nella regione degli agrumi (The tamarillo: a fruit crop that can be grown in citrus regions). *Italia Agricola* 113:96–99.

Dadlani, S.A. and K.P.S. Chandal. 1970. The little grown tree tomato. *Indian Horticulture* 14(2):13–14.

Dawes and Pringle, 1983. (See above under Fruits.)

Fletcher, W.A. 1979. *Growing Tamarillos*. Ministry of Agriculture and Fisheries Bulletin No. 307. Wellington, New Zealand.

Morton, J.F. 1982. The tree tomato, or "tamarillo," a fast-growing early-fruiting small tree for subtropical climates. *Proceedings of the Florida State Horticultural Society* 95:81–85.

Orihuela, M. n.d. *Tomate Andino; Manual Práctico para su Cultivo y Uso*. Centro de Estudios Rurales Andinos "Bartolomé de las Casas," Horacio Urteaga 452, Jesús Maria, Lima 11. 18 pp.

Sale, P.R. 1983a. *Tamarillos: General Requirements, Varieties, and Propagation for Commercial Production*. Aglink Horticultural Produce and Practice No. 296. (Pamphlet available from Ministry of Agriculture and Fisheries Media Services, Private Bag, Wellington, New Zealand.)

Sale, P.R. 1983b. *Tamarillos: Orchard Management, Planting, Training, Pruning, Nutrition, Harvesting*. Aglink Horticultural Produce and Practice No. 297. (Available as above.)

Sale, P.R. 1984. *Tamarillos; Pests and Diseases*. Aglink Horticultural Produce and Practice No. 298. (Available as above.)

NUTS

Quito Palm

Barry, B., Jr. 1960. The Ecuadorian relative of the Chilean wine palm. *Principes* 4:146–148.

Balslev, H. and A. Barfod. 1987. Ecuadorian palms—an overview. *Opera Botanica* 92:17–35.

Cárdenas, M. 1970. Palm forests of the Bolivian high Andes. *Principes* 14:50–54.

Fullington, J.C. 1987. *Parajubaea*—an unsurpassed palm for cool, mild areas. *Principes* 31(4):172–176. (Reprints available from author; see Research Contacts.)

Moore, H.E. 1970. Palm brief: *Parajubaea cocoides*. *Principes* 14:27.

Moraes R., M. and A. Henderson. In press. The genus *Parajubaea* (Palmae). Mimeo. 13 pp. (For information on availability, contact the authors; see Research Contacts.)

Walnuts

Endt, R. 1988. *A Walnut from the New World* Juglans neotropica, *a Notable New Introduction for New Zealand*. Mimeo. 2 pp. (Copies available from author; see Research Contacts.)

Manning, W.E. 1960. The genus *Juglans* in South America and the West Indies. *Brittonia* 12:1–26.

Pretell C., J., D. Ocaña V., R. Jon L., and E. Barahona C. 1985. Nogal. Pages 77–82 in *Apuntes Sobre Algunas Especies Forestales Natives de la Sierra Peruana*. Proyecto FAO/Holanda/INFOR (GCP/PER/027/NET), Lima, Peru.

Centers of Andean Crop Research

Listed below are South American organizations with known involvement in the development of Andean crops. (Appendix C lists individuals working with Andean crop research and promotion.) These institutions are likely to have information available on topics such as agronomy, biology, germplasm, nutrition, processing, and local knowledge of at least some of the crops discussed in this report. Also included are a few private organizations that work with these native crops.

In brackets at the end of some addresses we have noted the department from which we have received information. More organizations are becoming involved with these crops all the time, and this list, therefore, should not be considered definitive.

Argentina

Centro Nacional de Investigaciones Científicas y Técnicas, Centro de Estudios Farmacológicos y de Princípios Naturales (CEFAPRIN), Serrano 665, Buenos Aires 1414, Buenos Aires
Estación Experimental, 6326 Anguil, La Pampa
Instituto Nacional de Tecnología Agropecuaria (INTA), Centro de Investigaciones de Recursos Naturales, Castelar 1712, Buenos Aires
Instituto Nacional de Tecnología Agropecuaria (INTA), Estación Experimental Agropecuaria Santiago del Estero, Independencia No. 341, Casilla de Correo 268, Santiago del Estero 4200
Universidad Nacional de Córdoba, Casilla de Correo 495, Córdoba 5000 [Facultad de Ciencias Exactas, Físicas y Naturales]
Universidad Nacional de La Plata, La Plata 1900 [Facultad de Ciencias Naturales y Museo]
Universidad Nacional de Tucumán, Miguel Lillo 205, San Miguel de Tucumán [Fundación e Instituto Miguel Lillo]
Universidad de Buenos Aires, Facultad de Agronomía, Avenida San Martín 4453, Buenos Aires 1417, Buenos Aires [Cátedra de Botánica]

Bolivia

Centro de Investigaciones Fitotécnicas Pairumani, Casilla 3861, Cochabamba
Flores Montesillo, Casilla 053, Cochabamba
Herbario Nacional de Bolivia, Cajón Postal 20–127, La Paz
Instituto Boliviano de Tecnología Agropecuaria (IBTA), Calle Colombia, Casilla 2631, Cochabamba
Instituto Boliviano de Tecnología Agropecuaria (IBTA), Estación Experimental Patacamaya, Casilla Postal 5783, La Paz
Instituto de Ecología, Casilla 20127, La Paz
Instituto Interamericano de Cooperación para la Agricultura (IICA), Oficina del IICA en Bolivia, Casilla 6057, La Paz

Brazil

Empresa Brasileira de Pesquisa Agropecuária (EMBRAPA), Caixa Postal 07.0218, Brasilia, D.F. 70.359
Empresa Brasileira de Pesquisa Agropecuária (EMBRAPA), Empresa Goiâna de Pesquisa Agropecuária (EMGOPA), Caixa Postal 49, Goiâna, Goias 74.000
Empresa Brasileira de Pesquisa Agropecuária (EMBRAPA), Empresa de Assistência Técnica e Extensão Rural do Estado Minas Gerais (EMATER), Praça Barão de Santa Cecília No. 68 - 1 Andar, Carandaí, Minas Gerais 36.280
Instituto Nacional de Pesquisas da Amazônia (INPA), Estrada do Aleixo, Caixa Postal 478, Manaus 69.000, Amazonas
Universidade Estadual Paulista "Julio de Mesquita Filho," Campus de Rio Claro-UNESP, Caixa Postal 178, Rio Claro, São Paulo 13.500 [Departamento Botânica, Instituto de Biosciéncias]
Universidade Estadual Paulista, Campus de Jaboticabal, Jaboticabal [Facultad de Ciencias Agrarias e Veterinarias]
Universidade Federal de Viçosa (UFV), Viçosa, Minas Gerais 36.570 [Departamento de Fitotécnia]
Universidade Federal do Rio de Janeiro, Ilha de Cidade Universitária, Rio de Janeiro [Instituto de Nutrição]
Universidade do Estado do São Paulo (UNESP), Caixa Postal 237, Botucatu, São Paulo 18.600 [Departamento Horticultura]

Chile

Campex - Semillas Baer, Estación Gorbea, Casilla 87, Temuco
Corporación Nacional Forestal (CONAF), Región Metropolitana, Eliodoro Yáñez 1810, Santiago
Food & Agriculture Organization of the United Nations (FAO), Avenida Santa Maria 6700, Casilla 10095, Santiago
Pontificia Universidad Católica de Chile, Vic. Mackenna 4860, Casilla 114–D, Santiago [Laboratorio de Botánica, Facultad de Ciencias Biológicas]
Universidad Austral de Chile, Casilla 567, Valdivia [Institución Producción y Sanidad Vegetal]
Universidad Católica, Casilla 6177, Santiago [Facultad de Agronomía]
Universidad Católica de Valparaíso, Avenida Brasil 2950, Casilla 4059, Valparaíso [Escuela de Agronomía]
Universidad de Chile, Avenida Bernardo O'Higgins 1058, Casilla 10–D, Santiago [Departamento de Botánica, Facultad de Ciencias]
Universidad de Concepción, Casilla 237, Concepción

Colombia

Centro Internacional de Agricultura Tropical (CIAT), Apartado Aéreo 6713, Cali
Consejo Internacional de Recursos Fitogenéticos (CIRF-IBPGR), c/o CIAT, Apartado Aéreo 6713, Cali
Corporación de Araracuara, Apartado Aéreo 034174, Bogotá, D.E.
Federación Horto-Fruticola, Calle 37 No. 15–49, Bogotá
Instituto Colombiano Agropecuario (ICA), Apartado Aéreo 100, Rio Negro, Antioquia
Instituto Colombiano Agropecuario (ICA), Centro Nacional de Investigación (CNI), Apartado Aéreo 233, Palmira, Valle
Instituto Colombiano Agropecuario (ICA) - Tibaitata, Apartado Aéreo 151123, El Dorado, Bogotá
International Development Research Centre (IDRC), Oficina Regional para la América Latina y el Caribe, Apartado Aéreo 53016, Bogotá
Jardín Botánico de Colombia, Carrera 66A No. 56–84, Bogotá
Servicio Nacional de Aprendizaje (SENA), Regional Boyacá, Calle 65 No. 11–70, Apartado Aéreo 9801, Bogotá
Universidad Nacional de Colombia, Ciudad Universitaria, Apartado Aéreo 14.490, Bogotá [Instituto de Ciencias Naturales]
Universidad Nacional de Colombia (UNPALMIRA), Apartado Aéreo 237, Palmira, Valle [Facultad de Ciencias Agropecuarias, Departamento de Agricultura]
Universidad de Antioquia, Apartado Aéreo 1226, Ciudad Universitaria, Medellín, Antioquia [Departamento de Biología]
Universidad de Caldas, Apartado Aéreo 275, Manizales [Facultad de Agronomía]
Universidad de Nariño, Carrera 25. Mp/18–109, Pasto, Nariño [Facultad de Agronomía]

Ecuador

Centro Andino de Acción Popular (CAAP), Apartado 173–B, Quito
Centro de Arte y Acción Popular (CAAP), Chahuarpungo, Pichincha
Corporación Ambiente y Desarrollo (AMDE), PO Box 632, Ambato
Escuela Superior Politécnica de Chimborazo (ESPOCH), Casilla 4703, Riobamba [Facultad de Ingeniería Agronómica]
Instituto Interamericano de Cooperación para la Agricultura (IICA), Apartado 201–A, Quito
Instituto Nacional de Investigaciones Agropecuarias (INIAP), Estación Experimental del Austro, Apartado 554, Cuenca
Instituto Nacional de Investigaciones Agropecuarias (INIAP), Avenida Eloy Alfaro y Amazonas, Casilla 2600, Quito
Instituto Nacional de Investigaciones Agropecuarias (INIAP), Estación Experimental "Santa Catalina," Casilla Postal 340, Quito [Programa de Cultivos Andinos]
Latinreco S.A., Centro Nestlé, Casilla Postal 6053–CCI, Quito, Cumbaya
Museo Ecuatoriano de Ciencias Naturales, Casilla 8976, Quito [Herbario Nacional]
Universidad Central del Ecuador, Ciudadela Universitaria, Apartado 166, Quito [Facultad de Ciencias Agrícolas]
Universidad Nacional de Loja, Casilla Letra B, Loja [Jardín Botánico "Reinaldo Espinosa," Facultad de Ciencias Agricolas]
Universidad Técnica de Ambato, Avenida Colombia y Chile s/n, Casilla Postal 334, Ambato [Facultad de Ciencia]

Peru

Centro Internacional de Agricultura Tropical (CIAT), Apartado 14–0185, Lima 14
Centro Internacional de la Papa (CIP), Apartado 5969, Lima
Centro de Medicina Andina, Jirón Ricardo Palma No. 5, Santa Mónica, Apartado 711, Cusco
Comisión de Coordinación de Tecnología Andina (CCTA), Sede Institucional, Avenida Tullumayo 465, Apartado 477, Cusco [Centro de Estudios Rurales Andinos "Bartolomé de las Casas"]
Convenio Perú-Alemania para Cultivos Andinos-GTZ (CORDE), Avenida Sol 817, Casilla Postal 807, Cusco
Estación Experimental Agropecuaria Santa Ana, Apartado 411, Huancayo
Gentec - Grobman Genotécnica S.A., Oficina 702, Avenida República de Panamá 3563, San Isidro, Lima
Grupo de Investigación y Extensión de Tecnología Popular, MINKA-Grupo "Talpuy," Apartado No. 222, Avenida Centenario 589, San Carlos, Huancayo
Instituto Nacional Forestal y de Fauna (INFOR), Investigación y Capacitación Forestal y de Fauna (CENFOR VII), Huancayo
Instituto Nacional Forestal y de Fauna (INFOR), Componente Forestal de la Microregión de Juliaca, Juliaca, Puno
Instituto Nacional de Investigación Agraria y Agroindustrial (INIAA), Avenida Guzman Blanco 309, Lima 1 [Programa de Investigación de Recursos Genéticos (PRONARGEN)]
Instituto Nacional de Investigación Agraria y Agroindustrial (INIAA), Estación Experimental Baños del Inca, Julio Guerrero No. 121, Urb. Cajamarca, Cajamarca [Programa Nacional de Cultivos Andinos (PNCA)]
Instituto Nacional de Investigación Agraria y Agroindustrial (INIAA), Puno
Instituto Nacional de Investigación y Promoción Agropecuaria (INIPA), Apartado 110697, Lima 11 [Programa Nacional de Cultivos Andinos (PISA)]
Instituto Rural Vallegrande, Apartado 70, Cañete
Museo de Historia Natural "Javier Prado," Avenida Arenales 1256, Lima
Estación Experimental Yurimaguas, Proyecto Agroforestal, Programa de Suelos Tropicales, Yurimaguas, Loreto [North Carolina State University]
Universidad Nacional Agraria (UNA), Apartado 456, La Molina, Lima 100 [Facultad de Agronomía, Laboratoria de Biotecnología, Departamento de Fitotécnia]
Universidad Nacional Mayor de San Marcos de Lima, Avenida República de Chile 295, Casilla 454, Lima [Colegio Real]
Universidad Nacional Técnica de Cajamarca, Apartado 16, Cajamarca
Universidad Nacional Técnica del Altiplano, Ciudad Universitaria, Casilla 291, Puno
Universidad Nacional de Huánuco "Hermilio Valdizán," Jirón Dos de Mayo 680, Apartado 278, Huánuco [Facultad de Ciencias Agrarias]
Universidad Nacional de San Antonio Abad del Cusco (UNSAAC), Avenida de la Cultura s/n, Apartado 367, Cusco [Centro de Investigación en Cultivos Andinos (CICA)]

Universidad Nacional de San Cristóbal de Huamanga, Apartado 220, Ayacucho
Universidad Nacional de Trujillo, San Martín 380, Apartado 315, Trujillo [Facultad de Ciencias Biológicas]
Universidad Nacional del Centro del Perú, Apartado 138, Huancayo
Universidad San Martín de Porres, Calle Bolívar 348, Miraflores, Lima [Instituto de Estudios de Bromatología y Nutrición Andina]
Universidad del Pacifico, Jesús Maria, Avenida Salaverry 2020, Apartado 4683, Lima 11

Venezuela

Fondo Nacional de Investigaciones Agropecuarias (FONAIAP), Ministerio de Agricultura y Cría, Centro Simón Bolívar, Torre Norte, Piso 14, Caracas 1010
Universidad Central de Venezuela, Apartado 4579, Maracay, Estado Aragua 2101 [Facultad de Agronomía]

Appendix C

Research Contacts

With few exceptions, each of the research contacts listed below has contributed information to this report based on their personal expertise and experience. Most have agreed to cooperate with readers who contact them. Without the collaboration of the 600 individuals listed here, together with more than 50 other contributors, this report would not have been possible.

ACHIRA

Andean Region

Carlos Arbizú Avellaneda, Programa de Investigaciones en Cultivos Andinos, Universidad Nacional de San Cristóbal de Huamanga, Ayacucho, Peru

Fernando N. Barrantes Del Aguila, Programa de Cultivos Andinos, Facultad de Ciencias Agrarias, Universidad Nacional de San Cristóbal de Huamanga, Apartado 220, Ayacucho, Peru (germplasm, pathology)

Augusto Cardich, Facultad de Ciencias Naturales y Museo, Universidad Nacional de La Plata, La Plata 1900, Argentina (geography and climate)

César del Carpio Merino, Departamento de Manejo Forestal, Universidad Nacional Agraria (UNA), Centro de Datos Para La Conservación, Apartado 456, La Molina, Lima 100, Peru

Fondo Nacional de Investigaciones Agropecuarias (FONAIAP), Ministerio de Agricultura y Cría, Centro Simón Bolívar, Torre Norte, Piso 14, Caracas 1010, Venezuela

Michael Hermann, Program on Andean Root and Tuber Crops, International Board for Plant Genetic Resources (IBPGR), Universidad Nacional Agraria (UNA), Escuela de Post-Grado, Apartado 456, La Molina, Lima 100, Peru

Joy C. Horton Hofmann, PO Box 492, Loja, Ecuador

Zósimo Huamán Cueva, Genetic Resources Department, Centro Internacional de la Papa (CIP), Apartado 5969, Lima, Peru

Instituto Nacional de Investigaciones Agropecuarias (INIAP), Estación Experimental "Santa Catalina," Casilla Postal 340, Quito, Ecuador

Carlos Nieto C., Programa de Cultivos Andinos, Instituto Nacional de Investigaciones Agropecuarias (INIAP), Estación Experimental "Santa Catalina," Casilla Postal 340, Quito, Ecuador

Raúl Ríos E., Centro de Investigaciones Fitotécnicas Pairumani, Casilla 3861, Cochabamba, Bolivia

348 LOST CROPS OF THE INCAS

Juan Risi Carbone, Oficina Regional para la América Latina y el Caribe, International Development Research Centre (IDRC), Apartado Aéreo 53016, Bogotá, Colombia (cropping systems)

Juan Solano Lazo, Estación Experimental del Austro, Instituto Nacional de Investigaciones Agropecuarias (INIAP), Apartado 554, Cuenca, Ecuador

Other Countries

Gregory J. Anderson, Biological Sciences, Room 312, U-43, University of Connecticut, 75 North Eagleville Road, Storrs, Connecticut 06268, USA (ethnobotany and taxonomy)

Vichitr Benjasil, Field Crops Research Institute, Department of Agriculture, Bangkhen, Bangkok 10900, Thailand

Opas Boonseng, Banmai Samrong Field Crop Experiment Station, Department of Agriculture, Shikhiu, Nakhon Rachasima, Thailand

Ricardo Bressani, Instituto de Nutrición de Céntro América y Panamá (INCAP), Carretera Roosevelt Zona 11, Apartado Postal 1188, Guatemala City, Guatemala (nutrition)

Leslie Brownrigg, 7408 16th Avenue, Takoma Park, Maryland 20912, USA (indigenous systems)

Ramón de la Peña, College of Tropical Agriculture and Human Resources, University of Hawaii, Kauai Branch Station, 7370–A Kuamoo Road, Kapaa, Hawaii 96746, USA

Dick and Annemarie Endt, Landsendt Subtropical Fruits, 108 Parker Road, Oratia, Auckland 7, New Zealand

W. Hardy Eshbaugh, Department of Botany, Miami University, Oxford, Ohio 45056, USA

Michel Fanton, Seed Savers' Network, PO Box 24, Nimbin, New South Wales 2480, Australia

Daniel W. Gade, Department of Geography, Old Mill Building, University of Vermont, Burlington, Vermont 05405, USA

B.W.W. Grout, Faculty of Science, Department of Biological Sciences, Plymouth Polytechnic, Drake Circus, Plymouth, Devon PL4 8AA, UK

Ray Henkel, Department of Geography, Arizona State University, Tempe, Arizona 85287, USA (pioneer colonization)

International Board for Plant Genetic Resources (IBPGR), FAO Headquarters, Via delle Terme di Caracalla, Rome 00100, Italy (germplasm information)

Alan M. Kapuler, Peace Seeds, 2385 S.E. Thompson, Corvallis, Oregon 97333, USA

T.N. Khoshoo, Tata Energy Research Institute (teri), 7 Jor Bagh, New Delhi 110 003, India (triploids)

Steven R. King, Latin America Science Program, The Nature Conservancy, 1815 North Lynn Street, Arlington, Virginia 22209, USA (ethnobotany)

Tetsuo Koyama, Asiatic Programs, New York Botanical Garden, Bronx, New York 10458–9980, USA (polyploids)

J. Le Dividich, Station de Recherches Porcines, Institut National de la Recherche Agronomique (INRA), Saint-Gilles, L'Hermitage 35590, France (animal metabolism and nutrition)

Peter Lloyd, Department of Primary Industries, GPO 46, Brisbane, Queensland 4001, Australia

Paul J.M. Maas, Institute of Systematic Botany, Rijksuniversiteit te Utrecht, Vakgroep bijzondere Plantkunde, Heidelberglaan 2, PO Box 80.102, Utrecht 3508 PTC, Netherlands (taxonomy)

Franklin W. Martin, 2305 East Second, Lehigh Acres, Florida 33936, USA

K. Nishiyama, Faculty of Agriculture, Tokyo University of Agriculture and Technology, 3–8–1 Harumi-cho, Fuchu-shi, Tokyo 183, Japan

Mitsunori Oka, Department of Genetic Resources I, National Institute of Agrobiological Resources (NIAR), Kannondai, Tsukuba, Ibaraki-ken 305, Japan (growth characteristics)

Kevin Patterson, Division of Horticulture and Processing, Department of Scientific and Industrial Research (DSIR), Private Bag, Auckland, New Zealand

Martin L. Price, Educational Concerns for Hunger Organization (ECHO), 17430 Durrance Road, North Fort Myers, Florida 33917, USA (limited germplasm available)

Chris Rollins, Fruit and Spice Park, 24801 S.W. 187th Avenue, Homestead, Florida 33031, USA

Arne Rousi, Department of Biology, University of Turku, Turku SF-20500, Finland (physiology, virus-free culture)

Setijati Sastrapradja, Pusat Penelitian dan Pengembangan Bioteknologi, Centre for Research in Biotechnology, Gedung Kusnoto, Jalan Juanda 18, Bogor, Indonesia
Robin Saunders, Agricultural Research Service (ARS), U.S. Department of Agriculture (USDA), Western Regional Research Center, 800 Buchanan Street, Albany, California 94710, USA (processing starches)
Richard E. Schultes, Botanical Museum, Harvard University, 26 Oxford Street, Cambridge, Massachusetts 02138, USA (ethnobotany)
Seung Jin Kim, Department of Horticulture, Yonam Junior College of Livestock and Horticulture, Sung Whan PO Box 148, Chon Won, Choong Nam-Do 330–81, South Korea (germplasm)
Steven Spangler, Exotica Rare Fruit Nursery, 2508–B East Vista Way, PO Box 160, Vista, California 92083, USA
H.-D. Tscheuschner, Sektion Verarbeitungs- und Verfahrenstechnik, Technische Universität Dresden, Mommsenstrasse 13, Dresden 8027, East Germany (starches)
Le nqoc Tu, Faculty of Food Technology, Centre of Biotechnology, Polytechnic University of Hanoi, Hanoi, Vietnam (starches)
Donald Ugent, Department of Botany, University of Southern Illinois, Carbondale, Illinois 62901, USA
Watana Watananonta, Banmai Samrong Field Crop Experiment Station, Department of Agriculture, Shikhiu, Nakhon Rachasima, Thailand

AHIPA

Andean Region

Heinz Brücher, Finca Condorhuasi, Casilla de Correo 131, Mendoza 5500, Argentina (useful plants)
Raúl O. Castillo T., Instituto Nacional de Investigaciones Agropecuarias (INIAP), Estación Experimental "Santa Catalina," Casilla Postal 340, Quito, Ecuador (tissue culture, virus-free germplasm)
Michael Hermann, Peru (see Achira)
Zósimo Huamán Cueva, Peru (see Achira)
Instituto Nacional de Investigaciones Agropecuarias (INIAP), Ecuador (see Achira)
Juan Risi Carbone, Colombia (cropping systems) (see Achira)
Julio Valladolid Rivera, Oficina de Investigación, Departamento de Agronomía, Universidad Nacional de San Cristóbal de Huamanga, Apartado 243, Ayacucho, Peru

Other Countries

Ariel Azael, Institut Interaméricain de Cooperation pour l'Agriculture (IICA), Première Impasse Lavaud 14, PO Box 2020, Port-au-Prince, Haiti
Leslie Brownrigg, USA (indigenous systems) (see Achira)
D.J. Cotter, Department of Agronomy and Horticulture, New Mexico State University, Las Cruces, New Mexico 88003, USA
W. Hardy Eshbaugh, USA (see Achira)
Christine Franquemont, Latin American Studies Program, Cornell University, Uris Hall 190, Ithaca, New York 14853, USA (Quechua plant use)
R.E. Gomez, Plant and Pest Management Service, U.S. Department of Agriculture (USDA), South Agricultural Building, Washington, DC 20250, USA
B.W.W. Grout, UK (see Achira)
Jesse Jaynes, Department of Biochemistry, Louisiana State University, Baton Rouge, Louisiana 70803–1464, USA (gene splicing)
Steven R. King, USA (ethnobotany) (see Achira)

Avigdor Orr, Product and Package Development, DNA Plant Technology Corporation, 2611 Branch Pike, Cinnaminson, New Jersey 08077, USA

John Palmer, Division of Crop Research, New Crops Section, Department of Scientific and Industrial Research (DSIR), Private Bag, Christchurch, New Zealand

C.A. Schroeder, Department of Biology, University of California, Los Angeles, California 90024, USA

Richard E. Schultes, USA (ethnobotany) (see Achira)

Marten Sørensen, Botanisk Laboratorium, Kobenhavns Universitet, Gothersgade 140, Copenhagen DK-1123, Denmark

Calvin R. Sperling, Germplasm Services Laboratory, Agricultural Research Service (ARS), U.S. Department of Agriculture (USDA), Building 001, Room 321, Beltsville Agricultural Research Center (BARC-West), Beltsville, Maryland 20705, USA

Louis Trap, Box 59003, Mangere-Bridge, Auckland, New Zealand

ARRACACHA

Andean Region

Osmar Alves Carriji, Empresa Brasileira de Pesquisa Agropecuária (EMBRAPA)/CNPH, Caixa Postal 07.0218, Brasilia, D.F. 70.359, Brazil

Waldir Aparecido Marouelli, Empresa Brasileira de Pesquisa Agropecuária (EMBRAPA)/ CNPG, Caixa Postal 07.0218, Brasilia, D.F. 70.359, Brazil

Francisco Luiz Araujo Camara, Empresa Goiâna de Pesquisa Agropecuária (EMGOPA), Empresa Brasileira de Pesquisa Agropecuária (EMBRAPA), Caixa Postal 49, Goiânia, Goiâs 74.000, Brazil

Carlos Arbizú Avellaneda, Peru (see Achira)

Fernando N. Barrantes Del Aguila, Peru (germplasm, pathology) (see Achira)

Augusto Cardich, Argentina (geography and climate) (see Achira)

Vicente W.D. Casali, Departamento de Fitotécnia 185, Universidade Federal de Viçosa (UFV), Viçosa, Minas Gerais 36.570, Brazil

Raúl O. Castillo T., Ecuador (tissue culture, virus-free germplasm) (see Ahipa)

Fernando L. de Bastos Freire, Empresa de Assistência Técnica e Extensão Rural do Estado Minas Gerais (EMATER), Empresa Brasileira de Pesquisa Agropecuária (EMBRAPA), Praça Barão de Santa Cecília 68 - 1 Andar, Carandaí, Minas Gerais 36.280, Brazil

César del Carpio Merino, Peru (see Achira)

Luis Eduardo Mora, Facultad de Agronomía, Universidad de Nariño, Carrera 25, Mp/ 18– 109, Pasto, Nariño, Colombia

Rolando Egúsquiza, Programa de Papa, Universidad Nacional Agraria (UNA), Apartado 456, La Molina, Lima 100, Peru

Rolando Estrada J., Laboratorio de Recursos Genéticos y Biotecnologías, Universidad Nacional Mayor de San Marcos, Facultad de Ciencias Biológicas, Apartado 170138, Lima, Peru (somoclonal and virus-free germplasm)

Fondo Nacional de Investigaciones Agropecuarias (FONAIAP), Venezuela (see Achira)

Santiago D. Franco Pebe, Estación Experimental Baños del Inca, Programa Nacional de Cultivos Andinos (PNCA), Instituto Nacional de Investigación Agraria y Agroindustrial (INIAA), Julio Guerrero #121, Urb. Cajamarca, Cajamarca, Peru

Tomas Guerrero, Alemania 221, Quito, Ecuador

Michael Hermann, Peru (see Achira)

Fabio Higuita Muñoz, Hortilizas y Frutales, Instituto Colombiano Agropecuario (ICA), Apartado Aéreo 151123, Bogotá, Colombia

Andrés Miguel Hlatky Hernández, Facultad de Ingeniería Agronómica, Escuela Superior Politécnica de Chimborazo (ESPOCH), Casilla 4703, Riobamba, Ecuador

Miguel Holle, Programa de Investigación de Recursos Genéticos (PRONARGEN), Instituto Nacional de Investigación Agraria y Agroindustrial (INIAA), Avenida Guzman Blanco 309, Lima 1, Peru

Zósimo Huamán Cueva, Peru (see Achira)

Instituto de Ciencias Naturales, Universidad Nacional de Colombia, Ciudad Universitaria, Apartado Aéreo 14.490, Bogotá, Colombia
Instituto Nacional de Investigaciones Agropecuarias (INIAP), Ecuador (information, germplasm) (see Achira)
Latinreco S.A., Centro Nestlé, Casilla Postal 6053–CCI, Quito, Cumbaya, Ecuador
Sady Majino Bernardo, Facultad de Ciencias Agrarias, Universidad Nacional de Huánuco "Hermilio Valdizán," Jirón Dos de Mayo 680, Apartado 278, Huánuco, Peru
Reinaldo Monteiro, Departamento Botánica, Instituto de Biosciéncias, Universidade Estadual Paulista "Julio de Mesquita Filho," Campus de Rio Claro-UNESP, Caixa Postal 178, Rio Claro, São Paulo 13.500, Brazil
Laura Muñoz Espín, Programa de Cultivos Andinos, Instituto Nacional de Investigaciones Agropecuarias (INIAP), Estación Experimental "Santa Catalina," Casilla Postal 340, Quito, Ecuador
Carlos Nieto C., Ecuador (see Achira)
José Otocar Reina Barth, Facultad de Ciencias Agropecuarias, Departamento de Ciencias Sociales, Universidad Nacional de Colombia (UNPALMIRA), Apartado Aéreo 237, Palmira, Valle, Colombia (small-scale production)
Raúl Ríos E., Bolivia (see Achira)
Juan Risi Carbone, Colombia (cropping systems) (see Achira)
Jose Fernando Romero Cañizares, Facultad de Ingeniería Agronómica, Escuela Superior Politécnica de Chimborazo (ESPOCH), Panamericana Sur Km. 1, Casilla 4703, Riobamba, Ecuador
Isidoro Sánchez Vega, Los Fresnos 191, Apartado 55, Cajamarca, Peru
Peter E. Schmiediche, Genetic Resources Department, Centro Internacional de la Papa (CIP), Apartado 5969, Lima, Peru
Juan Solano Lazo, Ecuador (see Achira)
Julio Valladolid Rivera, Peru (see Ahipa)
Rebeca Vega de Rojas, Corporación Ambiente y Desarrollo (AMDE), PO Box 632, Ambato, Ecuador
Carlos Adolfo Vimos Naranjo, Programa de Cultivos Andinos, Instituto Nacional de Investigaciones Agropecuarias (INIAP), Estación Experimental "Santa Catalina," Casilla Postal 340, Quito, Ecuador
Antonio C.W. Zanin, Departamento Horticultura-FCA, Universidade do Estado do São Paulo (UNESP), Caixa Postal 237, Botucatu, São Paulo 18.600, Brazil

Other Countries

James Affolter, Botanical Garden, University of California, Berkeley, California 94720, USA (taxonomy and literature)
Gregory J. Anderson, USA (ethnobotany and taxonomy) (see Achira)
Vichitr Benjasil, Thailand (see Achira)
Lynn A. Bohs, Pringle Herbarium, Marsh Life Science Building, University of Vermont, Burlington, Vermont 05405, USA
Ricardo Bressani, Guatemala (nutrition) (see Achira)
Leslie Brownrigg, USA (indigenous systems) (see Achira)
Alan A. Brunt, Plant Pathology Department, AFRC Institute of Horticultural Research, Worthing Road, Littlehampton, West Sussex BN17 6LP, UK (disease- and virus-free germplasm)
Norberto Colón Ferrer, Agricultural Extension Service, University of Puerto Rico, Apartado 10, Barranquitos, Puerto Rico 00618, USA
Reinaldo Del Valle, Jr., Department of Agronomy and Soil, Agriculture Research Station, University of Puerto Rico, College of Agriculture, PO Box 21360, Rio Piedras, Puerto Rico 00928, USA (yield research)
R. Delhey, Institut für Nutzpflanzenforschung, Technische Universität Berlin, Fachbereich 15–Ostbau, Albrecht-Thaer-Weg 3, Berlin 33, West Germany (virus screening)
Michel Fanton, Australia (see Achira)
International Board for Plant Genetic Resources (IBPGR), Italy (germplasm information) (see Achira)

Jesse Jaynes, USA (gene splicing) (see Ahipa)

C. Antonio Jiménez Aparicio, Departamento de Graduados e Investigación en Alimentos, Instituto Politécnico Nacional (IPN), Escuela Nacional de Ciencias Biológicas (ENCB), Caixa Postal 02860, PO Box 26–350, México 07738, D.F., Mexico

Timothy A. Johns, School of Dietetics and Human Nutrition, Macdonald College of McGill University, Ste. Anne de Bellevue, Quebec H9X 1C0, Canada (ethnobotany, chemical ecology)

Steven R. King, USA (ethnobotany) (see Achira)

Mildred Mathias, c/o Department of Botany, University of California, Los Angeles, California 90024, USA (Umbelliferae)

Teddy E. Morelock, Department of Horticulture, University of Arkansas, 316 Plant Science, Fayetteville, Arkansas 72701, USA

Avigdor Orr, USA (see Ahipa)

John Palmer, New Zealand (see Ahipa)

Hugh Popenoe, International Program in Agriculture, University of Florida, 3028 McCarty Hall, Gainesville, Florida 32611, USA

Charles M. Rick, Department of Vegetable Crops, University of California, Davis, California 95616, USA

Richard E. Schultes, USA (ethnobotany) (see Achira)

Calvin R. Sperling, USA (see Ahipa)

MACA

Andean Region

Santiago Erik Antúñez de Mayolo R., Universidad Nacional Mayor de San Marcos, Apartado 18–5469, Lima 18, Peru (pre-Columbian plants, geography, and climate)

Carlos Arbizú Avellaneda, Peru (see Achira)

Augusto Cardich, Argentina (geography and climate) (see Achira)

Raúl O. Castillo T., Ecuador (tissue culture, virus-free germplasm) (see Ahipa)

César del Carpio Merino, Peru (see Achira)

Rolando Estrada J., Peru (somoclonal and virus-free germplasm) (see Arracacha)

Michael Hermann, Peru (see Achira)

Cipriano Mantari Camargo, Jirón Humbolt 645 - Chilca, Huancayo, Peru (botanical seed)

Ramiro Matos Mendieta, Colegio Real, Universidad Nacional Mayor de San Marcos, Jirón Andahuaylas 348, Avenida República de Chile 295, Casilla 454, Lima, Peru

Eloy Munive Jáuregui, Facultad de Agronomía, Universidad Nacional del Centro del Perú, Ciudad Universitaria, Huancayo, Peru

Carlos Nieto C., Ecuador (see Achira)

Francisco Rhon Dávila, Centro Andino de Acción Popular (CAAP), Apartado 173–B, Quito, Ecuador

Juan Risi Carbone, Colombia (cropping systems) (see Achira)

Emilio Rojas Mendoza, Avenida La Paz 857, Lima 18, Miraflores, Peru

Lauro Toribio Baltazar, Programa de Cultivos Andinos, Universidad Nacional del Centro del Peru, Estación Experimental Agropecuaria "El Mantaro," Carretera Central Km. 34, El Mantaro, Huancayo, Peru

Julio Valladolid Rivera, Peru (see Ahipa)

Noemí Zuñiga Lopez, Estación Experimental Agropecuaria Santa Ana, Apartado 411, Huancayo, Peru (tissue culture, virus-free germplasm)

Other Countries

Gregory J. Anderson, USA (ethnobotany and taxonomy) (see Achira)

Helio H.C. Bastien, 77 Zacatecas, San Miguel de Allende Guanajuato 37700, Mexico

W. Hardy Eshbaugh, USA (see Achira)

Christine Franquemont, USA (Quechua plant use) (see Ahipa)
Daniel W. Gade, USA (see Achira)
International Board for Plant Genetic Resources (IBPGR), Italy (germplasm information)
(see Achira)
Jesse Jaynes, USA (gene splicing) (see Ahipa)
Timothy A. Johns, Canada (ethnobotany, chemical ecology) (see Arracacha)
Ron Kadish, 1980 Hobart Drive, Camarillo, California 93010, USA (agronomy)
Steven R. King, USA (ethnobotany) (see Achira)
Jorge León, Apartado 480, San Pedro, Monte de Oca, San José, Costa Rica
Dag Olav Ovstedal, Milde Arboretum, The Norwegian Arboretum-ARBOHA, The University
of Bergen, Store Milde N-5067, Norway
Tej Partap, Mountain Farming Systems Division, International Centre for Integrated Mountain
Development (ICIMOD), GPO Box 3226, Kathmandu, Nepal (mountain agriculture,
genetic resources)
Deborah Pearsall, Department of Anthropology, University of Missouri, Columbia, Missouri
65211, USA (ethnobotanical archaeology)
Charles M. Rick, USA (see Arracacha)
John W. Rick, Department of Archeology, Stanford University, Stanford, California 94305,
USA
Richard E. Schultes, USA (ethnobotany) (see Achira)
A.A. Shah, Institut für Biochemie, Biologische Bundesanstalt, Messeweg 11–12, Braunschweig
D-3300, West Germany
Calvin R. Sperling, USA (see Ahipa)
Hermann Stegemann, Institut für Biochemie, Biologische Bundesanstalt, Messeweg 11–12,
Braunschweig D-3300, West Germany

MASHUA

Andean Region

Segundo Alandia, c/o Instituto Boliviano de Tecnología Agropecuaria (IBTA), Calle Colombia,
Casilla 2631, Cochabamba, Bolivia (pathology)
Osmar Alves Carriji, Brazil (see Arracacha)
Carlos Arbizú Avellaneda, Peru (see Achira)
Oscar Blanco Galdos, Institute of Andean Crops, Universidad Nacional de San Antonio
Abad del Cusco (UNSAAC), Apartado 921, Cusco 206, Peru
Heinz Brücher, Argentina (useful plants) (see Ahipa)
Anibal del Carpio Farfán, Programa Regional de Cultivos Andinos, Estación Experimental
Agropecuaria Andenes, Instituto Nacional de Investigación Agraria y Agroindustrial
(INIAA), Avenida Pachacutec No. 609, Casilla 807, Cusco, Peru
César del Carpio Merino, Peru (see Achira)
Rolando Egúsquiza, Peru (see Arracacha)
Rolando Estrada J., Peru (somoclonal and virus-free germplasm) (see Arracacha)
Fondo Nacional de Investigaciones Agropecuarias (FONAIAP), Venezuela (see Achira)
Santiago D. Franco Pebe, Peru (germplasm) (see Arracacha)
Michael Hermann, Peru (see Achira)
Zósimo Huamán Cueva, Peru (see Achira)
Instituto Interamericano de Cooperación para la Agricultura (IICA), Conservación y Manejo
de Recursos Naturales Renovables, Oficina del IICA en Bolivia, Casilla 6057, La Paz,
Bolivia
Instituto Nacional de Investigaciones Agropecuarias (INIAP), Ecuador (see Achira)
Cipriano Mantari Camargo, Peru (botanical seed) (see Maca)
Ulrich Mohr, Andean Crops for Human Nutrition, Convenio Perú-Alemania para Cultivos
Andinos-GTZ (CORDE), Avenida Sol 817, Casilla Postal 807, Cusco, Peru
David Morales V., Germplasm Section, Instituto Boliviano de Tecnología Agropecuaria
(IBTA), Cajón Postal 5783, La Paz, Bolivia

Eloy Munive Jáuregui, Peru (see Maca)

Laura Muñoz Espín, Ecuador (in vitro germplasm conservation) (see Arracacha)

Carlos Nieto C., Ecuador (see Achira)

Eduardo Peralta I., Programa de Cultivos Andinos, Instituto Nacional de Investigaciones Agropecuarias (INIAP), Estación Experimental "Santa Catalina," Casilla Postal 340, Quito, Ecuador

Raúl Ríos E., Bolivia (see Achira)

Juan Risi Carbone, Colombia (cropping systems) (see Achira)

Carlos Roersch, Centro de Medicina Andina, Jirón Ricardo Palma No. 5, Santa Mónica, Apartado 711, Cusco, Peru

Jose Fernando Romero Cañizares, Ecuador (see Arracacha)

Basilio Salas, Department of Plant Pathology, North Dakota State University, Fargo, North Dakota 58105, USA (Universidad Nacional del Altiplano, Puno, after 1990)

Luis Salazar, Centro Internacional de la Papa (CIP), Apartado 5969, Lima, Peru (virology)

Isidoro Sánchez Vega, Peru (see Arracacha)

Raúl Santana Paucar, MINKA-Grupo "Talpuy," Grupo de Investigación y Extensión de Tecnología Popular, Avenida Centenario 589 San Carlos, Apartado No. 222, Huancayo, Peru

Juan Seminario Cunya, Universidad Nacional Técnica de Cajamarca, Apartado 16, Cajamarca, Peru

Juan Solano Lazo, Ecuador (see Achira)

Lauro Toribio Baltazar, Peru (see Maca)

Julio Valladolid Rivera, Peru (see Ahipa)

Julio Valle, Corporación Ambiente y Desarrollo (AMDE), PO Box 632, Ambato, Ecuador (soil science)

Carlos Adolfo Vimos Naranjo, Ecuador (see Arracacha)

Jeffrey White, Programa de Frijoles, Centro Internacional de Agricultura Tropical (CIAT), Apartado Aéreo 6713, Cali, Colombia

Other Countries

Gregory J. Anderson, USA (ethnobotany and taxonomy) (see Achira)

Leslie Brownrigg, USA (indigenous systems) (see Achira)

Alan A. Brunt, UK (disease- and virus-free germplasm) (see Arracacha)

R. Delhey, West Germany (virus screening) (see Arracacha)

Deutsche Gesellschaft für Technische Zusammenarbeit (GTZ), 1 bei Frankfurt/Main, Dag-Hammarskjöld-weg 1, Postfach 5180, Eschborn D-6236, West Germany

Dick and Annemarie Endt, New Zealand (see Achira)

W. Hardy Eshbaugh, USA (see Achira)

Christine Franquemont, USA (Quechua plant use) (see Ahipa)

B.W.W. Grout, UK (see Achira)

Ron Hurov, PO Box 1596, Chula Vista, California 92012, USA

International Board for Plant Genetic Resources (IBPGR), Italy (germplasm information) (see Achira)

Jesse Jaynes, USA (gene splicing) (see Ahipa)

Timothy A. Johns, Canada (ethnobotany, chemical ecology) (see Arracacha)

Paula Jokela, Department of Biology, University of Turku, Turku SF-20500, Finland (physiology, virus-free culture)

Steven R. King, USA (ethnobotany) (see Achira)

Jess R. Martineau, NPI, University Research Park, 417 Wakara Way, PO Box 8049, Salt Lake City, Utah 84108, USA (biotechnology)

John Palmer, New Zealand (see Ahipa)

Michael N. Pearson, Department of Botany, University of Auckland, Private Bag, Auckland, New Zealand (virus-free germplasm)

Leene Pietilä, Department of Biology, University of Turku, Turku SF-20500, Finland (physiology, virus-free culture)

F.M. Quin, Koninklijk Instituut voor de Tropen (KIT) (Royal Tropical Institute), Mauritskade 63, Amsterdam 1092–AD, Netherlands

Charles M. Rick, USA (see Arracacha)
Warren G. Roberts, The University Arboretum, University of California, Davis, California 95616, USA
Arne Rousi, Finland (physiology, virus-free culture) (see Achira)
Robin Saunders, USA (processing starches) (see Achira)
L. Schilde-Rentschler, Arbeitsgruppe Prof. Dr. H. Ninnemann, Medizinisch-Naturwissenschaftliches Forschungszentrum, der Universität, Ob dem Himmelreich 7, Tübingen D-7400, West Germany
Richard E. Schultes, USA (ethnobotany) (see Achira)
Seung Jin Kim, South Korea (germplasm) (see Achira)
A.A. Shah, West Germany (see Maca)
Calvin R. Sperling, USA (see Ahipa)
Hermann Stegemann, West Germany (see Maca)
W.R. Sykes, Division of Botany, Department of Scientific and Industrial Research (DSIR), Private Bag, Christchurch, New Zealand

MACA

Andean Region

Raúl O. Castillo T., Ecuador (tissue culture, virus-free germplasm) (see Ahipa)
J. Chicaisa, Cochasquí, Tabacundo, Ecuador
Leopoldo Chontasi, Centro de Arte y Acción Popular (CAAP), Chahuarpungo, Pichincha, Ecuador (grower)
R. Chorlango, Cubinche, Tabacundo, Ecuador
Lola F. de Montenegro, Instituto de Estudios de Bromatología y Nutrición Andina, Universidad San Martín de Porres, Calle Bolívar 348, Miraflores, Lima, Peru (nutritional analysis)
César del Carpio Merino, Peru (see Achira)
Rolando Estrada J., Peru (somoclonal and virus-free germplasm) (see Arracacha)
Santiago D. Franco Pebe, Peru (germplasm) (see Arracacha)
Michael Hermann, Peru (see Achira)
Instituto Interamericano de Cooperación para la Agricultura (IICA), Bolivia (see Mashua)
Carlos Nieto C., Ecuador (see Achira)
Julio Rea, Casilla 21956, La Paz, Bolivia
Juan Rodríguez Cueva, Estación Experimental Baños del Inca, Programa Nacional de Cultivos Andinos (PNCA), INIAA, Julio Guerrero #121, Urb. Cajamarca, Cajamarca, Peru
Isidoro Sánchez Vega, Peru (see Arracacha)
Juan Seminario Cunya, Peru (see Mashua)
Mario E. Tapia Núñez, Programa Nacional de Cultivos Andinos, Instituto Nacional de Investigación y Promoción Agropecuaria (PISA - INIPA), Apartado 110697, Lima 11, Peru
Carlos Adolfo Vimos Naranjo, Ecuador (see Arracacha)

Other Countries

W. Hardy Eshbaugh, USA (see Achira)
José T. Esquinas-Alcázar, Seed Service (AGPS), Plant Production and Protection Service, Food and Agriculture Organization of the United Nations (FAO), Via delle Terme di Caracalla, Rome 00100, Italy
Jesse Jaynes, USA (gene splicing) (see Ahipa)
Steven R. King, USA (ethnobotany) (see Achira)
Jorge León, Costa Rica (see Maca)
Lyman Smith, c/o Department of Botany, National Museum of Natural History, Smithsonian Institution, Washington, DC 20560, USA (taxonomy)

OCA

Andean Region

Jaime Alba Aldunate, Asociación Integral de Organizaciones Agropecuarias, Casilla No. 7759, La Paz, Bolivia

Carlos A. Alvarez, Departamento de Biología, Universidad Nacional Agraria (UNA), Apartado 456, La Molina, Lima 100, Peru

Santiago Erik Antúñez de Mayolo R., Peru (pre-Columbian plants, geography, and climate) (see Maca)

Carlos Arbizú Avellaneda, Peru (see Achira)

Fernando N. Barrantes Del Aguila, Peru (germplasm, pathology) (see Achira)

Oscar Blanco Galdos, Peru (see Mashua)

Augusto Cardich, Argentina (geography and climate) (see Achira)

Raúl O. Castillo T., Ecuador (tissue culture, virus-free germplasm) (see Ahipa)

Andres Contreras M., Instituto de Producción Vegetal, Universidad Austral de Chile, Casilla 567, Valdivia, Chile (germplasm)

Hernán Cortés Bravo, Departamento Academicoide Agricultura, Universidad Nacional de San Antonio Abad del Cusco (UNSAAC), Apartado 921, Cusco, Peru

Daniel de Azkue, Centro Nacional de Investigaciones Científicas y Técnicas, Centro de Estudios Farmacológicos y de Principios Naturales (CEFAPRIN), Serrano 665, Buenos Aires 1414, Buenos Aires, Argentina (cytology)

Mery de Quitón, Instituto Boliviano de Tecnología Agropecuaria (IBTA), Calle Colombia, Casilla 2631, Cochabamba, Bolivia (pathology)

Anibal del Carpio Farfán, Peru (see Mashua)

César del Carpio Merino, Peru (see Achira)

Rolando Estrada J., Peru (somoclonal and virus-free germplasm) (see Arracacha)

Leonard Field, Centro Andino de Acción Popular (CAAP), Apartado 173–B, Quito, Ecuador

Fondo Nacional de Investigaciones Agropecuarias (FONAIAP), Venezuela (see Achira)

Santiago D. Franco Pebe, Peru (germplasm) (see Arracacha)

Humberto Gandarillas, Flores Montesillo, Casilla 053, Cochabamba, Bolivia

Alexander Grobman, Gentec - Grobman Genotécnica S.A., Oficina 702, Avenida República de Panamá 3563, San Isidro, Lima, Peru (germplasm collection)

Michael Hermann, Peru (see Achira)

Instituto Interamericano de Cooperación para la Agricultura (IICA), Bolivia (see Mashua)

Instituto Nacional de Investigaciones Agropecuarias (INIAP), Ecuador (see Achira)

Luis A. Jiménez Monroy, Universidad Nacional Técnica del Altiplano, Ciudad Universitaria, Casilla 291, Puno, Peru

Sady Majino Bernardo, Peru (see Arracacha)

Cipriano Mantari Camargo, Peru (botanical seed) (see Maca)

Arturo José Martínez, Centro Nacional de Investigaciones Científicas y Técnicas, Centro de Estudios Farmacológicos y de Principios Naturales (CEFAPRIN), Serrano 665, Buenos Aires 1414, Buenos Aires, Argentina (cytology)

Ulrich Mohr, Peru (see Mashua)

David Morales V., Bolivia (see Mashua)

Eloy Munive Jáuregui, Peru (see Maca)

Laura Muñoz Espín, Ecuador (in vitro germplasm conservation) (see Arracacha)

Carlos Nieto C., Ecuador (see Achira)

Eduardo Peralta I., Ecuador (see Mashua)

Julio Rea, Bolivia (indigenous food plants) (see Mauka)

Francisco Rhon Dávila, Ecuador (see Maca)

Raúl Ríos E., Bolivia (see Achira)

Juan Risi Carbone, Colombia (cropping systems) (see Achira)

Carlos Roersch, Peru (see Mashua)

Emilio Rojas Mendoza, Peru (see Maca)

Jose Fernando Romero Cañizares, Ecuador (see Arracacha)

Basilio Salas, USA (see Mashua)

Francisco Salas, Departamento de Tecnología y Productos Agropecuarios (TAPA), Universidad
Nacional Agraria (UNA), Apartado 456, La Molina, Lima 100, Peru
Luis Salazar, Peru (virology) (see Mashua)
Peter E. Schmiediche, Peru (see Arracacha)
Juan Seminario Cunya, Peru (see Mashua)
Juan Solano Lazo, Ecuador (see Achira)
Lauro Toribio Baltazar, Peru (see Maca)
Julio Valladolid Rivera, Peru (see Ahipa)
Julio Valle, Ecuador (soil science) (see Mashua)
Rebeca Vega de Rojas, Ecuador (see Arracacha)
Carlos Adolfo Vimos Naranjo, Ecuador (see Arracacha)
Jeffrey White, Colombia (see Mashua)

Other Countries

Gregory J. Anderson, USA (ethnobotany and taxonomy) (see Achira)
Helio H.C. Bastien, Mexico (see Maca)
Vichitr Benjasil, Thailand (see Achira)
Ricardo Bressani, Guatemala (nutrition) (see Achira)
William M. Brown, Department of Plant Pathology and Weed Science, Colorado State
University, Fort Collins, Colorado 80523, USA
Alan A. Brunt, UK (disease- and virus-free germplasm) (see Arracacha)
G. Burge, Levin Horticultural Research Centre, Ministry of Agriculture and Fisheries (MAF),
Private Bag, Levin, New Zealand
Sam Campbell, Botanical Garden, University of California, Berkeley, California 94720, USA
(germplasm)
R. Delhey, West Germany (virus screening) (see Arracacha)
Deutsche Gesellschaft für Technische Zusammenarbeit (GTZ), West Germany (see Mashua)
Dick and Annemarie Endt, New Zealand (see Achira)
W. Hardy Eshbaugh, USA (see Achira)
Michel Fanton, Australia (see Achira)
Stanley Gershoff, School of Nutrition, Tufts University, Medford, Massachusetts 02155, USA
B.W.W. Grout, UK (see Achira)
Peter Halford, Almadale Road, RD 7, Feilding, New Zealand
Einar Hellborn, Såningsvägen 86, Järfälla S-17545, Sweden (greenhouse grower of tropical
tubers)
Efraim Hernández X., Centro de Botánico, Institución Enseñanza e Investigación en Ciencias
Agrícolas, Colegio de Postgraduados, Montecillos, Chapingo, Estado de México 56230,
Mexico
Robert W. Hoopes, Frito-Lay, Incorporated, 4295 Tenderfoot Road, Rhinelander, Wisconsin
54501, USA (genetics, breeding, and pathology)
International Board for Plant Genetic Resources (IBPGR), Italy (germplasm information)
(see Achira)
Jesse Jaynes, USA (gene splicing) (see Ahipa)
Timothy A. Johns, Canada (ethnobotany, chemical ecology) (see Arracacha)
Paula Jokela, Finland (physiology, virus-free culture) (see Mashua)
Wayne Jones, NPI, University Research Park, 417 Wakara Way, PO Box 8049, Salt Lake
City, Utah 84108, USA (virus screening)
Ron Kadish, USA (agronomy) (see Maca)
Steven R. King, USA (ethnobotany) (see Achira)
Alicia Lourteig, Laboratoire de Phanérogamie, Muséum National d'Histoire Naturelle, 16,
rue Buffon, Paris 75005, France (taxonomy)
Jess R. Martineau, USA (biotechnology) (see Mashua)
Cesar Morales, Mundo Latino, 83 Broadway, Passaic, New Jersey 07055, USA (importation
to U.S.)
W.A. Nekkel, 122 Maple Lane, Covington, Virginia 24426, USA
Avigdor Orr, USA (see Ahipa)
W.A. Osborne, Manuwatu Fresh Produce, RD 10, Palmerston, New Zealand

Dag Olav Ovstedal, Norway (see Maca)
John Palmer, New Zealand (see Ahipa)
Tej Partap, Nepal (mountain agriculture, genetic resources) (see Maca)
Michael N. Pearson, New Zealand (virus-free germplasm) (see Mashua)
Charles Peters, Institute of Economic Botany, New York Botanical Garden, Bronx, New York 10458–9980, USA
Leene Pietilä, Finland (physiology, virus-free culture) (see Mashua)
Hugh Popenoe, USA (see Arracacha)
F.M. Quin, Netherlands (see Mashua)
Charles M. Rick, USA (see Arracacha)
John W. Rick, USA (see Maca)
John M. Riley, *Solanaceae Newsletter*, 3370 Princeton Court, Santa Clara, California 95051, USA (general information, germplasm)
Arne Rousi, Finland (physiology, virus-free culture) (see Achira)
L. Schilde-Rentschler, West Germany (see Mashua)
C.A. Schroeder, USA (see Ahipa)
Richard E. Schultes, USA (ethnobotany) (see Achira)
Seung Jin Kim, South Korea (germplasm) (see Achira)
A.A. Shah, West Germany (see Maca)
Calvin R. Sperling, USA (see Ahipa)
Hermann Stegemann, West Germany (see Maca)
W.R. Sykes, New Zealand (see Mashua)
Richard Valley, 3328 SE Kelly, Portland, Oregon 97202, USA (grower)
Guillermo Veliz, Peimco Natural Foods, 15216 Hartsook Street, Sherman Oaks, California 91403, USA (importation to U.S.)
Alejo von der Pahlen, Plant Production and Protection Division, Food and Agriculture Organization of the United Nations (FAO), Via delle Terme di Caracalla, Rome 00100, Italy
Benjamin H. Waite, Plant Pathology, Office of Agriculture, Bureau for Science and Technology, Agency for International Development (USAID), Washington, DC 20523, USA
R.B. Wynn-Williams, New Crops Section, Division of Crop Research, Department of Scientific and Industrial Research (DSIR), Private Bag, Christchurch, New Zealand

POTATOES

Andean Region

Segundo Alandia, Bolivia (pathology) (see Mashua)
Guillerma Aníbal Albornoz Pazmiño, Facultad de Ciencias Agrícolas, Universidad Central del Ecuador, Obispo Miguel Solier No. 152, Quito, Ecuador (germplasm)
Carlos A. Alvarez, Peru (see Oca)
Santiago Erik Antúñez de Mayolo R., Peru (pre-Columbian plants, geography, and climate) (see Maca)
Juan Astorga N., Departamento de Agricultura, Universidad Nacional Técnica del Altiplano, Ciudad Universitaria, Casilla 291, Puno, Peru (bitter potatoes)
Israel Aviles Pérez, Acción Rural Agrícola de Desarrollo Organizado (ARADO), Avenida Barrieutos 2347, Casilla 1710, Cochabamba, Bolivia
Antonio Bacigalupo, Food and Agriculture Organization of the United Nations (FAO), Avenida Santa Maria 6700, Casilla 10095, Santiago, Chile
Heinz Brücher, Argentina (see Ahipa)
Augusto Cardich, Argentina (geography and climate) (see Achira)
Centro Internacional de la Papa (CIP), Apartado 5969, Lima, Peru
Andres Contreras M., Chile (germplasm) (see Oca)
Rolando Egúsquiza, Peru (see Arracacha)
Nelson Estrada Ramos, Instituto Colombiano Agropecuario (ICA) - Tibaitata, Apartado Aéreo 151123, El Dorado, Bogotá, Colombia

Wilma Freire, Nutrition Unit, Center for Planning and Social Research (CEPLAES), Casilla 6127-CCI, Quito, Ecuador (nutrition)

Humberto Gandarillas, Bolivia (see Oca)

Zósimo Huamán Cueva, Peru (see Achira)

Instituto Interamericano de Cooperación para la Agricultura (IICA), Bolivia (see Mashua)

Instituto Nacional de Investigaciones Agropecuarias (INIAP), Ecuador (see Achira)

Ingo Junge Rodewald, Clasificador 26, Concepción, Chile (processing, nutrition)

Latinreco S.A., Ecuador (see Arracacha)

Lúis E. López Jaramilló, Consejo Internacional de Recursos Fitogenéticos (CIRF-IBPGR), c/o Centro Internacional de Agricultura Tropical (CIAT), Apartado Aéreo 6713, Cali, Colombia (Solanaceae)

Cipriano Mantari Camargo, Peru (botanical seed) (see Maca)

Humberto Mendoza, Centro Internacional de la Papa (CIP), Apartado 5969, Lima, Peru

David Morales V., Bolivia (see Mashua)

Carlos M. Ochoa, Department of Taxonomy, Centro Internacional de la Papa (CIP), Apartado 5969, Lima, Peru

Katsuo A. Okada, Instituto de Genética, Instituto Nacional de Tecnología Agropecuaria (INTA), Centro de Investigaciones de Recursos Naturales, Castelar 1712, Buenos Aires, Argentina (wild species of tuber-bearing solana)

Ramiro Ortega Dueñas, Programa de Papa, Centro de Investigación en Cultivos Andinos, Universidad Nacional de San Antonio Abad del Cusco (UNSAAC), Apartado 1006, K'ayra, Cusco, Peru (bitter potatoes)

Victor Otazú Monzón, Centro Internacional de la Papa (CIP), Apartado 5969, Lima, Peru (pathology)

Raúl Ríos E., Bolivia (see Achira)

Carlos Roersch, Peru (see Mashua)

Luis Salazar, Peru (virology) (see Mashua)

Raúl Santana Paucar, Peru (see Mashua)

Peter E. Schmiediche, Peru (tuberous solana) (see Arracacha)

María Scurrah, Breeding and Genetics Department, Centro Internacional de la Papa (CIP), Apartado 5969, Lima, Peru (bitter potatoes, nematodes)

Juan Solano Lazo, Ecuador (see Achira)

Lauro Toribio Baltazar, Peru (see Maca)

Cesar C. Vargas Calderón, Facultad de Botánica, Universidad del Cusco, Cusco, Peru

Jeffrey White, Colombia (see Mashua)

Other Countries

Gregory J. Anderson, USA (ethnobotany and taxonomy) (see Achira)

Vichitr Benjasil, Thailand (see Achira)

Meridith Bonierbale, Department of Plant Breeding and Biometry, Cornell University, Ithaca, New York 14853, USA (genetic mapping)

William M. Brown, USA (see Oca)

Stephen Brush, Department of Applied Behavioral Sciences, University of California, Davis, California 95616, USA (farming systems)

Manfred Dambroth, Bundesforschungsanstalt für Landwirtschaft, Institut für Pflanzenbau und Pflanzenzüchtung, Federal Research Centre of Agriculture (FAL), Bundesallee 50, Braunschweig D-3300, West Germany

W.G. D'Arcy, Missouri Botanical Garden, PO Box 299, St. Louis, Missouri 63166–0299, USA

L.M.W. Dellaert, Stichting voor Plantenveredeling (SVP), PO Box 117, Wageningen NL-6700–AC, Netherlands

Luis Destefano, Department of Biochemistry, Louisiana State University, Baton Rouge, Louisiana 70803–1464, USA (gene splicing, nutritional genes)

Craig Dremann, Redwood City Seed Company, PO Box 361, Redwood City, California 94064, USA

Dick and Annemarie Endt, New Zealand (see Achira)

360 LOST CROPS OF THE INCAS

Jan Engels, International Board for Plant Genetic Resources (IBPGR), c/o Pusa Campus, New Delhi 110 012, India
William Charles Evans (Department of Pharmacy, University of Nottingham), Buddlehayes, Southleigh, Colyton, Devon EX13 6JH, UK
Michel Fanton, Australia (see Achira)
Jorge A. Galindo, Centro de Fitopatología, Colegio de Postgraduados, Institución Enseñanza e Investigación en Ciencias Agricolas, Caixa Postal 56230, Montecillos, Estado de México, Mexico
Zaccheaus Oyesiju Gbile, Forestry Research Institute of Nigeria, PO Box 12747, G.P.O. Ibadan, Ibadan, Oyo, Nigeria (Phaseolus)
D.R. Glendinning, Scottish Crop Research Institute (SCRI), Pentlandfield, Roslin, Midlothian EH25 9RF, UK
Heiner Goldbach, Faculty of Plant Nutrition, Technische Universität München, Weihenstephan, Freising 12, D-8050, West Germany (seed storage and germination)
Paul Grun, Department of Biology, 208 Mueller Laboratory, Pennsylvania State University, University Park, Pennsylvania 16802, USA (cytology)
Robert E. Hanneman, Jr., Interregional Potato Introduction Station, University of Wisconsin, Route 2, Sturgeon Bay, Wisconsin 54235, USA
J.G. Hawkes, c/o Department of Geological Sciences, University of Birmingham, PO Box 363, Edgbaston, Birmingham B15 2TT, UK
Frank Haynes, Department of Horticulture, North Carolina State University, Raleigh, North Carolina 27695, USA
Kathleen Haynes, Agricultural Research Service (ARS), U.S. Department of Agriculture (USDA), Building 011, GH 13, Beltsville Agricultural Research Center (BARC-West), Beltsville, Maryland 20705, USA
J.H.Th. Hermsen, Institute for Agricultural Plant Breeding, Agricultural University, Lawickse Allee 166, Wageningen NL-6700 AJ, Netherlands (genetics)
Roel Hoekstra, German-Dutch Potato Collection, Institut für Pflanzenbau und Pflanzenzüchtung, Federal Research Centre of Agriculture (FAL), Bundesallee 50, Braunschweig D-3300, West Germany
Robert W. Hoopes, USA (genetics, breeding, and pathology) (see Oca)
Tom Hughes, National Potato Museum, 704 North Carolina Avenue, SE, Washington, DC 20003, USA (information)
International Board for Plant Genetic Resources (IBPGR), Italy (germplasm information) (see Achira)
Jesse Jaynes, USA (gene splicing) (see Ahipa)
Timothy A. Johns, Canada (ethnobotany, chemical ecology) (see Arracacha)
Wayne Jones, USA (virus screening) (see Oca)
Alan M. Kapuler, USA (see Achira)
Steven R. King, USA (ethnobotany) (see Achira)
Janet S. Luis, Northern Philippine Root Crops Research and Training Center, Benguet State University, La Trinidad, Benguet, Philippines
Cyrus McKell, NPI, University Research Park, 417 Wakara Way, PO Box 8049, Salt Lake City, Utah 84108, USA
Cesar Morales, USA (importation to U.S.) (see Oca)
Steve Neal, Route 1, Norwood, Missouri 65717, USA
John S. Niederhauser, 2474 Camino Valle Verde, Tucson, Arizona 85715, USA
Dag Olav Ovstedal, Norway (see Maca)
Tej Partap, Nepal (mountain agriculture, genetic resources) (see Maca)
Dov Pasternak, Boyko Institute, Ernst David Bergmann Campus, Ben-Gurion University of the Negev, PO Box 1025, Beer-Sheva 84110, Israel (Solanum phureja germplasm)
Stan Peloquin, Department of Horticulture, University of Wisconsin, Madison, Wisconsin 53706, USA
Robert Plaisted, Department of Plant Breeding and Biometry, Cornell University, Ithaca, New York 14853, USA
Jürgen Reckin, Experimentalgarten Finowfurt, Altenhofer Weg 1, Werbellin DDR-1303, East Germany
John M. Riley, USA (general information, germplasm) (see Oca)
H. Ross, Dompfaffenweg 33, Köln 30, D-5000, West Germany

APPENDIX C

David J. Sammons, Department of Agronomy, University of Maryland, College Park, Maryland 20742, USA

Robin Saunders, USA (processing starches) (see Achira)

Ewa Sawicka, Potato Research Institute, Mlochów, Poland (potato genetics)

Shao Qiquan, Genetic Transformation Laboratory and Genetic Resources of Plants, Academia Sinica Institute of Genetics, Building 917, De-sheng-Men-Wai, Bei-sha-Tan, Beijing 100012, China

N.W. Simmonds, School of Agriculture, University of Edinburgh, West Mains Road, Edinburgh, Midlothian, EH9 3JG, UK

Nigel Smith, Department of Geography, University of Florida, Gainesville, Florida 32611, USA

H. David Thurston, Department of Plant Pathology, Cornell University, Ithaca, New York 14853, USA

Donald Ugent, USA (see Achira)

Louis J.M. van Soest, New Potential Crops Project, Centrum voor Genetische Bronnen, Droevendaalsesteeg 1, Postbus 224, Wageningen 6700 AE, Netherlands

Alejo von der Pahlen, Italy (see Oca)

Merle Weaver, Agricultural Research Service (ARS), U.S. Department of Agriculture (USDA), Western Regional Research Center, 800 Buchanan Street, Albany, California 94710, USA

Raymon Webb, Agricultural Research Service (ARS), U.S. Department of Agriculture (USDA), Building 011, GH 13, Beltsville Agricultural Research Center (BARC-West), Beltsville, Maryland 20705, USA

Karl Zimmerer, Department of Geography, University of North Carolina, Chapel Hill, North Carolina 27514, USA

ULLUCO

Andean Region

Carlos A. Alvarez, Peru (see Oca)

Santiago Erik Antúñez de Mayolo R., Peru (pre-Columbian plants, geography, and climate) (see Maca)

Carlos Arbizú Avellaneda, Peru (see Achira)

Fernando N. Barrantes Del Aguila, Peru (germplasm, pathology) (see Achira)

Augusto Cardich, Argentina (geography and climate) (see Achira)

Raúl O. Castillo T., Ecuador (tissue culture, virus-free germplasm) (see Ahipa)

Anibal del Carpio Farfán, Peru (see Mashua)

César del Carpio Merino, Peru (see Achira)

Rolando Egúsquiza, Peru (see Arracacha)

Edgar Ivan Estrada, Facultad de Ciencias Agropecuarias, Departamento de Agricultura, Universidad Nacional de Colombia (UNPALMIRA), Apartado Aéreo 237, Palmira, Valle, Colombia

Rolando Estrada J., Peru (somoclonal and virus-free germplasm) (see Arracacha)

Leonard Field, Ecuador (see Oca)

Fondo Nacional de Investigaciones Agropecuarias (FONAIAP), Venezuela (see Achira)

Santiago D. Franco Pebe, Peru (germplasm) (see Arracacha)

Humberto Gandarillas, Bolivia (see Oca)

Juan Gaviria R., Grupo Botanica, Facultad de Ciencias, Departamento de Biología, Universidad de Los Andes, La Hechicera, Mérida 5101, Estado Mérida, Venezuela

Michael Hermann, Peru (see Achira)

Instituto Interamericano de Cooperación para la Agricultura (IICA), Bolivia (see Mashua)

Instituto Nacional de Investigaciones Agropecuarias (INIAP), Ecuador (see Achira)

Luis A. Jiménez Monroy, Peru (see Oca)

Sady Majino Bernardo, Peru (see Arracacha)

Cipriano Mantari Camargo, Peru (botanical seed) (see Maca)

Ulrich Mohr, Peru (see Mashua)
David Morales V., Bolivia (see Mashua)
Eloy Munive Jáuregui, Peru (see Maca)
Laura Muñoz Espín, Ecuador (in vitro germplasm conservation) (see Arracacha)
Carlos Nieto C., Ecuador (see Achira)
Eduardo Peralta I., Ecuador (see Mashua)
Francisco Rhon Dávila, Ecuador (see Maca)
Raúl Ríos E., Bolivia (see Achira)
Juan Risi Carbone, Colombia (cropping systems) (see Achira)
Carlos Roersch, Peru (see Mashua)
Emilio Rojas Mendoza, Peru (see Maca)
Jose Fernando Romero Cañizares, Ecuador (see Arracacha)
Basilio Salas, USA (see Mashua)
Francisco Salas, Peru (see Oca)
Luis Salazar, Peru (virology) (see Mashua)
Isidoro Sánchez Vega, Peru (see Arracacha)
Raúl Santana Paucar, Peru (see Mashua)
Peter E. Schmiediche, Peru (see Arracacha)
Juan Seminario Cunya, Peru (see Mashua)
Juan Solano Lazo, Ecuador (see Achira)
Lauro Toribio Baltazar, Peru (see Maca)
Julio Valladolid Rivera, Peru (see Ahipa)
Julio Valle, Ecuador (soil science) (see Mashua)
Carlos Adolfo Vimos Naranjo, Ecuador (see Arracacha)

Other Countries

Gregory J. Anderson, USA (ethnobotany and taxonomy) (see Achira)
Ricardo Bressani, Guatemala (nutrition) (see Achira)
William M. Brown, USA (see Oca)
Alan A. Brunt, UK (disease- and virus-free germplasm) (see Arracacha)
R. Delhey, West Germany (virus-free culture) (see Arracacha)
Department of Primary Industry, Director of Crop Research, PO Box 417, Konedobu, Papua
    New Guinea
Deutsche Gesellschaft für Technische Zusammenarbeit (GTZ), West Germany (see Mashua)
Dick and Annemarie Endt, New Zealand (see Achira)
W. Hardy Eshbaugh, USA (see Achira)
B.W.W. Grout, UK (see Achira)
International Board for Plant Genetic Resources (IBPGR), Italy (germplasm information)
    (see Achira)
Jesse Jaynes, USA (gene splicing) (see Ahipa)
Timothy A. Johns, Canada (ethnobotany, chemical ecology) (see Arracacha)
Paula Jokela, Finland (physiology, virus-free culture) (see Mashua)
Wayne Jones, USA (virus screening) (see Oca)
Steven R. King, USA (ethnobotany) (see Achira)
Terttu Lempiäinen, Department of Biology, University of Turku, Turku SF-20500, Finland
    (physiology, virus-free culture)
Jess R. Martineau, USA (biotechnology) (see Mashua)
Cesar Morales, USA (importation to U.S.) (see Oca)
John Palmer, New Zealand (see Ahipa)
Leene Pietilä, Finland (physiology, virus-free culture) (see Mashua)
Hugh Popenoe, USA (see Arracacha)
F.M. Quin, Netherlands (see Mashua)
Jürgen Reckin, East Germany (see Potatoes)
Charles M. Rick, USA (see Arracacha)
John W. Rick, USA (see Maca)
Arne Rousi, Finland (physiology, virus-free culture) (see Achira)
L. Schilde-Rentschler, West Germany (see Mashua)

Richard E. Schultes, USA (ethnobotany) (see Achira)
Seung Jin Kim, South Korea (germplasm) (see Achira)
A.A. Shah, West Germany (see Maca)
Calvin R. Sperling, USA (see Ahipa)
Hermann Stegemann, West Germany (see Maca)
W.R. Sykes, New Zealand (see Mashua)
Guillermo Veliz, USA (importation to U.S.) (see Oca)
Alejo von der Pahlen, Italy (see Oca)
Benjamin H. Waite, USA (see Oca)
Karl Zimmerer, USA (see Potatoes)

YACON

Andean Region

Carlos Arbizú Avellaneda, Peru (see Achira)
Fernando N. Barrantes Del Aguila, Peru (germplasm, pathology) (see Achira)
Oscar Blanco Galdos, Peru (see Mashua)
Augusto Cardich, Argentina (geography and climate) (see Achira)
Raúl O. Castillo T., Ecuador (tissue culture, virus-free germplasm) (see Ahipa)
César del Carpio Merino, Peru (see Achira)
Rolando Egúsquiza, Peru (see Arracacha)
Fondo Nacional de Investigaciones Agropecuarias (FONAIAP), Venezuela (see Achira)
Santiago D. Franco Pebe, Peru (germplasm) (see Arracacha)
Michael Hermann, Peru (see Achira)
Joy C. Horton Hofmann, Ecuador (see Achira)
Zósimo Huamán Cueva, Peru (see Achira)
Instituto Nacional de Investigaciones Agropecuarias (INIAP), Ecuador (see Achira)
Sady Majino Bernardo, Peru (see Arracacha)
Carlos Nieto C., Ecuador (see Achira)
Raúl Ríos E., Bolivia (see Achira)
Isidoro Sánchez Vega, Peru (see Arracacha)
Julio Valladolid Rivera, Peru (see Ahipa)
Carlos Adolfo Vimos Naranjo, Ecuador (see Arracacha)

Other Countries

Gregory J. Anderson, USA (ethnobotany and taxonomy) (see Achira)
Vichitr Benjasil, Thailand (see Achira)
Dick and Annemarie Endt, New Zealand (see Achira)
Stephen Facciola, 1870 Sunrise Drive, Vista, California 92084, USA (useful plants, germplasm)
Daniel W. Gade, USA (see Achira)
International Board for Plant Genetic Resources (IBPGR), Italy (germplasm information) (see Achira)
Timothy A. Johns, Canada (ethnobotany, chemical ecology) (see Arracacha)
Alan M. Kapuler, USA (see Achira)
Steven R. King, USA (ethnobotany) (see Achira)
Richard McCain, Quail Mountain Nursery, 14310 Campagna Way, Watsonville, California 95076, USA (horticulture)
D.J. Manners, Brewing and Biological Sciences, Heriot-Watt University, Chambers Street, Edinburgh EH1 1HX, Scotland, UK (inulin)
Jess R. Martineau, USA (biotechnology) (see Mashua)
Avigdor Orr, USA (see Ahipa)
John Palmer, New Zealand (see Ahipa)
Hugh Popenoe, USA (see Arracacha)

Martin L. Price, USA (limited germplasm available) (see Achira)

Harold Robinson, National Museum of Natural History, NHB-166, Smithsonian Institution, Washington, DC 20560, USA (taxonomy)

Elizabeth Schneider, 215 East 80th, New York, New York 10021, USA (food preparation)

Richard E. Schultes, USA (ethnobotany) (see Achira)

Seung Jin Kim, South Korea (germplasm) (see Achira)

Steven Spangler, USA (see Achira)

John Swift, Swift's Subtropicals, 3698 Clark Valley, Los Osos, California 93402, USA

W.R. Sykes, New Zealand (see Mashua)

Donald Ugent, USA (see Achira)

James R. Wells, Cranbrook Institute of Science, 500 Lone Pine Road, PO Box 801, Bloomfield Hill, Michigan 48013, USA (taxonomy)

KANIWA

Andean Region

Jaime Alba Aldunate, Bolivia (see Oca)

Alipio Canahua, Instituto Nacional de Investigación y Promoción Agropecuaria (INIPA), Casilla 172, Puno, Peru

Augusto Cardich, Argentina (geography and climate) (see Achira)

César del Carpio Merino, Peru (see Achira)

Humberto Gandarillas, Bolivia (see Oca)

Lionel O. Giusti, Fundación e Instituto Miguel Lillo, Universidad Nacional de Tucumán, Miguel Lillo 205, San Miguelde Tucumán, Argentina (systematics)

Luz Gomez, Programa de Cereales, Universidad Nacional Agraria (UNA), Apartado 456, La Molina, Lima 100, Peru

Alexander Grobman, Peru (germplasm collection) (see Oca)

Instituto Interamericano de Cooperación para la Agricultura (IICA), Bolivia (see Mashua)

Ingo Junge Rodewald, Chile (processing, nutrition) (see Potatoes)

Maximo Libermann Cruz, Facultad de Ciencias, Instituto de Ecología, Casilla 20127, La Paz, Bolivia (salt tolerance)

Cipriano Mantari Camargo, Peru (see Maca)

Ulrich Mohr, Peru (see Mashua)

David Morales V., Bolivia (see Mashua)

Angel Mujica, Instituto Nacional de Investigación y Promoción Agropecuaria (INIPA), Casilla 172, Puno, Peru

Maria de Lourdes Peñaloza Izurieta, Facultad de Ciencia e Ingeniería en Alimentos, Universidad Técnica de Ambato, Avenida Colombia y Chile s/n, Casilla Postal 334, Ambato, Ecuador (nutritional analysis)

Francisco Rhon Dávila, Ecuador (see Maca)

Raúl Ríos E., Bolivia (see Achira)

Juan Risi Carbone, Colombia (see Achira)

Carlos Roersch, Peru (see Mashua)

Emilio Rojas Mendoza, Peru (see Maca)

Basilio Salas, USA (see Mashua)

Mario E. Tapia Núñez, Peru (see Mauka)

Lauro Toribio Baltazar, Peru (see Maca)

Cesar C. Vargas Calderon, Peru (see Potatoes)

Eulogio Zanabria, Departamento de Agronomía, Universidad Nacional Técnica del Altiplano, Ciudad Universitaria, Casilla 291, Puno, Peru (pathology)

Other Countries

José Arze Borda, Departamento de Cultivos, Centro Agronómica Tropical de Investigación y Enseñanza (CATIE), Turrialba 7170, Costa Rica

Emigdio Ballón, Talavaya Center, Tesuque Drive, Box 2, Espanola, New Mexico 87532, USA

Ricardo Bressani, Guatemala (nutrition) (see Achira)
Rolf Carlsson, Department of Plant Physiology and Food Botany, University of Lund, PO
    Box 7007, Lund S-22007, Sweden (leaf protein, plant chemistry, and nutrition)
Miguel Carmen, FAO-Luanda, Angola, c/o Myllytie 16 as 24, Oulu 90500, Finland
Deutsche Gesellschaft für Technische Zusammenarbeit (GTZ), West Germany (see Mashua)
Craig Dremann, USA (see Potatoes)
Eden Foods, Inc., 701 Tecumseh Road, Clinton, Michigan 49236, USA (importation,
    processing, and distribution)
W. Hardy Eshbaugh, USA (see Achira)
Jorge D. Etchevers B., Centro de Edafología, Sección Fertilidad de Suelos, Institución
    Enseñanza e Investigación en Ciencias Agrícolas, Colegio de Postgraduados, Montecillos,
    Chapingo, Estado de México 56230, Mexico
Christine Franquemont, USA (Quechua plant use) (see Ahipa)
Daniel W. Gade, USA (see Achira)
N.W. Galwey, Department of Applied Biology, University of Cambridge, Pembroke Street,
    Cambridge CB2 3DX, UK
Stephen L. Gorad, 1929 Walnut Street #3, Boulder, Colorado 80302, USA (export/import)
B.W.W. Grout, UK (see Achira)
Barry Hammel, Department of Botany, Duke University, Durham, North Carolina 27706, USA
International Board for Plant Genetic Resources (IBPGR), Italy (germplasm information)
    (see Achira)
Duane Johnson, Department of Agronomy, Colorado State University, Fort Collins, Colorado
    80523, USA (production)
Ron Kadish, USA (agronomy) (see Maca)
Alan M. Kapuler, USA (see Achira)
Steven R. King, USA (ethnobotany) (see Achira)
John McCamant, Sierra Blanca Associates, 2560 South Jackson, Denver, Colorado 80210,
    USA
Cyrus McKell, USA (see Potatoes)
Jerry L. McLaughlin, Department of Medicine Chemistry and Pharmacy, Purdue University,
    West Lafayette, Indiana 47907, USA (toxicology)
Tej Partap, Nepal (mountain agriculture, genetic resources) (see Maca)
Martin L. Price, USA (limited germplasm available) (see Achira)
Robert Reid, International Board for Plant Genetic Resources (IBPGR), FAO Headquarters,
    Via delle Terme di Caracalla, Rome 00100, Italy
Alfredo Sánchez-Marroquin, San Miguel de Proyectos Agropecuarios, S.P.R. de R.S., Miami
    40, México 03810, D.F., Mexico
Richard E. Schultes, USA (ethnobotany) (see Achira)
Forest Shomer, Abundant Life Seed Foundation, PO Box 772, Port Townsend, Washington
    98368, USA
N.W. Simmonds, UK (see Potatoes)
P. Uotila, Botanical Museum, University of Helsinki, Unioninkalū 44, Helsinki SF-00170,
    Finland (herbarium, taxonomic review)
Richard Valley, USA (grower) (see Oca)
Benjamin H. Waite, USA (see Oca)
Hugh D. Wilson, Department of Biology, Texas A&M University, College Station, Texas
    77843, USA (systematics)
Rebecca Theurer Wood, PO Box 30, Crestone, Colorado 81131, USA

KIWICHA

Andean Region

Segundo Alandia, Bolivia (pathology) (see Mashua)
J. Andrango, Programa de Cultivos Andinos, Instituto Nacional de Investigaciones Agropecuarias
    (INIAP), Estación Experimental "Santa Catalina," Casilla Postal 340, Quito, Ecuador

Antonio Bacigalupo, Chile (see Potatoes)

Fernando N. Barrantes Del Aguila, Peru (germplasm, pathology) (see Achira)

Angelica Campana Sierra, Banco Germoplasma Cultivos Andinos, Instituto Nacional de Investigaciones Agropecuarias, Estación La Molina, Apartado 2791, Lima, Peru (native amaranths)

Andres Contreras M., Chile (germplasm) (see Oca)

Guillermo Covas, Estación Experimental, 6326 Anguil, La Pampa, Argentina

Daniel G. Debouck, Unidad de Recursos Genéticas, Centro Internacional de Agricultura Tropical (CIAT), Apartado Aéreo 6713, Cali, Colombia

Mabrouk A. El-Sharkawy, Centro Internacional de Agricultura Tropical (CIAT), Apartado Aéreo 6713, Cali, Colombia (physiology, photosynthesis)

Fondo Nacional de Investigaciones Agropecuarias (FONAIAP), Venezuela (see Achira)

Santiago D. Franco Pebe, Peru (germplasm) (see Arracacha)

Juan Gastó, Pontificia Universidad Católica de Chile, Vic. Mackenna 4860, Casilla 114–D, Santiago, Chile

Luz Gomez, Peru (see Kaniwa)

Armando T. Hunziker, Facultad de Ciencias Exactas, Físicas y Naturales, Museo Botánico, Universidad Nacional de Córdoba, Casilla de Correo 495, Córdoba 5000, Argentina

Joaquin Hurtado Arcila, Facultad de Agronomía, Universidad de Caldas, Apartado Aéreo No. 275, Manizales, Colombia

Nicanor Ibáñez-Herrera, Facultad de Ciencias Biológicas, Universidad Nacional de Trujillo, San Martín 380, Apartado 315, Trujillo, Peru

Instituto Interamericano de Cooperación para la Agricultura (IICA), Bolivia (see Mashua)

Instituto Nacional de Investigaciones Agropecuarias (INIAP), Ecuador (see Achira)

Michael J. Koziol, Latinreco S.A., Centro Nestlé, Casilla Postal 6053–CCI, Quito, Cumbaya, Ecuador

Francisco Javier López Macias, Facultad de Agronomía, Universidad de Caldas, Apartado Aéreo No. 275, Manizales, Colombia

Cipriano Mantari Camargo, Peru (see Maca)

Mario Alfredo Marcial Grijalva, Facultad de Ciencia e Ingeniería en Alimentos, Universidad Técnica de Ambato, Avenida Colombia y Chile s/n, Casilla Postal 334, Ambato, Ecuador (nutritional analysis)

Ulrich Mohr, Peru (see Mashua)

David Morales V., Bolivia (see Mashua)

Alexander M. Mueller, Integrated Rural Development Project, Corporación de Araracuara, Apartado Aéreo 034174, Bogotá, D.E., Colombia

Gladys Navas de Alvarado, Facultad de Ciencia e Ingeniería de Alimentos, Universidad Técnica de Ambato, Avenida Colombia y Chile s/n, Casilla Postal 334, Ambato, Ecuador (nutritional analysis)

Carlos Nieto C., Ecuador (see Achira)

Jaime Pacheco Navarro, Centro de Investigaciones de Cultivos Andinos (CICA), Universidad Nacional del Cusco, Apartado Postal 774, Cusco, Peru

Felipe Portocarrero S., Universidad del Pacifico, Jesús Maria, Avenida Salaverry 2020, Apartado 4683, Lima 11, Peru

Raúl Ríos E., Bolivia (see Achira)

Juan Risi Carbone, Colombia (cropping systems) (see Achira)

Carlos Roersch, Peru (see Mashua)

Emilio Rojas Mendoza, Peru (see Maca)

Isidoro Sánchez Vega, Peru (see Arracacha)

Raúl Santana Paucar, Peru (see Mashua)

Juan Seminario Cunya, Peru (see Mashua)

Juan Solano Lazo, Ecuador (see Achira)

Luis Sumar Kalinowski, Centro de Investigaciones de Cultivos Andinos (CICA), Universidad Nacional del Cusco, Apartado Postal 774, Cusco, Peru

Lauro Toribio Baltazar, Peru (see Maca)

Roberto Ugás C., Programa de Investigación en Hortalizas, Departamento de Horticultura, Universidad Nacional Agraria (UNA), Apartado 456, La Molina, Lima 100, Peru (leafy vegetables)

Cesar C. Vargas Calderon, Peru (see Potatoes)

César Vásconez Sevilla, Facultad de Ciencia e Ingeniería en Alimentos, Universidad Técnica de Ambato, Avenida Colombia y Chile s/n, Casilla Postal 334, Ambato, Ecuador (nutritional analysis)
Carlos Adolfo Vimos Naranjo, Ecuador (see Arracacha)
Samuel von Rütte, Latinreco S.A., Centro Nestlé, Casilla Postal 6053–CCI, Quito, Cumbaya, Ecuador

Other Countries

Mohammed Abid Alsaidy, College of Agriculture, University of Baghdad, Abu Gharib, Iraq (amaranth germplasm)
Gregory J. Anderson, USA (ethnobotany) (see Achira)
Wayne Applegate, American Amaranth Institute, PO Box 216, Bricelyn, Minnesota 56097, USA
Arrowhead Mills, Inc., PO Box 2059, Hereford, Texas 79045, USA (processing)
Suzanne Ashworth, 5007 Del Rio Road, Sacramento, California 95822, USA (amaranth)
Michael Avishai, Jerusalem and University Botanical Garden, Hebrew University of Jerusalem, Givat Ram, Mt. Scopus, Jerusalem 91904, Israel (germplasm)
Emigdio Ballón, USA (see Kaniwa)
Robert Becker, Agricultural Research Service (ARS), U.S. Department of Agriculture (USDA), Western Regional Research Center, 800 Buchanan Street, Albany, California 94710, USA (milling)
Ricardo Bressani, Guatemala (nutrition) (see Achira)
William M. Brown, USA (see Oca)
Leslie Brownrigg, USA (indigenous systems) (see Achira)
T.A. Campbell, Agricultural Research Service (ARS), U.S. Department of Agriculture (USDA), Building 001, Room 340, Beltsville Agricultural Research Center (BARC-West), Beltsville, Maryland 20705, USA (industrial crops)
Rolf Carlsson, Sweden (leaf protein, plant chemistry, and nutrition) (see Kaniwa)
Centroamericana de Semillas, Apartado Postal 1960, Tegucigalpa, Honduras
George B. Chibiliti, Crop Science Department, School of Agricultural Sciences, University of Zambia, Lusaka Campus, PO Box 32379, Lusaka, Zambia (amaranth)
Abebe Demissie, Plant Genetic Resources Center, PO Box 30726, Addis Ababa, Ethiopia (amaranths)
Deutsche Gesellschaft für Technische Zusammenarbeit (GTZ), West Germany (see Mashua)
Soonthorn Duriyaprapan, Thailand Institute of Scientific and Technological Research (TISTR), 196 Phahonyothin Road, Bangkhen, Bangkok 9, Thailand (amaranth)
Daniel Early, Central Oregon Community College, NW College Way, Bend, Oregon 97701, USA (amaranth)
B.O. Eggum, Department of Animal Physiology and Biochemistry, National Institute of Animal Science, 25 Rolighedsvej, Frederiksberg C, DK-2500, Denmark (nutrition)
Marco Romilio Estrada Muy, Facultad de Agronomía, Universidad de San Carlos, Apartado 1545, Ciudad Universitaria, Guatemala 12, Guatemala
Michel Fanton, Australia (see Achira)
Hector E. Flores, Department of Plant Pathology, Pennsylvania State University, University Park, Pennsylvania 16802, USA (phytochemistry)
B.W.W. Grout, UK (see Achira)
V.K. Gupta, Department of Crop Science, University of Nairobi, Nairobi, Kenya (amaranths)
D.J. Hagedorn, Department of Plant Pathology, University of Wisconsin, 1630 Linden Drive, Madison, Wisconsin 53706, USA
Ray Henkel, USA (pioneer colonization) (see Achira)
J.L. Hudson, Seedsman, PO Box 1058, Redwood City, California 94064, USA
Ron Hurov, USA (see Mashua)
International Board for Plant Genetic Resources (IBPGR), Italy (germplasm information) (see Achira)
S.K. Jain, Agronomy and Range Science Department, University of California, Davis, California 95616, USA (genetics)

C. Antonio Jiménez Aparicio, Mexico (see Arracacha)

Duane Johnson, USA (production) (see Kaniwa)

B.D. Joshi, Regional Substation, National Bureau of Plant Genetic Resources, Phagli, Simla 171 021, India (germplasm)

J. Juribe Ruíz, Academia Mexicana de Ciencias Agrícolas, Cuautla No. 76, Casilla Postal 06140, México 11, D.F., Mexico

Promila Kapoor, Commonwealth Science Council, Commonwealth Secretariat, Marlborough House, Pall Mall, London SW1Y 5HX, UK (amaranths)

Alan M. Kapuler, USA (see Achira)

Charles Kauffman, Rodale Research Center, Route 1, Box 323, Kutztown, Pennsylvania 19530, USA (germplasm, agronomy)

Steven R. King, USA (ethnobotany) (see Achira)

Peter Kulakow, The Land Institute, Salina, Kansas 67401, USA (amaranth)

Chi Won Lee, New Crops Section, Department of Horticulture, Colorado State University, Ft. Collins, Colorado 80523, USA (halophytes)

James Lehmann, Department of Agronomy, Iowa State University, Ames, Iowa 50011, USA (amaranth)

Jess R. Martineau, USA (biotechnology) (see Mashua)

Aníbal B. Martínez Muñoz, Facultad de Agronomía, Universidad de San Carlos, Apartado 1545, Guatemala City, Guatemala

Gary Nabhan, Desert Botanical Gardens, 1202 North Galvin Parkway, Phoenix, Arizona 85008, USA

Lenis Nelson, Institute of Agriculture and Natural Resources, University of Nebraska, Lincoln, Nebraska 68583, USA (amaranth)

John Palmer, New Zealand (amaranth) (see Ahipa)

Tej Partap, Nepal (amaranths) (see Maca)

Michael N. Pearson, New Zealand (amaranth) (see Mashua)

Birthe Pedersen, Department of Biotechnology, Carlsberg Research Laboratory, 10 Gamle Carlsbergvej, Valby DK-2500, Denmark (nutrition)

Martin L. Price, USA (limited germplasm available) (see Achira)

Daniel H. Putnam, Center for Alternative Crops and Products, University of Minnesota, 305 Alderman Hall, St. Paul, Minnesota 55108, USA (amaranths)

Mickey Reed, Arizona State University, Tempe, Arizona 85287, USA (salinity)

Robert Reid, Italy (see Kaniwa)

Charles M. Rick, USA (see Arracacha)

Robert Robinson, Department of Agronomy and Plant Genetics, University of Minnesota, Borlaug Hall, 1991 Buford Circle, St. Paul, Minnesota 55108, USA (amaranth)

David J. Sammons, USA (amaranth) (see Potatoes)

Alfredo Sánchez-Marroquin, Mexico (amaranths) (see Kaniwa)

Jonathan Sauer, Department of Geography, University of California, Los Angeles, California 90024, USA

Robin Saunders, USA (processing starches) (see Achira)

Jurgeon Schaeffer, Department of Plant and Soil Science, Montana State University, Bozeman, Montana 59717, USA (amaranth)

Al Schneiter, Department of Agronomy, North Dakota State University, Fargo, North Dakota 58105, USA (amaranth)

Shao Qiquan, China (see Potatoes)

Forest Shomer, USA (see Kaniwa)

William Stegmeier, Experiment Station, Fort Hays State University, Hays, Kansas 67601, USA (amaranth)

Guadalupe Suárez Ramos, ITESM Unidad Querétaro, Apartado Postal 37, Querétaro, Querétaro, Mexico (amaranth)

Sun Hongliang, Institute of Crop Breeding and Cultivation, Chinese Academy of Agricultural Sciences (CAAS), 30 Baishiqiao Road, West Suburbs, Beijing 100081, China (amaranths)

W.R. Sykes, New Zealand (amaranth) (see Mashua)

Antonio Trinidad Santos, Centro de Edafología, Sección Fertilidad de Suelos, Institución Enseñanza e Investigación en Ciencias Agrícolas, Colegio de Postgraduados, Montecillos, Chapingo, Estado de México 56230, Mexico (amaranths)

Alejo von der Pahlen, Italy (amaranths) (see Oca)
Leon W. Weber, Amaranth Project, Rodale Research Center, Route 1, Box 323, Kutztown, Pennsylvania 19530, USA (germplasm, agronomy)
Rebecca Theurer Wood, USA (see Kaniwa)
R.B. Wynn-Williams, New Zealand (amaranth) (see Oca)
Yue Shaoxian, Institute of Crop Breeding and Cultivation, Chinese Academy of Agricultural Sciences (CAAS), 30 Baishiqiao Road, West Suburbs, Beijing 100081, China (amaranths)

QUINOA

Andean Region

Segundo Alandia, Bolivia (pathology) (see Mashua)
Jaime Alba Aldunate, Bolivia (see Oca)
Aquilino Alvarez, Programa de Quinua, Universidad Nacional de San Antonio Abad del Cusco (UNSAAC), Apartado 1006, K'ayra, Cusco, Peru
Milton Alvarez B., Latinreco S.A., Centro Nestlé, Casilla Postal 6053–CCI, Quito, Cumbaya, Ecuador
Jeanine Anderson, Asociación Perú-Mujer, Avenida Arica 755–B, Miraflores, Lima, Peru
Antonio Bacigalupo, Chile (see Potatoes)
Fernando N. Barrantes Del Aguila, Peru (germplasm, pathology) (see Achira)
Alejandro Bonifacio F., Estación Experimental Patacamaya, Instituto Boliviano de Tecnología Agropecuaria (IBTA), Casilla Postal 5783, La Paz, Bolivia (genetics)
Hernán Caballero, Instituto Interamericano de Cooperación para la Agricultura (IICA), Apartado 201–A, Quito, Ecuador
Augusto Cardich, Argentina (geography and climate) (see Achira)
Angel Cari Choquehuanca, Instituto Nacional de Investigación Agraria y Agroindustrial (INIAA), Puno, Peru
Raúl O. Castillo T., Ecuador (see Ahipa)
Andres Contreras M., Chile (germplasm) (see Oca)
Daniel G. Debouck, Colombia (see Kiwicha)
Anibal del Carpio Farfán, Peru (see Mashua)
William Edwardson, Oficina Regional para la América Latina y el Caribe, International Development Research Centre (IDRC), Apartado Aéreo 53016, Bogotá, Colombia
José Egüez, Instituto Nacional de Investigaciones Agropecuarias (INIAP), Estación Experimental del Austro, Apartado 554, Cuenca, Ecuador
Fondo Nacional de Investigaciones Agropecuarias (FONAIAP), Venezuela (see Achira)
Santiago D. Franco Pebe, Peru (germplasm) (see Arracacha)
Humberto Gandarillas, Bolivia (see Oca)
Juan Gaviria R., Venezuela (see Ulluco)
Lionel O. Giusti, Argentina (systematics) (see Kaniwa)
Luz Gomez, Peru (see Kaniwa)
Emilio González González, c/o División Agrícola Carozzi, AGROZZI, Camino Longitudinal Sur 5201, Km. 23, San Bernardo, Santiago, Chile
Ruperto Hidalgo A., Hospital Cantonal de Colta, Colta, Chimborazo, Ecuador (child nutrition)
Instituto Interamericano de Cooperación para la Agricultura (IICA), Bolivia (see Mashua)
Instituto Nacional de Investigaciones Agropecuarias (INIAP), Ecuador (see Achira)
Ingo Junge Rodewald, Chile (desaponization, processing, and nutrition) (see Potatoes)
Michael J. Koziol, Ecuador (see Kiwicha)
Johann Krug, Andean Crops for Human Nutrition, Convenio Perú-Alemania para Cultivos Andinos-GTZ (CORDE), Avenida Sol 817, Casilla Postal 807, Cusco, Peru ("life-support" crops)
Latinreco S.A., Ecuador (see Arracacha)
Craig Leon, Grupo Morisaenz, Casilla 625, Quito, Ecuador (indigenous agriculture)
Maximo Libermann Cruz, Bolivia (salt tolerance) (see Kaniwa)
Cipriano Mantari Camargo, Peru (see Maca)

Mario Alfredo Marcial Grijalva, Ecuador (nutritional analysis) (see Kiwicha)
Ulrich Mohr, Peru (see Mashua)
David Morales V., Bolivia (see Mashua)
Angel Mujica, Peru (see Kaniwa)
Eloy Munive Jáuregui, Peru (see Maca)
Gladys Navas de Alvarado, Ecuador (nutritional analysis) (see Kiwicha)
Carlos Nieto C., Ecuador (see Achira)
José Ochoa L., Programa de Cultivos Andinos, Instituto Nacional de Investigaciones
    Agropecuarias (INIAP), Estación Experimental "Santa Catalina," Casilla Postal 340, Quito,
    Ecuador
Victor Otazú Monzón, Peru (pathology) (see Potatoes)
Maria de Lourdes Peñaloza Izurieta, Ecuador (nutritional analysis) (see Kaniwa)
Eduardo Peralta I., Ecuador (see Mashua)
A. Reggiardo, Departamento de Tecnología y Productos Agropecuarios (TAPA), Universidad
    Nacional Agraria (UNA), Apartado 456, La Molina, Lima 100, Peru (processing)
Raúl Ríos E., Bolivia (see Achira)
Juan Risi Carbone, Colombia (see Achira)
Carlos Roersch, Peru (see Mashua)
Emilio Rojas Mendoza, Peru (see Maca)
Jose Fernando Romero Cañizares, Ecuador (see Arracacha)
Steve Sacks, Grupo Morisaenz, Casilla 625, Quito, Ecuador (landraces)
Basilio Salas, USA (see Mashua)
Raúl Santana Paucar, Peru (see Mashua)
Juan Seminario Cunya, Peru (see Mashua)
Juan Solano Lazo, Ecuador (see Achira)
Mario E. Tapia Núñez, Peru (see Mauka)
Lauro Toribio Baltazar, Peru (see Maca)
Julio Valle, Ecuador (soil science) (see Mashua)
Cesar C. Vargas Calderon, Peru (see Potatoes)
César Vásconez Sevilla, Ecuador (nutritional analysis) (see Kiwicha)
Carlos Adolfo Vimos Naranjo, Ecuador (see Arracacha)
Erik von Baer von Lochow, Estación Gorbea, Campex - Semillas Baer, Casilla 87, Temuco,
    Chile
Samuel von Rütte, Ecuador (see Kiwicha)
José Zaporta R., Hospital Cantonal de Colta, Colta, Chimborazo, Ecuador (child nutrition)

Other Countries

Sergey M. Alexanyan, N.I. Vavilov Institute of Plant Industry, Ulitsa Gertsena 44, Leningrad
    190 000, USSR
Arrowhead Mills, Inc., USA (processing) (see Kiwicha)
Ariel Azael, Haiti (see Ahipa)
Emigdio Ballón, USA (see Kaniwa)
Robert Becker, USA (milling) (see Kiwicha)
Nate Bower, Department of Chemistry, Colorado College, Colorado Springs, Colorado
    80903, USA
Ricardo Bressani, Guatemala (nutrition) (see Achira)
William M. Brown, USA (see Oca)
Rolf Carlsson, Sweden (leaf protein, plant chemistry, and nutrition) (see Kaniwa)
Miguel Carmen, Finland (see Kaniwa)
George Clark, Box 163, Rossburn, Manitoba R0J 1V0, Canada (grower)
D. J. Cotter, USA (see Ahipa)
Lynton J. Cox, Central Laboratory, Nestlé Products Technical Assistance Company, Avenue
    Nestlé 55, Vevey CH-1800, Switzerland
Ewen Coxworth, Biomass Resources, Saskatchewan Research Council, 15 Innovation
    Boulevard, Saskatoon, Saskatchewan S7N 2X8, Canada (salt tolerance)
Deutsche Gesellschaft für Technische Zusammenarbeit (GTZ), West German (see Mashua)
Eden Foods, Inc., USA (importation, processing, and distribution) (see Kaniwa)

W. Hardy Eshbaugh, USA (see Achira)
Jorge D. Etchevers B., Mexico (see Kaniwa)
Michel Fanton, Australia (see Achira)
Erich W. Forster, Suite 490, 109 Minna Street, San Francisco, California 94105, USA (desaponization)
Christine Franquemont, USA (Quechua plant use) (see Ahipa)
N.W. Galwey, UK (see Kaniwa)
Good Seed, PO Box 702, Tonasket, Washington 98855, USA
Stephen L. Gorad, USA (export/import) (see Kaniwa)
B.W.W. Grout, UK (see Achira)
V.L. Guzman, Everglades Research and Education Center - Belle Glade, Institute of Food and Agricultural Sciences, University of Florida, PO Box 8003, Belle Glade, Florida 33430, USA
Barry Hammel, USA (see Kaniwa)
Ray Henkel, USA (pioneer colonization) (see Achira)
Sandra Hernández, 165 East 112th #8A, New York, New York 10029, USA
Karl Herz, Food and Agriculture Organization of the United Nations (FAO), Via delle Terme di Caracalla, Rome 00100, Italy
Gretchen Hofmann, The Windstar Foundation, PO Box 178, Snowmass, Colorado 81654, USA
Robert W. Hoopes, USA (genetics, breeding, and pathology) (see Oca)
J.L. Hudson, USA (see Kiwicha)
International Board for Plant Genetic Resources (IBPGR), Italy (germplasm information) (see Achira)
International Development Research Centre (IDRC), International Development Research Centre (IDRC), PO Box 8500, Ottawa, Ontario K1G 3H9, Canada
Jesse Jaynes, USA (gene splicing) (see Ahipa)
C. Antonio Jiménez Aparicio, Mexico (see Arracacha)
Duane Johnson, USA (production, processing, and distribution) (see Kaniwa)
Alan M. Kapuler, USA (see Achira)
Jack A. Kernan, Biomass Resources, Saskatchewan Research Council, 15 Innovation Boulevard, Saskatoon, Saskatchewan S7N 2X8, Canada (salt tolerance)
Steven R. King, USA (ethnobotany) (see Achira)
Vladamir Krivchenko, N.I. Vavilov Institute of Plant Industry, Ulitsa Gertsena 44, Leningrad 190 000, USSR
Chi Won Lee, USA (halophytes) (see Kiwicha)
John McCamant, USA (see Kaniwa)
Cyrus McKell, USA (see Potatoes)
Jerry L. McLaughlin, USA (toxicology) (see Kaniwa)
Bruce Macler, Mail Stop 239–4, National Aeronautics and Space Administration (NASA), Ames Research Center, Moffett Field, California 94035, USA (biochemistry of plant growth in closed systems)
Malachite Small Farms, Gardner, Colorado 81040, USA
Jess R. Martineau, USA (biotechnology) (see Mashua)
Cesar Morales, USA (importation to U.S.) (see Oca)
Kapiton V. Novozhilov, Pushkin Institute of Plant Protection, Shosse Podbel'skogo 3, Leningrad 188 620, USSR
John Palmer, New Zealand (see Ahipa)
Tej Partap, Nepal (mountain agriculture, genetic resources) (see Maca)
Alan M. Paton, Division of Bacteriology, Department of Agriculture, Aberdeen University, 581 King Street, Aberdeen AB9 1UD, UK (fusion of nitrogen fixation genes)
Martin L. Price, USA (limited germplasm available) (see Achira)
Daniel H. Putnam, USA (see Kiwicha)
F.M. Quin, Netherlands (see Mashua)
Quinoa Corporation, 24248 Crenshaw Boulevard, PO Box 1039, Torrance, California 90505, USA (importation, processing, and distribution)
Jerry Rabe, Pillsbury Company, 311 Second Street Southeast, Minneapolis, Minnesota 55414, USA

Robert Reid, Italy (see Kaniwa)
Alfredo Sanchez-Marroquin, Mexico (see Kaniwa)
Robin Saunders, USA (processing starches) (see Achira)
Elizabeth Schneider, USA (food preparation) (see Yacon)
Richard E. Schultes, USA (ethnobotany) (see Achira)
Forest Shomer, USA (see Kaniwa)
N.W. Simmonds, UK (see Potatoes)
Richard Storey, Department of Biology, Colorado College, Colorado Springs, Colorado 80903, USA
W.R. Sykes, New Zealand (see Mashua)
Rick Torwalt, Leroy, Saskatchewan S0K 2P0, Canada (grower)
R. Uotila, Finland (herbarium, taxonomic review) (see Kaniwa)
Richard Valley, USA (grower) (see Oca)
Mary van Buren, P.O. Box 162, Cutchogue, New York 11935, USA
Louis J.M. van Soest, Netherlands (germplasm) (see Potatoes)
Guillermo Veliz, USA (importation to U.S.) (see Oca)
Hugh D. Wilson, USA (systematics) (see Kaniwa)
Rebecca Theurer Wood, USA (see Kaniwa)
R.B. Wynn-Williams, New Zealand (see Oca)
Karl Zimmerer, USA (see Potatoes)

BASUL

Andean Region

L.E. Acero Duarte, Apartado Aéreo 55332, Bogotá, Colombia
N. Barrera Marin, Ciencias Agropecuarias, Universidad Nacional, Apartado 237, Palmira, Colombia
Cecilia de Martinez, Laboratorio de Bioquímica, Departamento de Química, Universidad Nacional de Colombia, Bogotá, Colombia
Daniel G. Debouck, Colombia (see Kiwicha)
Estela Díaz, Laboratorio de Bioquímica, Departamento de Química, Universidad Nacional de Colombia, Bototá, Colombia
Marc Dourojeanni, Universidad Nacional Agraria (UNA), Centro de Datos Para La Conservación, Apartado 456, La Molina, Lima 100, Peru
Jaime Gaete Calderon, Región Metropolitana, Corporación Nacional Forestal (CONAF), Eliodora Yáñez 1810, Santiago, Chile
Instituto Nacional de Investigaciones Agropecuarias (INIAP), Ecuador (see Achira)
Shirley Keel, Universidad Nacional Agraria (UNA), Centro de Datos Para La Conservación, Apartado 456, La Molina, Lima 100, Peru (botany)
Alexander M. Mueller, Colombia (see Kiwicha)
C. Percy Nuñez Vargas, Proyecto Flora del Perú, Universidad Nacional de San Antonio Abad del Cusco (UNSAAC), Umanchata 136, Cusco, Peru
Isaac C. Peralta Vargas, Departamento Academia de Ciencias Biológicas, Universidad Nacional de San Antonio Abad del Cusco (UNSAAC), Apartado Corea 162, Cusco, Peru
Gerardo Pérez, Laboratorio de Bioquímica, Departamento de Química, Universidad Nacional de Colombia, Bogotá, Colombia
Manuel Ríos, Departmento de Manejo Forestal, Universidad Nacional Agraria (UNA), Apartado 456, La Molina, Lima 100, Peru (forestry)
José Gabriel Sánchez Vega, Los Sauces No. 207, Apartado 53, Cajamarca, Peru
Cesar C. Vargas Calderón, Peru (see Potatoes)

Other Countries

Rupert C. Barneby, New York Botanical Garden, Bronx, New York 10458-9980, USA
John Beer, Centro Agronómica Tropical de Investigación y Enseñanza (CATIE), Turrialba 7170, Costa Rica

Gilles Bourgeois, 2181 Navaho #608, Nepeau, Ottawa, Ontario K2C 3K3, Canada
James L. Brewbaker, Nitrogen Fixing Tree Association (NFTA), PO Box 680, Waimanalo, Hawaii 96795, USA
Leslie Brownrigg, USA (indigenous systems) (see Achira)
Curt Brubaker, Department of Botany, Iowa State University, Ames, Iowa 50011, USA (leaflet anatomy)
Centroamericana de Semillas, Honduras (see Kiwicha)
Nancy Glover, Nitrogen Fixing Tree Association (NFTA), PO Box 680, Waimanalo, Hawaii 96795, USA
Ralph J. Hervey, Departamento de Recursos Naturales Renovables Proyecto: Erythrina, Centro Agronómica Tropical de Investigación y Enseñanza (CATIE), Turrialba 7170, Costa Rica
Richard McCain, USA (horticulture) (see Yacon)
David Neill, Missouri Botanical Garden, c/o USAID/Quito, U.S. Agency for International Development (AID), Washington, DC 20523, USA
Charles Peters, USA (see Oca)
Michael Pilarski, Friends of the Trees, PO Box 1466, Chelan, Washington 98816, USA (reforestation and tree crops)
Antonio M. Pinchinat, Instituto Interamericano de Cooperación para la Agricultura (IICA), PO Box 1223, Castries, St. Lucia
Peter H. Raven, Missouri Botanical Garden, PO Box 299, St. Louis, Missouri 63166–0299, USA
Chris Rollins, USA (see Achira)
Ricardo O. Russo, School of Forestry and Environmental Studies, Yale University, 370 Prospect Street, New Haven, Connecticut 06511, USA
Pedro A. Sánchez, Tropical Soils Research Program, Department of Soil Science, North Carolina State University, Raleigh, North Carolina 27695, USA (integrated farming in humid tropics)
Eduardo C. Schröder, Department of Agronomy and Soils, College of Agricultural Sciences, University of Puerto Rico, Recinto Universitario de Mayagüez, PO Box 5000, Mayagüez, Puerto Rico 00708–5000, USA (rhizobia)
Richard E. Schultes, USA (ethnobotany) (see Achira)
James L. Zarucchi, Missouri Botanical Garden, PO Box 299, St. Louis, Missouri 63166–0299, USA

NUÑAS

Andean Region

Hernán Caballero, Ecuador (see Quinoa)
Augusto Cardich, Argentina (geography and climate) (see Achira)
Alfonso Cerrate Valenzuela, Universidad Nacional Agraria (UNA), Apartado 456, La Molina, Lima 100, Peru
Andres Contreras M., Chile (germplasm) (see Oca)
Jeremy Davis, c/o Centro Internacional de Agricultura Tropical (CIAT), Apartado Aéreo 6713, Cali, Colombia
Daniel G. Debouck, Colombia (see Kiwicha)
Guillermo Gálvez, Programa de Frijoles Andinas, Centro Internacional de Agricultura Tropical (CIAT), Apartado 14–0185, Lima 14, Peru
Miguel Holle, Peru (see Arracacha)
Instituto Nacional de Investigaciones Agropecuarias (INIAP), Ecuador (see Achira)
Julia Kornegay, Programa de Frijoles, Centro Internacional de Agricultura Tropical (CIAT), Apartado Aéreo 6713, Cali, Colombia
Maria del Carmen Menendez Sevillano, Facultad de Agronomía, Cátedra de Botánica, Universidad de Buenos Aires, Avenida San Martín 4453, Buenos Aires 1417, Buenos Aires, Argentina

Raúl Ríos E., Bolivia (see Achira)
Juan Risi Carbone, Colombia (cropping systems) (see Achira)
Jaime Rojas, Corporación Ambiente y Desarrollo (AMDE), PO Box 632, Ambato, Ecuador
José Gabriel Sánchez Vega, Peru (see Basul)
Raúl Santana Paucar, Peru (see Mashua)
Juan Seminario Cunya, Peru (see Mashua)
Rut Margarita Solari, Instituto de Genética, Instituto Nacional de Tecnología Agropecuaria (INTA), Centro de Investigaciones de Recursos Naturales, Castelar 1712, Buenos Aires, Argentina
R. Sotomayor, c/o Universidad Nacional Agraria (UNA), Apartado 456, La Molina, Lima 100, Peru
Joseph Tohme, Centro Internacional de Agricultura Tropical (CIAT), Apartado Aéreo 6713, Cali, Colombia
L. Trugo, Instituto de Química, Universidade Federal do Rio de Janeiro, C.T. Bloco A 5 Andar, Ilha do Fundão, Rio de Janeiro, RJ, Brazil (nutrition)
Jeffrey White, Colombia (see Mashua)

Other Countries

Fred A. Bliss, Department of Pomology, University of California, Davis, California 95616, USA
Ricardo Bressani, Guatemala (nutrition) (see Achira)
Felix Carmarena, Faculté des Sciences Agronomiques de L'État, Gembloux 5800, Belgium (Phaseolus)
Alfonso Delgado Salinas, Herbario Nacional, Instituto de Biología, Universidad Nacional Autónoma de México (UNMA), Apartado Postal 70–367, México 04510, D.F., Mexico
Abebe Demissie, Ethiopia (Phaseolus) (see Kiwicha)
Craig Dremann, USA (see Potatoes)
Jan Engels, West Germany (see Potatoes)
W. Hardy Eshbaugh, USA (see Achira)
Sharon Fleming, Department of Nutritional Sciences, University of California, Berkeley, California 94720, USA
Christine Franquemont, USA (Quechua plant use) (see Ahipa)
Zaccheaus Oyesiju Gbile, Nigeria (Phaseolus) (see Potatoes)
Paul Gepts, Department of Agronomy, University of California, Davis, California 95616, USA
B.W.W. Grout, UK (see Achira)
Leland W. Hudson, Regional Plant Introduction Station, U.S. Department of Agriculture (USDA), Washington State University, Pullman, Washington 99163, USA
International Board for Plant Genetic Resources (IBPGR), Italy (germplasm information) (see Achira)
C. Antonio Jiménez Aparicio, Mexico (see Arracacha)
Lawrence Kaplan, Department of Biology, University of Massachusetts, Harbor 15 Campus, Boston, Massachusetts 02125, USA (archeobotany)
Charles Kauffman, USA (germplasm) (see Kiwicha)
Steven R. King, USA (ethnobotany) (see Achira)
Cal F. Konzak, Department of Agronomy and Soils, Washington State University, Pullman, Washington 99164–6420, USA
Janet Long-Solís, Apartado Postal 41 593, México 11000, D.F., Mexico
Carol Mackey, Department of Anthropology, California State University, 18111 Nordhoff Street, Northridge, California 91330, USA
Jess R. Martineau, USA (biotechnology) (see Mashua)
Tej Partap, Nepal (mountain agriculture, genetic resources) (see Maca)
Hugh Popenoe, USA (see Arracacha)
Martin L. Price, USA (see Achira)
L. Quagliotti, Institute of Plant Breeding and Seed Production, Via Giuria 15, Turin, I-10126, Italy
Jürgen Reckin, East Germany (see Potatoes)

Shao Qiquan, China (Phaseolus) (see Potatoes)
Stephen Spaeth, Grain Legume Genetic and Physiological Research, Agricultural Research
    Service (ARS), U.S. Department of Agriculture (USDA), 215 Johnson Hall, Washington
    State University, Pullman, Washington 99164–6421, USA
Karl Zimmerer, USA (see Potatoes)

TARWI

Andean Region

Segundo Alandia, Bolivia (pathology) (see Mashua)
Jaime Alba Aldunate, Bolivia (see Oca)
Fernando N. Barrantes Del Aguila, Peru (germplasm, pathology) (see Achira)
Oscar Blanco Galdos, Peru (see Mashua)
Heinz Brücher, Argentina (useful plants) (see Ahipa)
Augusto Cardich, Argentina (geography and climate) (see Achira)
Alfonso Cerrate Valenzuela, Peru (see Nuñas)
Andres Contreras M., Chile (germplasm) (see Oca)
Daniel G. Debouck, Colombia (see Kiwicha)
Fondo Nacional de Investigaciones Agropecuarias (FONAIAP), Venezuela (see Achira)
Santiago D. Franco Pebe, Peru (germplasm) (see Arracacha)
Alexander Grobman, Peru (hybrid seed, germplasm collection) (see Oca)
Instituto Interamericano de Cooperación para la Agricultura (IICA), Bolivia (see Mashua)
Instituto Nacional de Investigaciones Agropecuarias (INIAP), Ecuador (see Achira)
Ingo Junge Rodewald, Chile (processing, nutrition) (see Potatoes)
Michael J. Koziol, Ecuador (see Kiwicha)
Johann Krug, Peru ("life-support" crops) (see Quinoa)
Latinreco S.A., Ecuador (see Arracacha)
Lúis E. López Jaramilló, Colombia (see Potatoes)
Martha Beatriz Meléndez Ibarra, Facultad de Ciencia e Ingeniería en Alimentos, Universidad
    Técnica de Ambato, Avenida Colombia y Chile s/n, Casilla Postal 334, Ambato, Ecuador
    (nutritional analysis)
Ulrich Mohr, Peru (see Mashua)
Reinaldo Monteiro, Brazil (Brazilian lupins) (see Arracacha)
David Morales V., Bolivia (see Mashua)
Angel Mujica, Peru (see Kaniwa)
Gladys Navas de Alvarado, Ecuador (nutritional analysis) (see Kiwicha)
Carlos Nieto C., Ecuador (see Achira)
Victor Otazú Monzón, Peru (pathology) (see Potatoes)
Julio Rea, Bolivia (indigenous food plants) (see Mauka)
Raúl Ríos E., Bolivia (see Achira)
Juan Risi Carbone, Colombia (cropping systems) (see Achira)
Jose Fernando Romero Cañizares, Ecuador (see Arracacha)
Basilio Salas, USA (see Mashua)
José Gabriel Sánchez Vega, Peru (see Basul)
Patricia Santamaría Freire, Facultad de Ciencia e Ingeniería en Alimentos, Universidad
    Técnica de Ambato, Avenida Colombia y Chile s/n, Casilla Postal 334, Ambato, Ecuador
    (nutritional analysis)
Raúl Santana Paucar, Peru (see Mashua)
Juan Seminario Cunya, Peru (see Mashua)
Juan Solano Lazo, Ecuador (see Achira)
Mario E. Tapia Núñez, Peru (see Mauka)
Lauro Toribio Baltazar, Peru (see Maca)
L. Trugo, Brazil (nutrition) (see Nuñas)
Cesar C. Vargas Calderon, Peru (see Potatoes)

Ramiro Velastegui, Corporación Ambiente y Desarrollo (AMDE), PO Box 632, Ambato, Ecuador (pathology)
Dietrich von Baer, Universidad de Concepción, Casilla 237, Concepción, Chile
Erik von Baer von Lochow, Chile (see Quinoa)

Other Countries

Rudolf F.W. Binsack, Consultative Group on International Agricultural Research, CGIAR, The World Bank, 1818 H Street, NW, Washington, DC 20433, USA (plant production)
Lynn A. Bohs, USA (see Arracacha)
Gilles Bourgeois, Canada (see Basul)
Ricardo Bressani, Guatemala (nutrition) (see Achira)
William M. Brown, USA (see Oca)
Eduardo Busquets, Section 211, Deutsche Gesellschaft für Technische Zusammenarbeit (GTZ), 1 bei Frankfurt/Main, Dag-Hammarskjöld-weg 1, Postfach 5180, Eschborn D-6236, West Germany
Daniel Cohen, PO Box 401, Davis, California 95616, USA
Manfred Dambroth, West Germany (see Potatoes)
Deutsche Gesellschaft für Technische Zusammenarbeit (GTZ), West Germany (see Mashua)
Christine Franquemont, USA (Quechua plant use) (see Ahipa)
Daniel W. Gade, USA (see Achira)
R. Gross, Section 412, Deutsche Gesellschaft für Technische Zusammenarbeit (GTZ), 1 bei Frankfurt/Main, Dag-Hammarskjöld-weg 1, Postfach 5180, Eschborn D-6236, West Germany
B.W.W. Grout, UK (see Achira)
Chaia Clara Heyn, Department of Botany, Institute of Life Sciences, Hebrew University of Jerusalem, Givat Ram, Mt. Scopus, Jerusalem 91904, Israel
G.D. Hill, University College of Agriculture, Lincoln College, Canterbury, New Zealand
J.L. Hudson, USA (see Kiwicha)
Ron Hurov, USA (see Mashua)
International Board for Plant Genetic Resources (IBPGR), Italy (germplasm information) (see Achira)
C. Antonio Jiménez Aparicio, Mexico (see Arracacha)
Tadeusz Kazimierski, Institute of Plant Genetics, Polish Academy of Sciences, Strzeszynska 34, Poznan 60–479, Poland
Friedhelm Koch, Department IC-ATAV, Degussa AG, ZN Wolfgang, PO Box 1345, Hanau D-6450, West Germany (nutrition)
F.B. Lucas, Facultad de Química, Universidad Nacional Autónoma de México (UNMA), Apartado Postal 70–367, Mexico 04510, D.F., México
Barbara Lynch, 115 Kelvin Place, Ithaca, New York 14850, USA
R. Marquard, Agrarwissenschaften, Institut für Pflanzenbau und Pflanzenzüchtung, Ludwigstrasse 23, Giessen D-2300, West Germany (nutrition)
Jess R. Martineau, USA (biotechnology) (see Mashua)
John Palmer, New Zealand (see Ahipa)
Tej Partap, Nepal (mountain agriculture, genetic resources) (see Maca)
John Pate, Department of Botany, University of Western Australia, Nedlands, Western Australia 6009, Australia (physiology)
Uzi Plitman, Department of Botany, Hebrew University of Jerusalem, Givat Ram, Mt. Scopus, Jerusalem 91904, Israel (Lupinus)
Martin L. Price, USA (limited germplasm available) (see Achira)
Lech Ratajczak, Department of Biology, University of Poznan, Fredry 10, Poznan 61–701, Poland
Peter Roemer, Institut für Pflanzenbau und Pflanzenzüchtung I, Justus-Liebig-Universität Giessen, Fachbereich 17, Ludwigstrasse 23, Giessen D-6300, West Germany (breeding)
Robin Saunders, USA (processing starches) (see Achira)
Shao Qiquan, China (see Potatoes)
Forest Shomer, USA (see Kaniwa)
Richard Storey, USA (see Quinoa)

Guillermo Veliz, USA (importation to U.S.) (see Oca)
Alejo von der Pahlen, Italy (see Oca)
M. Wink, Institut für Pharmazie, Johannes Gutenberg Universität, Saarstrasse 21, Postfact
    3980, Mainz 1, D-6500 West Germany (nutrition)
Wanda Wojciechowska, Institute of Plant Genetics, Polish Academy of Sciences, Strzeszynska
    34, Poznan 60–479, Poland (pollination)

PEPPERS

Andean Region

Cesar Barberan, Corporación Ambiente y Desarrollo (AMDE), PO Box 632, Ambato, Ecuador
Ricardo Bellón, Facultad de Agronomía, Cátedra de Botánica, Universidad de Buenos Aires,
    Avenida San Martín 4453, Buenos Aires 1417, Buenos Aires, Argentina
Augusto Cardich, Argentina (geography and climate) (see Achira)
Andres Contreras M., Chile (germplasm) (see Oca)
Mario Crespo, Centro de Investigaciones Fitotécnicas Pairumani, Casilla 3861, Cochabamba,
    Bolivia
Carmen E.B. de Rojas, Facultad de Agronomía, Universidad Central de Venezuela, Apartado
    4579, Maracay, Estado Aragua 2101, Venezuela
Francisco Delgado de la Flor B., Programa de Investigación en Hortalizas, Universidad
    Nacional Agraria (UNA), Apartado 456, La Molina, Lima 100, Peru (germplasm
    information)
José Díaz Flores, Universidad Nacional de San Cristóbal de Huamanga, Apartado 220,
    Ayacucho, Peru
Fondo Nacional de Investigaciones Agropecuarias (FONAIAP), Venezuela (see Achira)
Juan Gaviria R., Venezuela (see Ulluco)
Alexander Grobman, Peru (hybrid seed, germplasm collection) (see Oca)
Miguel Holle, Peru (see Arracacha)
Instituto Interamericano de Cooperación para la Agricultura (IICA), Bolivia (see Mashua)
Instituto Nacional de Investigaciones Agropecuarias (INIAP), Ecuador (see Achira)
Juan Jaramilló Vasquez, Sección Hortalizas, Instituto Colombiano Agropecuario (ICA),
    Centro Nacional de Investigación (CNI), Apartado Aéreo 233, Palmira, Valle, Colombia
    (germplasm)
Mario Lobo, Sección Hortalizas, Vegetable Crops Program, Instituto Colombiano Agropecuario
    (ICA), Apartado Aéreo 100, Rio Negro, Antioquia, Colombia
Lúis E. López Jaramilló, Colombia (Solanaceae) (see Potatoes)
Raúl Ríos E., Bolivia (see Achira)
Juan Risi Carbone, Colombia (cropping systems) (see Achira)
Raul Salazar C., Sección Frutales, Instituto Colombiano Agropecuario (ICA), Apartado Aéreo
    233, Palmira, Valle, Colombia
Raúl Santana Paucar, Peru (see Mashua)
Cesar C. Vargas Calderon, Peru (see Potatoes)
Ramiro Velastegui, Ecuador (pathology) (see Tarwi)

Other Countries

Jórge Arce, Unidad de Recursos Genéticos, Centro Agronómica Tropical de Investigación
    y Enseñanza (CATIE), Turrialba 7170, Costa Rica (germplasm)
Ariel Azael, Haiti (see Ahipa)
Lynn A. Bohs, USA (see Arracacha)
Paul Bosland, Department of Agronomy and Horticulture, New Mexico State University, Las
    Cruces, New Mexico 88003, USA
Ricardo Bressani, Guatemala (nutrition) (see Achira)

H.H. Bryan, Institute of Food and Agricultural Sciences (IFAS), University of Florida, Tropical Research and Education Center (TREC), Homestead, Florida 33031, USA

D.J. Cotter, USA (see Ahipa)

Gabor Csilleery, Research Institute for Vegetable Crops, Station Budateeteeny, PO Box 95, Budapest, Hungary (hybridization and mutation, germplasm)

W.G. D'Arcy, USA (see Potatoes)

R. Dumas de Vaulx, Station d'Amélioration des Plantes Maraîchères, Centre de Recherches Agronomiques d'Avignon, Institut National de la Recherche Agronomique (INRA), 84140 Montfavet, Avignon, France

Abebe Demissie, Ethiopia (see Kiwicha)

William Doty, Indoor Citrus and Rare Fruit Society, 176 Coronado Avenue, Los Altos, California 94022, USA (germplasm)

Craig Dremann, USA (see Potatoes)

Jan Engels, West Germany (see Potatoes)

W. Hardy Eshbaugh, USA (taxonomy) (see Achira)

Jorge A. Galindo, Mexico (see Potatoes)

Zaccheaus Oyesiju Gbile, Nigeria (see Potatoes)

Heiner Goldbach, West Germany (seed storage and germination) (see Potatoes)

V.L. Guzman, USA (see Quinoa)

Charles B. Heiser, Jr., Department of Biology, Jordan Hall 138, Indiana University, Bloomington, Indiana 47405, USA

J.L. Hudson, USA (see Kiwicha)

International Board for Plant Genetic Resources (IBPGR), Italy (germplasm information) (see Achira)

C. Antonio Jiménez Aparicio, Mexico (see Arracacha)

Alan M. Kapuler, USA (see Achira)

C. Todd Kennedy, 452 South Westminster, Los Angeles, California 90020, USA

José A. Laborde Cancino, Centro de Investigaciones Agrícolas de El Bajío (CIAB), Unidad de Recursos Genéticos, Instituto Nacional de Investigaciones Agrícolas (INIA), Apartado Postal 112, Celaya, Guanajuato, Mexico

Jorge León, Costa Rica (see Maca)

W.H. Lindhout, Institute for Horticultural Plant Breeding, Instituut voor de Veredeling van Tuinbouwgewassen (IVT), Postbus 16, Wageningen NL-6700 A, Netherlands

Janet Long-Solís, Mexico (see Nuñas)

Cyrus McKell, USA (see Potatoes)

Cesar Morales, USA (importation to U.S.) (see Oca)

Gary Nabhan, USA (see Kiwicha)

Roy M. Nakayama, Department of Agronomy and Horticulture, New Mexico State University, Las Cruces, New Mexico 88003, USA

Yasuo Ohta, 608–102, Takezono 3 chome, Tsukuba, Ibaraki-ken 305, Japan

Barbara Pickersgill, Department of Agricultural Botany, University of Reading, Whiteknights, Reading RG6 2AS, UK

E. Pochard, Station d'Amélioration des Plantes Maraîchères, Centre de Recherches Agronomiques d'Avignon, Institut National de la Recherche Agronomique (INRA), 84140 Montfavet, Avignon, France

L. Quagliotti, Italy (see Nuñas)

Rare Fruit Council International, Inc., 13609 Old Cutler Road, Miami, Florida 33158, USA

Chris Rollins, USA (see Achira)

Elizabeth Schneider, USA (food preparation) (see Yacon)

Shao Qiquan, China (see Potatoes)

Paul G. Smith, c/o Department of Vegetable Crops, University of California, Davis, California 95616, USA

Guillermo Veliz, USA (importation to U.S.) (see Oca)

Victor Villadolid, Unidad de Recursos Genéticos, Centro Agronómica Tropical de Investigación y Enseñanza (CATIE), Turrialba 7170, Costa Rica

Benigno Villalon, Texas Agricultural Experiment Center at Weslaco, 2415 East Highway 83, Weslaco, Texas 78596, USA

Alejo von der Pahlen, Italy (see Oca)

Raymon Webb, USA (see Potatoes)
Harold F. Winters, 10717 Kinloch Road, Silver Spring, Maryland 20903, USA

SQUASHES AND THEIR RELATIVES

Andean Region

María Inés Báez, Estación Experimental Agropecuaria Santiago del Estero, Instituto Nacional
    de Tecnología Agropecuaria (INTA), Independencia 341, Casilla de Correo 268, Santiago
    del Estero 4200, Argentina
Cesar Barberan, Ecuador (see Peppers)
Fernando N. Barrantes Del Aguila, Peru (germplasm, pathology) (see Achira)
Andres Contreras M., Chile (germplasm) (see Oca)
Bernardo Eraso Silva, Federación Horto-Fruticola, Calle 37 No. 15–49, Bogotá, Colombia
Alexander Grobman, Peru (germplasm collection) (see Oca)
Instituto Interamericano de Cooperación para la Agricultura (IICA), Bolivia (see Mashua)
Instituto Nacional de Investigaciones Agropecuarias (INIAP), Ecuador (see Achira)
Juan Jaramilló Vasquez, Colombia (germplasm) (see Peppers)
Ingo Junge Rodewald, Chile (processing, nutrition) (see Potatoes)
Ana Maria Miante Alzogaray, Facultad de Agronomía, Cátedra de Botánica, Universidad de
    Buenos Aires, Avenida San Martín 4453, Buenos Aires 1417, Buenos Aires, Argentina
Patricio Montaldo Bustos, Institución Producción y Sanidad Vegetal, Universidad Austral de
    Chile, Casilla 567, Valdivia, Chile (pre-Columbian food plants)
Alexander M. Mueller, Colombia (see Kiwicha)
Carlos Nieto C., Ecuador (see Achira)
Raúl Ríos E., Bolivia (see Achira)
Víctor Rodríguez, Instituto Nacional de Investigaciones Agropecuarias (INIAP), Estación
    Experimental "Santa Catalina," Casilla Postal 340, Quito, Ecuador
Raul Salazar C., Colombia (see Peppers)
Raúl Santana Paucar, Peru (see Mashua)
Juan Solano Lazo, Ecuador (see Achira)
Jorge Soria V., Avenida La Republica 1754, Quito, Ecuador
César Vásconez Sevilla, Ecuador (nutritional analysis) (see Kiwicha)

Other Countries

Thomas Andres, Department of Plant Breeding and Biometry, Cornell University, Ithaca,
    New York 14853, USA
Ricardo Bressani, Guatemala (nutrition) (see Achira)
Neil Delroy, Western Australia Department of Agriculture, Baron-Hay Court, South Perth,
    Western Australia 6151, Australia
Dick and Annemarie Endt, New Zealand (see Achira)
Jan Engels, West Germany (see Potatoes)
José T. Esquinas-Alcázar, Italy (see Mauka)
Michel Fanton, Australia (see Achira)
Zaccheaus Oyesiju Gbile, Nigeria (see Potatoes)
Heiner Goldbach, West Germany (seed storage and germination) (see Potatoes)
Doug Grant, Pukekohe Research Station, Department of Scientific and Industrial Research
    (DSIR), Cronin Road, RD 1, Pukekohe, New Zealand
J.L. Hudson, USA (see Kiwicha)
International Board for Plant Genetic Resources (IBPGR), Italy (germplasm information)
    (see Achira)
Alan M. Kapuler, USA (see Achira)
Jorge León, Costa Rica (see Maca)
Janet Long-Solís, Mexico (see Nuñas)

Barry McGlasson, Division of Food Research, Centre for International Agricultural Research, Commonwealth Scientific and Industrial Research Organization (CSIRO), PO Box 52, North Ryde, New South Wales 2113, Australia

Franklin W. Martin, USA (see Achira)

Laura Merrick, Department of Plant and Soil Sciences, 105 Deering Hall, University of Maine, Orono, Maine 04469, USA

Henry M. Munger, New York State Agricultural Experiment Station, Cornell University, PO Box 462, Geneva, New York 14456, USA

Michael Nee, New York Botanical Garden, Bronx, New York 10458–9980, USA

Tej Partap, Nepal (mountain agriculture, genetic resources) (see Maca)

Martin L. Price, USA (limited germplasm available) (see Achira)

R.W. Robinson, New York State Agricultural Experiment Station, Cornell University, Geneva, New York 14456, USA

Chris Rollins, USA (see Achira)

W.R. Sykes, New Zealand (see Mashua)

Raymon Webb, USA (see Potatoes)

Thomas W. Whitaker, 2534 Ellentown Road, La Jolla, California 92037, USA

BERRIES

Andean Region

Mary Arroyo, Departamento de Botánica, Facultad de Ciencias, Universidad de Chile, Avenida Bernardo O'Higgins 1058, Casilla 10–D, Santiago, Chile (Ugni)

Antonio Bacigalupo, Chile (Ugni) (see Potatoes)

Luis Arcadio Becerra, Programa de Frutales, Federación Nacional de Cafeteros de Colombia, Calle 73 No. 8–13, Bogotá, Colombia

Augusto Cardich, Argentina (geography and climate) (see Achira)

Vicente W.D. Casali, Brazil (see Arracacha)

Charles R. Clement, Instituto Nacional de Pesquisas da Amazônia (INPA), Estrada do Aleixo, Caixa Postal 478, Manaus 69.000, Amazonas, Brazil (tropical fruits)

Bernardo Eraso Silva, Colombia (see Squashes)

Fondo Nacional de Investigaciones Agropecuarias (FONAIAP), Venezuela (see Achira)

Jaime Gaete Calderon, Chile (Ugni) (see Basul)

Juan Gaviria R., Venezuela (see Ulluco)

María Inés González, Fruticola y Forestal Sudamericana S.A., Arauco 974, Chillan, Chile

Emilio González González, Chile (Ugni) (see Quinoa)

Fabio Higuita Muñoz, Colombia (see Arracacha)

Instituto Nacional de Investigaciones Agropecuarias (INIAP), Ecuador (see Achira)

Miguel Jordan, Laboratorio de Botánica, Facultad de Ciencias Biológicas, Pontificia Universidad Católica de Chile, Vic. Mackenna 4860, Casilla 114–D, Santiago, Chile (Ugni)

Ingo Junge Rodewald, Chile (processing, nutrition) (see Potatoes)

Latinreco S.A., Ecuador (see Arracacha)

Pascual Londoño, COLFRUTAS, Calle 55 No. 49-50, Medellín, Antioquia, Colombia (trade association)

Lúis E. López Jaramilló, Colombia (see Potatoes)

Patricio Montaldo Bustos, Chile (pre-Columbian food plants) (see Squashes)

Eloy Munive Jáuregui, Peru (see Maca)

Jaime Osorio Bedoya, Hortilizas y Frutales, Instituto Colombiano Agropecuario (ICA), Apartado Aéreo 151123, Bogotá, Colombia

Raúl Ríos E., Bolivia (see Achira)

Raul Salazar C., Colombia (see Peppers)

Francisco Sánchez, Jardín Botánico de Colombia, Carrera 66A No. 56–84, Bogotá, Colombia

Servicio Nacional de Aprendizaje (SENA), Regional Boyacá, Calle 65 No. 11–70, Apartado Aéreo 9801, Bogotá, Colombia

Jorge Soria V., Ecuador (see Squashes)
Julio Cesar Toro Meza, Industrial Crops Division, Instituto Colombiano Agropecuario (ICA), Apartado Aéreo 233, Palmira, Valle, Colombia
Cesar C. Vargas Calderon, Peru (see Potatoes)
Erik von Baer von Lochow, Chile (Ugni) (see Quinoa)
Claudio Wernli K., c/o División Agrícola Carozzi, AGROZZI, Camino Longitudinal Sur 5201, Km. 23, San Bernardo, Santiago, Chile (Ugni)

Other Countries

Vichitr Benjasil, Thailand (see Achira)
Edward Cope, Bailey Horatorium, Cornell University, Ithaca, New York 14853, USA (taxonomy)
Peter Del Tredici, The Arnold Arboretum, Harvard University, The Arborway, Jamaica Plain, Massachusetts 02130, USA
Arlen D. Draper, c/o Small Fruits Lab, Agricultural Research Service (ARS), U.S. Department of Agriculture (USDA), Beltsville Agricultural Research Center (BARC-West), Beltsville, Maryland 20705, USA
Craig Dremann, USA (see Potatoes)
Michel Fanton, Australia (see Achira)
Gene Galletta, Small Fruits Laboratory, Agricultural Research Service (ARS), U.S. Department of Agriculture (USDA), Beltsville Agricultural Research Center (BARC-West), Beltsville, Maryland 20705, USA
Ray Gereau, Missouri Botanical Garden, PO Box 299, St. Louis, Missouri 63166–0299, USA (Rubus taxonomy)
Harvey K. Hall, Riwaka Research Station, Department of Scientific and Industrial Research (DSIR), Old Mill Road, RD 3, Motueka, New Zealand (Rubus)
Richard A. Hamilton, Department of Horticulture, College of Tropical Agriculture, University of Hawaii at Manoa, St. John's Plant Science Laboratory #102, 3190 Maile Way, Honolulu, Hawaii 96822, USA
Roy Hart, Riwaka Research Station, Department of Scientific and Industrial Research (DSIR), Old Mill Road, RD 3, Motueka, New Zealand (Vaccinium)
David Himelrick, 84 East Main Street, Silver Creek, New York 14136, USA
J.L. Hudson, USA (see Kiwicha)
Gojka Jelenkovic, Department of Horticulture and Forestry, Rutgers University, PO Box 2101, New Brunswick, New Jersey 08903, USA (Vaccinium)
Derek Leonard Jennings, Medway Fruits, 'Clifton' Honey Lane Otham, Maidstone, Kent ME13 4RJ, UK (Rubus)
C. Antonio Jiménez Aparicio, Mexico (see Arracacha)
G. Linsley-Noakes, Fruit and Fruit Technology Research Institute, Department of Agriculture and Water Supply, Private Bag X5013, Stellenbosch 7600, South Africa (Rubus, Vaccinium)
Professor Lüdders, Institut für Nutzpflanzenforschung, Technische Universität Berlin, Fachbereich 15–Ostbau, Albrecht-Thaer-Weg 3, Berlin 33, West Germany
Cyrus McKell, USA (see Potatoes)
Julia F. Morton, Morton Collectanea, University of Miami, PO Box 248204, Coral Gables, Florida 33124, USA (general information)
Tej Partap, Nepal (mountain agriculture, genetic resources) (see Maca)
Rare Fruit Council International, Inc., USA (see Peppers)
Jorge Rodríguez A., Centro de Fruticultura, Institución Enseñanza e Investigación en Ciencias Agrícolas, Colegio de Postgraduados, Montecillos, Chapingo, Estado de México 56230, Mexico (Rubus)
Chris Rollins, USA (see Achira)
Richard E. Schultes, USA (ethnobotany) (see Achira)
Steven Spangler, USA (see Achira)
Harry Jan Swartz, College of Agriculture, University of Maryland, College Park, Maryland 20742, USA

W.R. Sykes, New Zealand (Ugni) (see Mashua)
Victor A. Wynne, Haiti Seed Store - Wynne Farm, PO Box 15146, Petion-Ville, Haiti

CAPULI CHERRY

Andean Region

Emilio Barahona Chura, Componente Forestal de la Microregión de Juliaca, Instituto
    Nacional Forestal y de Fauna (INFOR), Juliaca, Puno, Peru
Vidal Bautista, Programa de Frutales, Facultad de Agronomía, Universidad Nacional Agraria
    (UNA), Apartado 456, La Molina, Lima 100, Peru
Hernán Caballero, Ecuador (see Quinoa)
José Calzada Benza, Cervecena San Juan, Pucallpa, Peru
Raúl O. Castillo T., Ecuador (see Ahipa)
S.A. Centauro, Apartado 1088, Lima 100, Peru
Bernardo Eraso Silva, Colombia (see Squashes)
Jorge Fabara, Corporación Ambiente y Desarrollo (AMDE), PO Box 632, Ambato, Ecuador
Andrés Miguel Hlatky Hernández, Ecuador (see Arracacha)
Miguel Holle, Peru (see Mashua)
Instituto Nacional de Investigaciones Agropecuarias (INIAP), Ecuador (see Achira)
Ricardo Jon Llap, Investigación y Capacitación Forestal y de Fauna (CENFOR VII), Instituto
    Nacional Forestal y de Fauna (INFOR), Huancayo, Peru
Patricio Montaldo Bustos, Chile (pre-Columbian food plants) (see Squashes)
Eloy Munive Jáuregui, Peru (see Maca)
Gladys Navas de Alvarado, Ecuador (nutritional analysis) (see Kiwicha)
Carlos Nieto C., Ecuador (see Achira)
David Ocaña Vidal, Investigación y Capacitación Forestal y de Fauna (CENFOR III), Instituto
    Nacional Forestal y de Fauna (INFOR), Huaraz, Peru
José Pretell Chiclote, Proyecto Reforestación Piloto en Cajamarca (CENFOR II), Instituto
    Nacional Forestal y de Fauna (INFOR), Cajamarca, Peru
Carlos Roersch, Peru (see Mashua)
Jaime Rojas, Ecuador (see Nuñas)
Raul Salazar C., Colombia (see Peppers)
Francisco Sánchez, Colombia (see Berries)
Isidoro Sánchez Vega, Peru (see Arracacha)
Juan Solano Lazo, Ecuador (see Achira)
Jorge Soria V., Ecuador (see Squashes)
César Vásconez Sevilla, Ecuador (nutritional analysis) (see Kiwicha)
Rebeca Vega de Rojas, Ecuador (see Arracacha)

Other Countries

Gilles Bourgeois, Canada (see Basul)
Ferdinando Cossio, Istituto Sperimentale di Frutticoltura, Amministrazione Provinciale di
    Verona, Via San Giacomo 25, Verona 37135, Italy
Peter Del Tredici, USA (germplasm) (see Berries)
Dick and Annemarie Endt, New Zealand (see Achira)
Gojka Jelenkovic, USA (see Berries)
Steven R. King, USA (ethnobotany) (see Achira)
Richard McCain, USA (horticulture) (see Yacon)
Rogers McVaugh, Department of Biology, University of North Carolina, Coker Hall
    CB# 3280, Chapel Hill, North Carolina 27599, USA (taxonomy)
Simon E. Malo, Escuela Agrícola Panamericana (Zamorano), Casilla Postal 93, Tegucigalpa,
    Honduras
Martin L. Price, USA (limited germplasm available) (see Achira)

Antonino Raimondo, Istituto de Coltivazioni Arboree, Università degli Studi - Palermo, c/o Università degli Studi, Piazza Marina 61, Palermo 90128, Italy
Rare Fruit Council International, Inc., USA (see Peppers)
Jorge Rodríguez A., Mexico (see Berries)
Chris Rollins, USA (see Achira)
Shao Qiquan, China (see Potatoes)
Claude Sweet, Horticulture Department T-6, MiraCosta College, One Bernard Drive, Oceanside, California 92056, USA
Louis Trap, New Zealand (see Ahipa)

CHERIMOYA

Andean Region

Antonio Bacigalupo, Chile (see Potatoes)
Vidal Bautista, Peru (see Capuli Cherry)
Hernán Caballero, Ecuador (see Quinoa)
Augusto Cardich, Argentina (geography and climate) (see Achira)
Charles R. Clement, Brazil (tropical fruits) (see Berries)
M. J. Durán, Departamento de Fitotécnia, Universidad Nacional Agraria (UNA), Apartado 456, La Molina, Lima 100, Peru
William Edwardson, Colombia (see Quinoa)
Bernardo Eraso Silva, Colombia (see Squashes)
Jorge Fabara, Ecuador (see Capuli Cherry)
Fondo Nacional de Investigaciones Agropecuarias (FONAIAP), Venezuela (see Achira)
Francisco Gardiazábal, Escuela de Agronomía, Universidad Católica de Valparaíso, Avenida Brasil 2950, Casilla 4059, Valparaíso, Chile
Instituto Nacional de Investigaciones Agropecuarias (INIAP), Ecuador (see Achira)
Miguel Jordan, Chile (cropping) (see Berries)
Juan León F., Programa Fruticultura, Estación Experimental Tumbaco, Instituto Nacional de Investigaciones Agropecuarias (INIAP), Avenida Eloy Alfaro y Amazonas, Casilla 2600, Quito, Ecuador (horticulture and germplasm)
Miguel Morán Robles, Laboratoria de Biotecnología, Departamento de Fitotécnia, Facultad de Agronomía, Universidad Nacional Agraria (UNA), Apartado 456, La Molina, Lima 100, Peru
Francisco Rhon Dávila, Ecuador (see Maca)
Raúl Ríos E., Bolivia (see Achira)
Emilio Rojas Mendoza, Peru (see Maca)
G. Rosenberg, Escuela de Agronomía, Universidad Católica de Valparaíso, Avenida Brasil 2950, Casilla 4059, Valparaíso, Chile
Isidoro Sánchez Vega, Peru (see Arracacha)
Fernando Santa Cruz-B., Casilla 73, La Cruz, Chile (grower)
Jorge Soria V., Ecuador (see Squashes)
Luis Torellis, Escuela de Agronomía, Universidad Católica de Valparaíso, Avenida Brasil 2950, Casilla 4059, Valparaíso, Chile
Julio Cesar Toro Meza, Colombia (see Berries)
Francisco A. Vivar C., Jardín Botánico "Reinaldo Espinosa," Facultad de Ciencias Agricolas, Universidad Nacional de Loja, Casilla Letra B, Loja, Ecuador

Other Countries

David Austen, Exotic Nurseries, Larmar Road, RD #1, Kaitaia, New Zealand
A.G. Banks, Maroochy Horticultural Research Station, Department of Primary Industries, PO Box 5083, Sunshine Coast Mail Center, Nambour, Queensland 4560, Australia (atemoya)

Robert Barnum, Possum Trot Nursery, 14955 S.W. 214th Street, Miami, Florida 33187, USA (cherimoya and atemoya)

Boxer-Lerner, 208 Hertzel Street, Rechovot 76270, Israel (propagation, germplasm)

Ricardo Bressani, Guatemala (nutrition) (see Achira)

Tony Brown, California Tropics, Rancho Bonita Vista Corporation, 6950 Casitas Pass Road, Carpinteria, California 93013, USA

Carl Campbell, Institute of Food and Agricultural Sciences (IFAS), University of Florida, Tropical Research and Education Center (TREC), Homestead, Florida 33031, USA

Tom Cowdell, Welcome Bay, Tauranga, New Zealand (grower, pollination)

Stuart N. Dawes, Division of Horticulture and Processing, Department of Scientific and Industrial Research (DSIR), Private Bag, Auckland, New Zealand

William Doty, USA (germplasm) (see Peppers)

Norman C. Ellstrand, Department of Botany and Plant Sciences, University of California, Riverside, California 92521–0124, USA (genetics and breeding)

Dick and Annemarie Endt, New Zealand (see Achira)

W. Hardy Eshbaugh, USA (see Achira)

José Farré, Estación Experimental "La Mayora," Universidad de Málaga, Algarrobo Costa, Málaga 29750, Spain (germplasm, agronomy, and pollination)

Ricardo Carlos F. França, Centro de Fruticultura Subtropical, Direcção Regional de Agricultura, Quebradas, São Martinho, Funchal 9000, Madeira, Portugal (germplasm)

Daniel W. Gade, USA (see Achira)

F. Gazit, The Volcani Center, Agricultural Research Organization, PO Box 6, Bet-Dagan 50250, Israel

Alan P. George, Maroochy Horticultural Research Station, Department of Primary Industries, PO Box 5083, Sunshine Coast Mail Center, Nambour, Queensland 4560, Australia (taxonomy, agronomy)

Heiner Goldbach, West Germany (seed storage and germination) (see Potatoes)

V.L. Guzman, USA (see Quinoa)

Richard A. Hamilton, USA (see Berries)

M.E. Hopping, Ruakura Soil and Plant Research Station, Ministry of Agriculture and Fisheries (MAF), Private Bag, Hamilton, New Zealand

Ron Kadish, USA (agronomy) (see Maca)

Alan M. Kapuler, USA (see Achira)

Professor Lüdders, West Germany (see Berries)

Paul J.M. Maas, Netherlands (taxonomy, genetics; bibliography available) (see Achira)

Richard McCain, USA (horticulture) (see Yacon)

Cyrus McKell, USA (see Potatoes)

Simon E. Malo, Honduras (see Capuli Cherry)

Kenny Maxfield, Route 62, Box 9, Moyie Springs, Idaho 83845, USA (cold tolerance)

Mauro Raúl Mendoza López, INIFAP, Campo Experimental Sierra Tarasca, Pátzcuaro, Michoacán, Mexico (cherimoya and atemoya)

Mike Moncur, Division for Forest Research, Centre for International Agricultural Research - CSIRO, Canberra, Australia (atemoya)

Julia F. Morton, USA (general information) (see Berries)

Hannah Nadel, Tropical Research and Education Center, Institute of Food and Agricultural Sciences, 18905 S.W. 280th Street, Homestead, Florida 33031, USA (pollination)

T.D. Pennington, c/o The Herbarium, Royal Botanic Gardens, Kew, Richmond, Surrey TW9 3AE, UK (taxonomy)

Antonio M. Pinchinat, St. Lucia (see Basul)

Martin L. Price, USA (limited germplasm available) (see Achira)

Rare Fruit Council International, Inc., USA (see Peppers)

Herman Real, Biological Institute of Tropical America (BIOTA), PO Box 2585, Menlo Park, California 94026–2585, USA (tropical ecosystems)

Jorge Rodríguez A., Mexico (see Berries)

G.M. Sanewski, Maroochy Horticultural Research Station, Department of Primary Industries, PO Box 5083, Sunshine Coast Mail Center, Nambour, Queensland 4560, Australia (atemoya)

George E. Schatz, Missouri Botanical Garden, PO Box 299, St. Louis, Missouri 63166–0299, USA (pollination)

C.A. Schroeder, USA (see Ahipa)
Steven Spangler, USA (see Achira)
Claude Sweet, USA (see Capuli Cherry)
W.R. Sykes, New Zealand (see Mashua)
Louis Trap, New Zealand (see Ahipa)
Richard Valley, USA (grower) (see Oca)
Benjamin H. Waite, USA (see Oca)

GOLDENBERRY

Andean Region

Segundo Alandia, Bolivia (pathology) (see Mashua)
David Baumann, Instituto Rural Vallegrande, Apartado 70, Cañete, Peru
Andres Contreras M., Chile (germplasm) (see Oca)
Orlando Cortés, Departamento Academicoide Agricultura, Universidad Nacional de San
    Antonio Abad del Cusco (UNSAAC), Apartado 921, Cusco, Peru
Hernán Cortés Bravo, Peru (see Oca)
Carmen E.B. de Rojas, Venezuela (see Peppers)
Juan Gaviria R., Venezuela (see Ulluco)
Fabio Higuita Muñoz, Colombia (see Arracacha)
Instituto Nacional de Investigaciones Agropecuarias (INIAP), Ecuador (see Achira)
Mario Lobo, Colombia (see Peppers)
Lúis E. López Jaramilló, Colombia (Solanaceae) (see Potatoes)
Miguel Morán Robles, Peru (see Cherimoya)
Modesto Soria V., Batate, Estado Tungurahua, Ecuador (grower)
Julio Cesar Toro Meza, Colombia (see Berries)
Ramiro Velastegui, Ecuador (pathology) (see Tarwi)

Other Countries

Gregory J. Anderson, USA (ethnobotany and taxonomy) (see Achira)
Suzanne Ashworth, USA (see Kiwicha)
John E. Averett, National Wildflower Research Center, 2600 FM 973 North, Austin, Texas
    78725, USA
Lynn A. Bohs, USA (see Arracacha)
Gilles Bourgeois, Canada (see Basul)
M. Bureau et fils, 5, rue Hutte, La Possonière, St. George sur Loire F-49170 France
    (nurserymen)
G. Burge, New Zealand (see Oca)
Carl Campbell, USA (see Cherimoya)
George B. Chibiliti, Zambia (see Kiwicha)
Alan Child, Ryburn, 30 Middle Street, Nafferton, Driffield, East Yorkshire YO2 50JS, UK
    (Solanaceae)
Roy Danforth, BP 1377, Bangui, Central African Republic
W.G. D'Arcy, USA (see Potatoes)
Stuart N. Dawes, New Zealand (see Cherimoya)
William Doty, USA (germplasm) (see Peppers)
Craig Dremann, USA (see Potatoes)
Dick and Annemarie Endt, New Zealand (see Achira)
Jan Engels, West Germany (see Potatoes)
Hector E. Flores, USA (tissue culture, phytochemistry) (see Kiwicha)
Heiner Goldbach, West Germany (seed storage and germination) (see Potatoes)
Richard A. Hamilton, USA (see Berries)
Doug Hammonds, Turners and Growers, Auckland, New Zealand (fruit export)

Hawaiian Fruit Preserving Company, Limited, PO Box 637, Kalaheo, Kauai, Hawaii 96741, USA (fruit products)
Don Hudson, Chewonki Foundation, Wiscasset, Maine 04578, USA
Nancy Jarris, University of California Extension Service, Suite 200, 2175 The Alameda, San Jose, California 95126, USA
Alan M. Kapuler, USA (see Achira)
Kenya Orchards Limited, Mau Hills, Machako District, Kenya (commercial production)
David J. Klinac, Ruakura Soil and Plant Research Station, Ministry of Agriculture and Fisheries (MAF), Private Bag, Hamilton, New Zealand
G. Linsley-Noakes, South Africa (see Berries)
Professor Lüdders, West Germany (see Berries)
Janet S. Luis, Philippines (see Potatoes)
Richard McCain, USA (horticulture) (see Yacon)
D.A. Miller, Horticultural Advisory Service Experimental Station, Longue Rue (Burnt Lane), St. Martin's, Guernsey, Channel Islands, UK
Henry Y. Nakasone, St. John's Plant Science Laboratory #102, Department of Horticulture, College of Tropical Agriculture, University of Hawaii at Manoa, 3190 Maile Way, Honolulu, Hawaii 96822, USA
Michael Nee, USA (see Squashes)
John P. Ogier, Exotics Group, Horticultural Advisory Service Experimental Station, Longue Rue (Burnt Lane), St. Martin's, Guernsey, Channel Islands, UK
John Palmer, New Zealand (see Ahipa)
Dov Pasternak, Israel (see Potatoes)
Kevin Patterson, New Zealand (see Achira)
Jean-Yves Peron, Department of Vegetable and Seed Crops, National Institute for Horticulture Science (E.N.I.T.A.H.), 2, Rue le Notre, Angers 49045, France (germplasm)
Martin L. Price, USA (limited germplasm available) (see Achira)
Rare Fruit Council International, Inc., USA (see Peppers)
John M. Riley, USA (general information, germplasm) (see Oca)
Chris Rollins, USA (see Achira)
Elizabeth Schneider, USA (food preparation) (see Yacon)
C.A. Schroeder, USA (see Ahipa)
Steven Spangler, USA (see Achira)
Janet Sullivan, Department of Botany, University of New Hampshire, Nesmith Hall, Durham, New Hampshire 03824–3597, USA
Claude Sweet, USA (see Capuli Cherry)
W.R. Sykes, New Zealand (see Mashua)
Loren Toomey, 24791 Belgreen Place, El Toro, California 92630, USA
Louis Trap, New Zealand (see Ahipa)
Benjamin H. Waite, USA (see Oca)
F.H. Wood, Ruakura Soil and Plant Research Station, Ministry of Agriculture and Fisheries (MAF), Private Bag, Hamilton, New Zealand

HIGHLAND PAPAYAS

Andean Region

Segundo Alandia, Bolivia (pathology) (see Mashua)
Victor M. Badillo, Facultad de Agronomía, Universidad Central de Venezuela, Apartado 4579, Maracay, Estado Aragua 2101, Venezuela (taxonomy)
Cesar Barberan, Ecuador (see Peppers)
Hernán Caballero, Ecuador (see Quinoa)
Saul E. Camacho B., Instituto Colombiano Agropecuario (ICA), Apartado Aéreo 233, Palmira, Valle, Colombia
Carmen E.B. de Rojas, Venezuela (see Peppers)
César del Carpio Merino, Peru (see Achira)

Bernardo Eraso Silva, Colombia (see Squashes)
Jorge Fabara, Ecuador (see Capuli Cherry)
Ramón Ferreyra, Museo de Historia Natural "Javier Prado," Avenida Arenales 1256, Lima, Peru
Fondo Nacional de Investigaciones Agropecuarias (FONAIAP), Venezuela (see Achira)
Juan Gaviria R., Venezuela (see Ulluco)
Joy C. Horton Hofmann, Ecuador (see Achira)
Instituto Nacional de Investigaciones Agropecuarias (INIAP), Ecuador (see Achira)
Miguel Jordan, Chile (cropping) (see Berries)
Mario Lobo, Colombia (*Carica crassifolia*) (see Peppers)
Miguel Morán Robles, Peru (see Cherimoya)
Laura Muñoz Espín, Ecuador (see Arracacha)
Raúl Ríos E., Bolivia (see Achira)
Víctor Rodríguez, Ecuador (see Squashes)
Jaime Rojas, Ecuador (see Nuñas)
Isidoro Sánchez Vega, Peru (see Arracacha)
Jorge Soria V., Ecuador (see Squashes)
Modesto Soria V., Ecuador (grower, hybrids) (see Goldenberry)
Julio Cesar Toro Meza, Colombia (see Berries)
Rebeca Vega de Rojas, Ecuador (tissue culture) (see Arracacha)
Ramiro Velastegui, Ecuador (pathology) (see Tarwi)

Other Countries

David Austen, New Zealand (hybrids) (see Cherimoya)
Robert Barnum, USA (see Cherimoya)
Gilles Bourgeois, Canada (see Basul)
G. Burge, New Zealand (see Oca)
Carl Campbell, USA (see Cherimoya)
Ferdinando Cossio, Italy (germplasm) (see Capuli Cherry)
John Couch, Strybing Arboretum, Golden Gate Park, 9th Avenue at Lincoln Way, San Francisco, California 94720, USA (germplasm)
Stuart N. Dawes, New Zealand (see Cherimoya)
William Doty, USA (germplasm) (see Peppers)
Dick and Annemarie Endt, New Zealand (see Achira)
W. Hardy Eshbaugh, USA (see Achira)
Michel Fanton, Australia (see Achira)
Enrique Forero, Missouri Botanical Garden, PO Box 299, St. Louis, Missouri 63166–0299, USA
Phil and Georgie Gardener, c/o P.O. Opua, Bay of Islands, New Zealand (hybrids)
F. Gazit, Israel (see Cherimoya)
Zaccheaus Oyesiju Gbile, Nigeria (see Potatoes)
Jane Harman, Division of Horticulture and Processing, Postharvest Physiology, Department of Scientific and Industrial Research (DSIR), Mount Albert Research Centre, 120 Mt. Albert Road, Private Bag, Auckland, New Zealand
Ernest P. Imle, 10802 Bornedale Drive, Adelphi, Maryland 20783, USA
Sherry Kitto, Department of Plant Science, University of Delaware, Newark, Delaware 19716–1303, USA (tissue culture)
Robert Knight, Agricultural Research Service (ARS), U.S. Department of Agriculture (USDA), 13601 Old Cutler Road, Miami, Florida 33158, USA
Richard E. Litz, University of Florida Research Station, University of Florida, Homestead, Florida 33030, USA (tissue culture)
Paul Loo, Plantek International, #04–01A, Block 14, Lee Maxwell, Science Park Drive, Singapore 0511, Singapore (tissue culture)
Richard McCain, USA (horticulture) (see Yacon)
Cyrus McKell, USA (see Potatoes)
D.A. Miller, UK (see Goldenberry)
Henry Y. Nakasone, USA (see Goldenberry)

John P. Ogier, UK (see Goldenberry)
John Palmer, New Zealand (see Ahipa)
Martin L. Price, USA (see Achira)
Rare Fruit Council International, Inc., USA (see Peppers)
C.A. Schroeder, USA (see Ahipa)
Shao Qiquan, China (see Potatoes)
Steven Spangler, USA (see Achira)
Claude Sweet, USA (see Capuli Cherry)
John Swift, USA (see Yacon)
W.R. Sykes, New Zealand (see Mashua)
Louis Trap, New Zealand (see Ahipa)
Francisco Vásquez, Tropical Agriculture Research Station, U.S. Department of Agriculture (USDA), PO Box 70, Mayagüez, Puerto Rico 00789, USA
Francis Zee, Hawaii Branch Station, Beaumont Agricultural Research Center, University of Hawaii, College of Tropical Agriculture, 461 West Lanakaula Street, Hilo, Hawaii 96720–4094, USA (clonal variation and propagation)

LUCUMA

Andean Region

Vidal Bautista, Peru (bibliography available) (see Capuli Cherry)
Jorge Bermudez, Departamento Fitotécnia, Universidad Nacional Agraria (UNA), Apartado 456, La Molina, Lima 100, Peru
José Calzada Benza, Peru (see Capuli Cherry)
César del Carpio Merino, Peru (see Achira)
Ramón Ferreyra, Peru (see Highland Papayas)
Wilma Freire, Ecuador (nutrition) (see Potatoes)
Jaime Gaete Calderon, Chile (see Basul)
Miguel Jordan, Chile (cropping) (see Berries)
Patricio Montaldo Bustos, Chile (pre-Columbian food plants) (see Squashes)
Miguel Morán Robles, Peru (see Cherimoya)
Laura Muñoz Espín, Ecuador (see Arracacha)
Emilio Rojas Mendoza, Peru (see Maca)
Raul Salazar C., Colombia (see Peppers)
Fernando Santa Cruz-B., Chile (grower) (see Cherimoya)
Cesar C. Vargas Calderón, Peru (see Potatoes)

Other Countries

Robert Barnum, USA (see Cherimoya)
Gilles Bourgeois, Canada (see Basul)
Jesús Axayacatl Cuevas Sánchez, Departamento de Fitotécnia, Universidad Autónoma Chapingo, Chapingo, Estado de México 56230, Mexico
Stuart N. Dawes, New Zealand (see Cherimoya)
Dick and Annemarie Endt, New Zealand (see Achira)
Michel Fanton, Australia (see Achira)
Richard A. Hamilton, USA (see Berries)
Professor Lüdders, West Germany (see Berries)
Simon E. Malo, Honduras (see Capuli Cherry)
T.D. Pennington, UK (taxonomy) (see Cherimoya)
C.A. Schroeder, USA (see Ahipa)
Steven Spangler, USA (see Achira)
Louis Trap, New Zealand (see Ahipa)

NARANJILLA (Lulo)

Andean Region

Carlos Bueno, Instituto Nacional de Pesquisas da Amazônia (INPA), Estrada do Aleixo, Caixa
    Postal 478, Manaus 69.000, Amazonas, Brazil
Hernán Caballero, Ecuador (see Quinoa)
Saul E. Camacho B., Colombia (see Highland Papayas)
Saul Comacho, Horticultura Moderna, Apartado Aéreo 20236, Cali, Colombia
Carmen E.B. de Rojas, Venezuela (see Peppers)
Edgar Ivan Estrada, Colombia (see Ulluco)
Michael Hermann, Peru (see Achira)
Miguel Holle, Peru (see Arracacha)
Joy C. Horton Hofmann, Ecuador (see Achira)
Instituto Nacional de Investigaciones Agropecuarias (INIAP), Ecuador (see Achira)
Mario Lobo, Colombia (see Peppers)
Lúis E. López Jaramilló, Colombia (Solanaceae) (see Potatoes)
Miguel Morán Robles, Peru (see Cherimoya)
Gladys Navas de Alvarado, Ecuador (nutritional analysis) (see Kiwicha)
José Otocar Reina Barth, Colombia (small-scale production) (see Arracacha)
Víctor Rodríguez, Ecuador (see Squashes)
Jorge Soria V., Ecuador (see Squashes)
Julio Cesar Toro Meza, Colombia (see Berries)
Rebeca Vega de Rojas, Ecuador (see Arracacha)
Ramiro Velastegui, Ecuador (pathology) (see Tarwi)

Other Countries

Robert Barnum, USA (see Cherimoya)
A. Benzioni, Boyko Institute, Ernst David Bergmann Campus, Ben-Gurion University of the
    Negev, PO Box 1025, Beer-Sheva 84110, Israel
Lynn A. Bohs, USA (see Arracacha)
Ricardo Bressani, Guatemala (nutrition) (see Achira)
Carl Campbell, USA (see Cherimoya)
Alan Child, UK (Solanaceae) (see Goldenberry)
W.G. D'Arcy, USA (see Potatoes)
Romulo Davide, College of Agriculture, University of the Philippines at Los Baños, Los
    Baños, Philippines (nematocides)
Stuart N. Dawes, New Zealand (see Cherimoya)
F.G. Dennis, Jr., Department of Horticulture, College of Agriculture, Michigan State University,
    Plant and Soil Sciences Building, East Lansing, Michigan 48824–1325, USA
William Doty, USA (germplasm) (see Peppers)
Craig Dremann, USA (see Potatoes)
Dick and Annemarie Endt, New Zealand (see Achira)
Jan Engels, West Germany (see Potatoes)
W. Hardy Eshbaugh, USA (see Achira)
William Charles Evans, UK (see Potatoes)
Heiner Goldbach, West Germany (seed storage and germination) (see Potatoes)
B.W.W. Grout, UK (see Achira)
Richard A. Hamilton, USA (see Berries)
Kathleen Haynes, USA (see Potatoes)
Charles B. Heiser, Jr., USA (see Peppers)
Ray Henkel, USA (pioneer colonization) (see Achira)
J.L. Hudson, USA (see Kiwicha)
Ron Hurov, USA (see Mashua)
Nancy Jarris, USA (see Goldenberry)
Alan M. Kapuler, USA (see Achira)

John (Jack) Kelly, Department of Horticulture, Michigan State University, East Lansing, Michigan 48824–1112, USA

Brian R. Kerry, Entomology and Nematology Department, Crop and Environmental Protection Division, Rothamsted Experimental Station, Harpenden, Hertsfordshire AL5 2JQ, UK

Richard E. Litz, USA (tissue culture) (see Highland Papayas)

Richard McCain, USA (horticulture) (see Yacon)

Franklin W. Martin, USA (see Achira)

S. Mendlinger, Boyko Institute, Ernst David Bergmann Campus, Ben-Gurion University of the Negev, PO Box 1025, Beer-Sheva 84110, Israel

Michael Nee, USA (see Squashes)

John Palmer, New Zealand (see Ahipa)

Dov Pasternak, Israel (see Potatoes)

Martin L. Price, USA (limited germplasm available) (see Achira)

Rare Fruit Council International, Inc., USA (see Peppers)

Herman Real, USA (tropical ecosystems) (see Cherimoya)

John M. Riley, USA (general information, germplasm) (see Oca)

Chris Rollins, USA (see Achira)

Elizabeth Schneider, USA (food preparation) (see Yacon)

C.A. Schroeder, USA (see Ahipa)

Richard E. Schultes, USA (ethnobotany) (see Achira)

Steven Spangler, USA (see Achira)

Claude Sweet, USA (see Capuli Cherry)

W.R. Sykes, New Zealand (see Mashua)

Louis Trap, New Zealand (see Ahipa)

Benjamin H. Waite, USA (see Oca)

Victor A. Wynne, Haiti (see Berries)

PACAY (Ice-Cream Beans)

Andean Region

Rodrigo Arce Rojas, Herbario Forestal, Facultad de Ciencias Forestales, Universidad Nacional Agraria (UNA), Centro de Datos Para La Conservación, Apartado 456, La Molina, Lima 100, Peru

Daniel G. Debouck, Colombia (see Kiwicha)

Marc Dourojeanni, Peru (see Basul)

Erick C.M. Fernandes, Tropical Soils Research Program, Department of Soil Science, North Carolina State University, Programa de Suelos Tropicales, Estación Experimental Yurimaguas, Yurimaguas, Loreto, Peru (integrated farming in humid tropics, germplasm)

Jaime Gaete Calderon, Chile (see Basul)

Patricia Gomez Andrade, Herbario Nacional, Museo Ecuatoriano de Ciencias Naturales, Casilla 8976 Sucre 7, Quito, Ecuador (fruit types)

Instituto Nacional de Investigaciones Agropecuarias (INIAP), Ecuador (see Achira)

Shirley Keel, Peru (botany) (see Basul)

Miguel Morán Robles, Peru (see Cherimoya)

Alexander M. Mueller, Colombia (see Kiwicha)

C. Percy Nuñez Vargas, Peru (see Basul)

Isaac C. Peralta Vargas, Peru (see Basul)

Carlos Reynel, Facultad de Ciencias Forestales, Universidad Nacional Agraria (UNA), Apartado 456, La Molina, Lima 100, Peru

Manuel Ríos, Peru (forestry) (see Basul)

Angel Salazar, Proyecto Agroforestal, North Carolina State University, Programa de Suelos Tropicales, Estación Experimental Yurimaguas, Yurimaguas, Loreto, Peru (germplasm)

José Gabriel Sánchez Vega, Peru (see Basul)

Lawrence Szott, Tropical Soils Research Program, Department of Soil Science, North Carolina State University, Programa de Suelos Tropicales, Estación Experimental Yurimaguas, Yurimaguas, Loreto, Peru (germplasm)

Other Countries

Robert Barnum, USA (see Cherimoya)

Jose R. Benites, Tropical Soils Research Program, Department of Soil Science, North Carolina State University, Raleigh, North Carolina 27695, USA

Lynn A. Bohs, USA (see Arracacha)

Gilles Bourgeois, Canada (see Basul)

James L. Brewbaker, USA (see Basul)

Leslie Brownrigg, USA (indigenous systems) (see Achira)

Gerardo Budowski, United Nations University for Peace, PO Box 199–1250, Escazú, Costa Rica

Centroamericana de Semillas, Honduras (see Kiwicha)

Dick and Annemarie Endt, New Zealand (see Achira)

Michel Fanton, Australia (see Achira)

Enrique Forero, USA (see Highland Papayas)

Zaccheaus Oyesiju Gbile, Nigeria (see Potatoes)

Nancy Glover, USA (see Basul)

Richard A. Hamilton, USA (see Berries)

Ralph J. Hervey, Costa Rica (see Basul)

Suzanne Koptur, Department of Biology, Florida International University, Miami, Florida 33199, USA

Jorge León, Costa Rica (see Maca)

Richard McCain, USA (horticulture) (see Yacon)

Simon E. Malo, Honduras (see Capuli Cherry)

Yosef Mizrahi, Institute of Applied Research and Department of Biology, Ben-Gurion University of the Negev, PO Box 1025, Beer-Sheva 84110, Israel

William L. Nelson, Pacific Tree Farms, 4301 Lynnwood Drive, Chula Vista, California 92010, USA

A. Nerd, Boyko Institute, Ernst David Bergmann Campus, Ben-Gurion University of the Negev, PO Box 1025, Beer-Sheva 84110, Israel

Dov Pasternak, Israel (see Potatoes)

Charles Peters, USA (see Oca)

Michael Pilarski, USA (reforestation and tree crops) (see Basul)

Antonio M. Pinchinat, St. Lucia (see Basul)

De Lourdes Rico-Arce, Botany Department c/405, General Herbarium, British Museum (Natural History), Cromwell Road, London SW7 5BD, UK (taxonomy)

Chris Rollins, USA (see Achira)

Pedro A. Sánchez, USA (integrated farming in humid tropics) (see Basul)

Eduardo C. Schröder, USA (rhizobia) (see Basul)

C.A. Schroeder, USA (see Ahipa)

Richard E. Schultes, USA (ethnobotany) (see Achira)

W.R. Sykes, New Zealand (see Mashua)

James L. Zarucchi, USA (see Basul)

PASSIONFRUITS

Andean Region

Hernán Caballero, Ecuador (see Quinoa)

Andres Contreras M., Chile (germplasm) (see Oca)

João Lúcio de Azevedo, BioPlanta do Brazil, Caixa Postal 1141, Campinas São Paulo 13.100, Brazil (mycorrhiza, tissue culture)
César del Carpio Merino, Peru (see Achira)
Bernardo Eraso Silva, Colombia (see Squashes)
Linda Albert de Escobar, Departamento de Biología, Universidad de Antioquia, Apartado Aéreo 1226, Ciudad Universitaria, Medellín, Antioquia, Colombia (germplasm, economic use, and taxonomy)
Jorge Fabara, Ecuador (see Capuli Cherry)
Fondo Nacional de Investigaciones Agropecuarias (FONAIAP), Venezuela (see Achira)
Juan Gaviria R., Venezuela (see Ulluco)
Instituto Nacional de Investigaciones Agropecuarias (INIAP), Ecuador (see Achira)
Pascual Londoño, Colombia (trade association) (see Berries)
Sady Majino Bernardo, Peru (see Arracacha)
Miguel Morán Robles, Peru (see Cherimoya)
C. Percy Nuñez Vargas, Peru (see Basul)
Margareta Maria Restrepo, Department of Biology, Universidad de Antioquia, Apartado Aéreo 1226, Ciudad Universitaria, Medellín, Antioquia, Colombia
Raúl Ríos E., Bolivia (see Achira)
Víctor Rodríguez, Ecuador (see Squashes)
Carlos Roersch, Peru (see Mashua)
Carlos Ruggiero, Facultad de Ciencias Agrarias e Veterinarias, Universidade Estadual Paulista, Campus de Jaboticabal, Jaboticabal, Brazil
Jorge Soria V., Ecuador (see Squashes)
Modesto Soria V., Ecuador (grower) (see Goldenberry)
Julio Cesar Toro Meza, Colombia (see Berries)

Other Countries

Robert Barnum, USA (see Cherimoya)
Gilles Bourgeois, Canada (see Basul)
Ricardo Bressani, Guatemala (nutrition) (see Achira)
Tony Brown, USA (see Cherimoya)
Ferdinando Cossio, Italy (see Capuli Cherry)
John Couch, USA (germplasm) (see Highland Papayas)
Stuart N. Dawes, New Zealand (see Cherimoya)
William Doty, USA (germplasm) (see Peppers)
Dick and Annemarie Endt, New Zealand (see Achira)
W. Hardy Eshbaugh, USA (see Achira)
Tunde Fatunla, Department of Plant Science, Obafemi Awolowo University, Ile-Ife, Nigeria (processing)
Zaccheaus Oyesiju Gbile, Nigeria (see Potatoes)
Heiner Goldbach, West Germany (seed storage and germination) (see Potatoes)
B.W.W. Grout, UK (see Achira)
Hawaiian Fruit Preserving Company, Limited, USA (fruit products) (see Goldenberry)
J.L. Hudson, USA (see Kiwicha)
C. Antonio Jiménez Aparicio, Mexico (see Arracacha)
Peter Moller Jorgensen, Botanical Institute, University of Aarhus, 68 Nordlandsvej, Risskov DK-8240, Denmark
Robert Knight, USA (see Highland Papayas)
G. Linsley-Noakes, South Africa (see Berries)
Professor Lüdders, West Germany (see Berries)
Richard McCain, USA (horticulture) (see Yacon)
John M. MacDougal, Missouri Botanical Garden, PO Box 299, St. Louis, Missouri 63166–0299, USA (taxonomy)
Simon E. Malo, Honduras (see Capuli Cherry)
Franklin W. Martin, USA (see Achira)
D.A. Miller, UK (see Goldenberry)

APPENDIX C393

Dewey Moore, Department of Plant Pathology, University of Wisconsin, 1630 Linden Drive, Madison, Wisconsin 53706, USA
Mónica Moraes, Botanical Institute, University of Aarhus, 68 Nordlandsvej, Risskov DK-8240, Denmark
Julia F. Morton, USA (general information) (see Berries)
J.S. Mugawara, Faculty of Agriculture, J.C.M. Odungu, Department of Agronomy, Makerere University, Kampala, Uganda (shipping)
Henry Y. Nakasone, USA (see Goldenberry)
John P. Ogier, UK (see Goldenberry)
John Palmer, New Zealand (see Ahipa)
Tej Partap, Nepal (mountain agriculture, genetic resources) (see Maca)
Antonio M. Pinchinat, St. Lucia (see Basul)
Martin L. Price, USA (limited germplasm available) (see Achira)
Greg J. Pringle, Division of Horticulture and Processing, Department of Scientific and Industrial Research (DSIR), Private Bag, Auckland, New Zealand
Rare Fruit Council International, Inc., USA (see Peppers)
Herman Real, USA (tropical ecosystems) (see Cherimoya)
Chris Rollins, USA (see Achira)
Pat R. Sale, Citrus and Subtropical Horticulture, Ministry of Agriculture and Fisheries (MAF), Private Bag, Tauranga, New Zealand
Elizabeth Schneider, USA (food preparation) (see Yacon)
C.A. Schroeder, USA (see Ahipa)
Richard E. Schultes, USA (ethnobotany) (see Achira)
Steven Spangler, USA (see Achira)
Claude Sweet, USA (see Capuli Cherry)
W.R. Sykes, New Zealand (see Mashua)
Louis Trap, New Zealand (see Ahipa)
Guillermo Veliz, USA (importation to U.S.) (see Oca)
Alejo von der Pahlen, Italy (see Oca)
Benjamin H. Waite, USA (see Oca)
Donald S.C. Wright, Division of New Crops Section, Crop Research, Department of Scientific and Industrial Research (DSIR), Private Bag, Christchurch, New Zealand
Victor A. Wynne, Haiti (see Berries)

PEPINO

Andean Region

Segundo Alandia, Bolivia (pathology) (see Mashua)
Antonio Bacigalupo, Chile (see Potatoes)
Cesar Barberan, Ecuador (see Peppers)
David Baumann, Peru (see Goldenberry)
Andres Contreras M., Chile (germplasm) (see Oca)
Carmen E.B. de Rojas, Venezuela (see Peppers)
Michael Hermann, Peru (see Achira)
Fabio Higuita Muñoz, Colombia (see Arracacha)
Miguel Holle, Peru (see Arracacha)
Instituto Nacional de Investigaciones Agropecuarias (INIAP), Ecuador (see Achira)
Christian Krarup H., Facultad de Agronomía, Universidad Católica, Casilla 6177, Santiago, Chile
Lúis E. López Jaramilló, Colombia (Solanaceae) (see Potatoes)
Sady Majino Bernardo, Peru (see Arracacha)
Miguel Morán Robles, Peru (see Cherimoya)
Laura Muñoz Espín, Ecuador (see Arracacha)
Raúl Ríos E., Bolivia (see Achira)
Víctor Rodríguez, Ecuador (see Squashes)

Juan Solano Lazo, Ecuador (see Achira)
Jorge Soria V., Ecuador (see Squashes)

Other Countries

Gregory J. Anderson, USA (ethnobotany and taxonomy) (see Achira)
Suzanne Ashworth, USA (see Kiwicha)
Robert Barnum, USA (see Cherimoya)
Rudolf F.W. Binsack, USA (plant production) (see Tarwi)
Lynn A. Bohs, USA (see Arracacha)
Gilles Bourgeois, Canada (see Basul)
Ricardo Bressani, Guatemala (nutrition) (see Achira)
Carl Campbell, USA (see Cherimoya)
Alan Child, UK (Solanaceae) (see Goldenberry)
Ferdinando Cossio, Italy (see Capuli Cherry)
James and Joy Crawshaws, Pepino Growers of New Zealand Ltd., Waimauku, Northland,
    New Zealand
W.G. D'Arcy, USA (see Potatoes)
Stuart N. Dawes, New Zealand (see Cherimoya)
William Doty, USA (germplasm) (see Peppers)
Dick and Annemarie Endt, New Zealand (see Achira)
Jan Engels, West Germany (see Potatoes)
Michel Fanton, Australia (see Achira)
Heiner Goldbach, West Germany (seed storage and germination) (see Potatoes)
B.W.W. Grout, UK (see Achira)
Keith R.W. Hammett, Division of Horticulture and Processing, Department of Scientific and
    Industrial Research (DSIR), Mount Albert Research Centre, 120 Mt. Albert Road, Private
    Bag, Auckland 3, New Zealand
Doug Hammonds, New Zealand (fruit export) (see Goldenberry)
Jane Harman, New Zealand (see Highland Papayas)
Kathleen Haynes, USA (see Potatoes)
Alan M. Kapuler, USA (see Achira)
G. Linsley-Noakes, South Africa (see Berries)
Janet S. Luis, Philippines (see Potatoes)
Richard McCain, USA (horticulture) (see Yacon)
Cyrus McKell, USA (see Potatoes)
Carol Mackey, USA (see Nuñas)
N.A. Martin, Division of Entomology, Department of Scientific and Industrial Research
    (DSIR), Private Bag, Auckland, New Zealand
Jess R. Martineau, USA (biotechnology) (see Mashua)
Donald Maynard, Gulf Coast Research and Education Center, University of Florida, 5007
    60th Street East, Bradenton, Florida 34203, USA
D.A. Miller, UK (see Goldenberry)
Michael Morley-Bunker, Pepino Project, Horticulture Department, Lincoln College, Canterbury,
    New Zealand
Michael Nee, USA (see Squashes)
John P. Ogier, UK (see Goldenberry)
John Palmer, New Zealand (see Ahipa)
Dov Pasternak, Israel (see Potatoes)
Jean-Yves Peron, France (germplasm) (see Goldenberry)
Martin L. Price, USA (limited germplasm available) (see Achira)
Rare Fruit Council International, Inc., USA (see Peppers)
Charles M. Rick, USA (see Arracacha)
John M. Riley, USA (general information, germplasm) (see Oca)
Chris Rollins, USA (see Achira)
Elizabeth Schneider, USA (food preparation) (see Yacon)
C.A. Schroeder, USA (see Ahipa)
Richard E. Schultes, USA (ethnobotany) (see Achira)

Steven Spangler, USA (see Achira)
Calvin R. Sperling, USA (see Ahipa)
Don Steenstra, Waikato Polytechnic, Tristram Street, Box 982, Hamilton, New Zealand
Claude Sweet, USA (see Capuli Cherry)
John Swift, USA (see Yacon)
W.R. Sykes, New Zealand (see Mashua)
Louis Trap, New Zealand (see Ahipa)
Noel Turner, Division of Basic Science, Department of Scientific and Industrial Research
    (DSIR), Private Bag, Wellington, New Zealand
Guillermo Veliz, USA (importation to U.S.) (see Oca)
Alejo von der Pahlen, Italy (see Oca)
Benjamin H. Waite, USA (see Oca)
Donald S.C. Wright, New Zealand (see Passionfruit)
R.B. Wynn-Williams, New Zealand (see Oca)

TAMARILLO (Tree Tomato)

Andean Region

Guillerma Aníbal Albornoz Pazmiño, Ecuador (germplasm) (see Potatoes)
Alonso Atarihuana C., Facultad de Ciencias Agrícolas, Universidad Central del Ecuador,
    Ciudadela Universitaria, Apartado 166, Quito, Ecuador (germplasm)
Antonio Bacigalupo, Chile (see Potatoes)
Marco Barahona, Programa Fruticultura, Estación Experimental Tumbaco, Instituto Nacional
    de Investigaciones Agropecuarias (INIAP), Avenida Eloy Alfaro y Amazonas, Casilla 2600,
    Quito, Ecuador
Hernán Caballero, Ecuador (see Quinoa)
Carmen E.B. de Rojas, Venezuela (see Peppers)
Jorge Fabara, Ecuador (see Capuli Cherry)
Fondo Nacional de Investigaciones Agropecuarias (FONAIAP), Venezuela (see Achira)
Alexander Grobman, Peru (germplasm collection) (see Oca)
Joy C. Horton Hofmann, Ecuador (*Cyphomandra cajanumensis*) (see Achira)
Instituto Nacional de Investigaciones Agropecuarias (INIAP), Ecuador (see Achira)
Mario Lobo, Colombia (see Peppers)
Lúis E. López Jaramilló, Colombia (Solanaceae) (see Potatoes)
Miguel Morán Robles, Peru (see Cherimoya)
Laura Muñoz Espín, Ecuador (see Arracacha)
Manuel Orihuela Herrera, Centro de Estudios Rurales Andinos "Bartolomé de las Casas,"
    Comisión de Coordinación de Tecnología Andina (CCTA), SEDE Institucional, Avenida
    Tullumayo 465, Apartado 477, Cusco, Peru
Raúl Ríos E., Bolivia (see Achira)
Víctor Rodríguez, Ecuador (see Squashes)
Isidoro Sánchez Vega, Peru (see Arracacha)
Jorge Soria V., Ecuador (see Squashes)
Modesto Soria V., Ecuador (grower) (see Goldenberry)
Julio Cesar Toro Meza, Colombia (see Berries)
Cesar C. Vargas Calderon, Peru (see Potatoes)
Ramiro Velastegui, Ecuador (pathology) (see Tarwi)

Other Countries

Gregory J. Anderson, USA (ethnobotany and taxonomy) (see Achira)
Ariel Azael, Haiti (see Ahipa)
Lynn A. Bohs, USA (taxonomy) (see Arracacha)
Ricardo Bressani, Guatemala (nutrition) (see Achira)

Carl Campbell, USA (see Cherimoya)
Alan Child, UK (Solanaceae, taxonomy) (see Goldenberry)
W.G. D'Arcy, USA (see Potatoes)
Stuart N. Dawes, New Zealand (see Cherimoya)
William Doty, USA (germplasm) (see Peppers)
Dick and Annemarie Endt, New Zealand (see Achira)
Jan Engels, West Germany (see Potatoes)
William Charles Evans, UK (see Potatoes)
Heiner Goldbach, West Germany (seed storage and germination) (see Potatoes)
B.W.W. Grout, UK (see Achira)
V.L. Guzman, USA (see Quinoa)
Richard A. Hamilton, USA (see Berries)
Doug Hammonds, New Zealand (fruit export) (see Goldenberry)
Jane Harman, New Zealand (see Highland Papayas)
Roy Hart, New Zealand (see Berries)
J.L. Hudson, USA (see Kiwicha)
Nancy Jarris, USA (see Goldenberry)
Alan M. Kapuler, USA (see Achira)
John Laurenson, New Zealand Tamarillo Growers Association, Inc., PO Box 258, Kerikeri,
    New Zealand
Professor Lüdders, West Germany (see Berries)
Janet S. Luis, Philippines (see Potatoes)
Richard McCain, USA (horticulture) (see Yacon)
Cyrus McKell, USA (see Potatoes)
N.A. Martin, New Zealand (see Pepino)
Michael Nee, USA (see Squashes)
John Palmer, New Zealand (see Ahipa)
Tej Partap, Nepal (mountain agriculture, genetic resources) (see Maca)
Dov Pasternak, Israel (see Potatoes)
Martin L. Price, USA (limited germplasm available) (see Achira)
Greg J. Pringle, New Zealand (see Passionfruit)
Rare Fruit Council International, Inc., USA (see Peppers)
Herman Real, USA (tropical ecosystems) (see Cherimoya)
Charles M. Rick, USA (see Arracacha)
John M. Riley, USA (general information, germplasm) (see Oca)
Chris Rollins, USA (see Achira)
Pat R. Sale, New Zealand (see Passionfruit)
Elizabeth Schneider, USA (food preparation) (see Yacon)
C.A. Schroeder, USA (see Ahipa)
Richard E. Schultes, USA (ethnobotany) (see Achira)
Steven Spangler, USA (see Achira)
Calvin R. Sperling, USA (see Ahipa)
Claude Sweet, USA (see Capuli Cherry)
W.R. Sykes, New Zealand (see Mashua)
Loren Toomey, USA (grower) (see Goldenberry)
Louis Trap, New Zealand (see Ahipa)
Benjamin H. Waite, USA (see Oca)
Donald S.C. Wright, New Zealand (see Passionfruit)
Victor A. Wynne, Haiti (see Berries)
R.B. Wynn-Williams, New Zealand (see Oca)

QUITO PALM

Andean Region

Stephan G. Beck, Herbario Nacional de Bolivia, Cajón Postal 20–127, La Paz, Bolivia
    (ecology)

José Calzada Benza, Peru (see Capuli Cherry)
Carmen E.B. de Rojas, Venezuela (see Peppers)
Instituto Interamericano de Cooperación para la Agricultura (IICA), Bolivia (see Mashua)
Instituto Nacional de Investigaciones Agropecuarias (INIAP), Ecuador (see Achira)
Miguel Morán Robles, Peru (see Cherimoya)

Other Countries

Ariel Azael, Haiti (see Ahipa)
Michael Balick, Institute of Economic Botany, New York Botanical Garden, Bronx, New York
    10458–9980, USA
H. Balslev, Botanical Institute, University of Aarhus, 68 Nordlandsvej, Risskov DK-8240,
    Denmark (Ecuadorean palms)
A. Barfod, Botanical Institute, University of Aarhus, 68 Nordlandsvej, Risskov DK-8240,
    Denmark (Ecuadorean palms)
Warren Dolby, 5331 Golden Gate Avenue, Oakland, California 94618, USA (limited
    germplasm)
John Dransfield, Royal Botanic Gardens, Kew, Richmond, Surrey TW9 3AE, UK
J. Garrin Fullington, PO Box 10411, Hilo, Hawaii 96721, USA (limited germplasm)
Andrew Henderson, New York Botanical Garden, Bronx, New York 10458–9980, USA
International Board for Plant Genetic Resources (IBPGR), Italy (germplasm information)
    (see Achira)
Dennis Johnson, DESFIL Coordinator, Tropical Research and Development, Inc., 624 Ninth
    Street, Sixth Floor, Washington, DC 20001, USA
Peter Moller Jorgensen, Denmark (see Passionfruit)
Mónica Moraes, Denmark (*Parajubaea torallyi*) (see Passionfruit)

WALNUTS

Andean Region

Emilio Barahona Chura, Peru (see Capuli Cherry)
Cesar Barberan, Ecuador (see Peppers)
Ramón Ferreyra, Peru (see Highland Papayas)
Jaime Gaete Calderon, Chile (see Basul)
Juan Gaviria R., Venezuela (see Ulluco)
Instituto Nacional de Investigaciones Agropecuarias (INIAP), Ecuador (see Achira)
Ricardo Jon Llap, Peru (see Capuli Cherry)
Laura Muñoz Espín, Ecuador (see Arracacha)
David Ocaña Vidal, Peru (see Capuli Cherry)
José Pretell Chiclote, Peru (see Capuli Cherry)
Raul Salazar C., Colombia (see Peppers)
K. Thelen, Food and Agriculture Organization of the United Nations (FAO), Avenida Santa
    Maria 6700, Casilla 10095, Santiago, Chile

Other Countries

Robert Barnum, USA (see Cherimoya)
Gerardo Budowski, Costa Rica (see Pacay)
Edilberto Camacho Vargas, Apartado 361, Guadalupe, Goicoechea, San José, Costa Rica
Dick and Annemarie Endt, New Zealand (see Achira)
Ernest P. Imle, USA (see Highland Papayas)
International Board for Plant Genetic Resources (IBPGR), Italy (germplasm information)
    (see Achira)

Gale McGranahan, Department of Pomology, University of California, Davis, California 95616, USA

Wayne F. Manning, 27 Brown Street, Lewisburg, Pennsylvania 17837, USA (taxonomy)

William L. Nelson, USA (see Pacay)

John Palmer, New Zealand (see Ahipa)

Tej Partap, Nepal (mountain agriculture, genetic resources) (see Maca)

Chris Rollins, USA (see Achira)

Richard E. Schultes, USA (ethnobotany) (see Achira)

Shao Qiquan, China (see Potatoes)

Steven Spangler, USA (see Achira)

Calvin R. Sperling, USA (see Ahipa)

Donald E. Stone, Department of Botany, Duke University, Durham, North Carolina 27706, USA

Appendix D

Biographical Sketches of Panel Members

HUGH L. POPENOE (*Chairman*) is professor of soils, agronomy, botany, and geography and director of the Center for Tropical Agriculture and International Programs (Agriculture) at the University of Florida, Gainesville. His principal research interest has been tropical agriculture and land use. His early work in shifting cultivation is one of the few contributions to knowledge of this system. He was born in Guatemala and has traveled and worked in most of the countries in the tropical areas of Latin America, Asia, and Africa. He is chairman of the Board of Trustees of the Escuela Agrícola Panamericana in Honduras, visiting lecturer on tropical public health at the Harvard School of Public Health, and is a fellow of the American Society of Agronomy, the American Geographical Society, and the International Soils Science Society. His father, Wilson Popenoe (1892-1975), was a plant explorer who traveled extensively through the Andes and was a pioneer in the promotion of the more extensive use of Andean fruits and other crops.

STEVEN R. KING is currently chief botanist for Latin America for The Nature Conservancy's Latin America Science Program. Prior to 1989, Dr. King was a research associate with the Committee on Managing Global Genetic Resources of the National Research Council's Board on Agriculture. He received his B.A. from the College of the Atlantic in 1980, and his M.S. from the City University of New York in 1986. In 1988, he received his Ph.D. from the City University of New York, where he worked with the Andean tuber crop complex (potatoes, oca, mashua, ulluco, and maca) and searched for the wild ancestors of modern cultigens. He has done field work throughout Latin America as well as in Southeast Asia. From 1986–1988 he was a fellow of The New York Botanical Garden's Institute of Economic Botany.

JORGE LEÓN is one of the world's foremost experts on Andean agriculture. A native of Costa Rica, he received his Ph.D. in botany from Washington University (under the famed economic botanist Edgar Anderson). He became botanist and head of the Plant Industry Department at IICA in Turrialba, Costa Rica, and then spent seven years heading the Andean Zone Research Program at IICA in Lima, Peru. Subsequently, he was with the FAO in Rome and was for many years chief of FAO's Crop Ecology and Genetic Resources Unit. After leaving FAO he was chief of the Genetic Resource Unit, Centro Agronómica Tropical de Investigación y Enseñanza (CATIE) until his retirement. Dr. León is a fellow of the Linnean Society and author of some 60 technical articles, 5 bulletins, 2 books, and 40 technical missions and consultations. His book *Plantas Alimenticias Andinas* is a classic survey of the native crops of the Andes.

LUIS SUMAR KALINOWSKI, Centro de Investigaciones de Cultivos Andinos, Universidad Nacional Técnica del Altiplano, Cuzco, Peru, received his degree of agricultural engineering from the University of Cuzco in 1961. Postgraduate studies were in phytogenics at the National University of Cuzco. From 1964 to 1975, he was associate professor of vegetative therapeutics at the University of Cuzco, while also serving with the Department of Agricultural Development of Cuzco Corporation (a quasi-governmental institution). He was concurrently an instructor at the University of Lima and the National Agricultural University at La Molina during 1974. Since 1975, Dr. Sumar has been a principal professor in the Agriculture Department of the University of Cuzco. He was made head of the department in 1981. Dr. Sumar has been involved with nutrition, pathology, genetic conservation, and plant improvement for many years, and has traveled worldwide as a plant collector and as a consultant. In 1982 he became the only civilian recipient of Peru's Gold Medal of the Order of the Sun, in recognition of his contributions to the nutritional well-being of the poor.

NOEL D. VIETMEYER, staff officer and technical writer for this study, is a senior program officer of the Board on Science and Technology for International Development. A New Zealander with a Ph.D. in organic chemistry from the University of California, Berkeley, he now works on innovations in science and technology that are important for the future of developing countries.

INDEX OF PLANTS

WILLIAM E. GORDON, Foreign Secretary, National Academy of Sciences, *ex officio*

Board on Science and Technology for International Development
Publications and Information Services (HA–476E)
Office of International Affairs
National Research Council
2101 Constitution Avenue, N.W.
Washington, D.C. 20418 USA

How to Order BOSTID Reports

BOSTID manages programs with developing countries on behalf of the U.S. National Research Council. Reports published by BOSTID are sponsored in most instances by the U.S. Agency for International Development. They are intended for distribution to readers in developing countries who are affiliated with governmental, educational, or research institutions, and who have professional interest in the subject areas treated by the reports.

BOSTID books are available from selected international distributors. For more efficient and expedient service, please place your order with your local distributor. (See list on back page.) Requestors from areas not yet represented by a distributor should send their orders directly to BOSTID at the above address.

Energy

33. **Alcohol Fuels: Options for Developing Countries.** 1983, 128pp. Examines the potential for the production and utilization of alcohol fuels in developing countries. Includes information on various tropical crops and their conversion to alcohols through both traditional and novel processes. ISBN 0–309–04160–0.

36. **Producer Gas: Another Fuel for Motor Transport.** 1983, 112pp. During World War II Europe and Asia used wood, charcoal, and coal to fuel more than a million gasoline and diesel vehicles. However, the technology has since been virtually forgotten. This report reviews producer gas and its modern potential. ISBN 0–309–04161–9.

56. **The Diffusion of Biomass Energy Technologies in Developing Countries.** 1984, 120pp. Examines economic, cultural, and political factors

that affect the introduction of biomass-based energy technologies in developing countries. It includes information on the opportunities for these technologies as well as conclusions and recommendations for their application. ISBN 0–309–04253–4.

Technology Options

14. More Water for Arid Lands: Promising Technologies and Research Opportunities. 1974, 153pp. Outlines little-known but promising technologies to supply and conserve water in arid areas. ISBN 0–309–04151–1.

21. Making Aquatic Weeds Useful: Some Perspectives for Developing Countries. 1976, 175pp. Describes ways to exploit aquatic weeds for grazing, and by harvesting and processing for use as compost, animal feed, pulp, paper, and fuel. Also describes utilization for sewage and industrial wastewater. ISBN 0–309–04153–X.

34. Priorities in Biotechnology Research for International Development: Proceedings of a Workshop. 1982, 261pp. Report of a workshop organized to examine opportunities for biotechnology research in six areas: 1) vaccines, 2) animal production, 3) monoclonal antibodies, 4) energy, 5) biological nitrogen fixation, and 6) plant cell and tissue culture. ISBN 0–309–04256–9.

61. Fisheries Technologies for Developing Countries. 1987, 167pp. Identifies newer technologies in boat building, fishing gear and methods, coastal mariculture, artificial reefs and fish aggregating devices, and processing and preservation of the catch. The emphasis is on practices suitable for artisanal fisheries. ISBN 0–309–04260–7.

Plants

25. Tropical Legumes: Resources for the Future. 1979, 331pp. Describes plants of the family Leguminosae, including root crops, pulses, fruits, forages, timber and wood products, ornamentals, and others. ISBN 0–309–04154–6.

37. Winged Bean: A High Protein Crop for the Tropics. 1981, 2nd edition, 59pp. An update of BOSTID's 1975 report of this neglected tropical legume. Describes current knowledge of winged bean and its promise. ISBN 0–309–04162–7.

47. Amaranth: Modern Prospects for an Ancient Crop. 1983, 81pp. Before the time of Cortez, grain amaranths were staple foods of the Aztec and Inca. Today this nutritious food has a bright future. The report discusses vegetable amaranths also. ISBN 0–309–04171–6.

53. Jojoba: New Crop for Arid Lands. 1985, 102pp. In the last 10 years, the domestication of jojoba, a little-known North American desert shrub, has been all but completed. This report describes the plant and its promise to provide a unique vegetable oil and many likely industrial uses. ISBN 0–309–04251–8.

63. Quality-Protein Maize. 1988, 100pp. Identifies the promise of a nutritious new form of the planet's third largest food crop. Includes chapters on the importance of maize, malnutrition and protein quality, experiences with quality-protein maize (QPM), QPM's potential uses in feed and food, nutritional qualities, genetics, research needs, and limitations. ISBN 0–309–04262–3.

64. Triticale: A Promising Addition to the World's Cereal Grains. 1989, 105pp. Outlines the recent transformation of triticale, a hybrid between wheat and rye, into a food crop with much potential for many marginal lands. Includes chapters on triticale's history, nutritional quality, breeding, agronomy, food and feed uses, research needs, and limitations. ISBN 0–309–04263–1.

67. Lost Crops of the Incas. 1989, approx. 415pp. The Andes is one of the seven major centers of plant domestication, but the world is largely unfamiliar with its native food crops. When the Conquistadores brought the potato to Europe, they ignored the other domesticated Andean crops — fruits, legumes, tubers, and grains that had been cultivated for centuries by the Incas. This book focuses on 30 of the "forgotten" Incan crops that show promise not only for the Andes, but for warm-temperate, subtropical, and upland tropical regions in many parts of the world. ISBN 0–309–04264-X

69. Saline Agriculture: Salt-Tolerant Plants for Developing Countries. 1989, approx. 150pp. The purpose of this report is to create greater awareness of salt-tolerant plants and the special needs they may fill in developing countries. Examples of the production of food, fodder, fuel, and other products are included. Salt-tolerant plants can use land and water unsuitable for conventional crops and can harness saline resources that are generally neglected or considered as impediments to, rather than opportunities for, development. ISBN 0-309-04266-6.

Innovations in Tropical Forestry

35. **Sowing Forests from the Air.** 1981, 64pp. Describes experiences with establishing forests by sowing tree seed from aircraft. Suggests testing and development of the techniques for possible use where forest destruction now outpaces reforestation. ISBN 0–309–04257–7.

40. **Firewood Crops: Shrub and Tree Species for Energy Production.** Volume II, 1983, 92pp. Examines the selection of species of woody plants that seem suitable candidates for fuelwood plantations in developing countries. ISBN 0–309–04164–3 (Vol. II).

41. **Mangium and Other Fast-Growing Acacias for the Humid Tropics.** 1983, 63pp. Highlights 10 acacia species that are native to the tropical rainforest of Australasia. That they could become valuable forestry resources elsewhere is suggested by the exceptional performance of *Acacia mangium* in Malaysia. ISBN 0–309–04165–1.

42. **Calliandra: A Versatile Small Tree for the Humid Tropics.** 1983, 56pp. This Latin American shrub is being widely planted by the villagers and government agencies in Indonesia to provide firewood, prevent erosion, provide honey, and feed livestock. ISBN 0–309–04166-X.

43. **Casuarinas: Nitrogen-Fixing Trees for Adverse Sites.** 1983, 118pp. These robust, nitrogen-fixing, Australasian trees could become valuable resources for planting on harsh, eroding land to provide fuel and other products. Eighteen species for tropical lowlands and highlands, temperate zones, and semiarid regions are highlighted. ISBN 0–309–04167–8.

52. **Leucaena: Promising Forage and Tree Crop in Developing Countries.** 1984, 2nd edition, 100pp. Describes a multi-purpose tree crop of potential value for much of the humid lowland tropics. Leucaena is one of the fastest growing and most useful trees for the tropics. ISBN 0–309–04250-X.

Managing Tropical Animal Resources

32. The Water Buffalo: New Prospects for an Underutilized Animal. 1981, 188pp. The water buffalo is performing notably well in recent trials in such unexpected places as the United States, Australia, and Brazil. Report discusses the animal's promise, particularly emphasizing its potential for use outside Asia. ISBN 0–309–04159–7.

44. Butterfly Farming in Papua New Guinea. 1983, 36pp. Indigenous butterflies are being reared in Papua New Guinea villages in a formal government program that both provides a cash income in remote rural areas and contributes to the conservation of wildlife and tropical forests. ISBN 0-309-04168-6.

45. Crocodiles as a Resource for the Tropics. 1983, 60pp. In most parts of the tropics, crocodilian populations are being decimated, but programs in Papua New Guinea and a few other countries demonstrate that, with care, the animals can be raised for profit while protecting the wild populations. ISBN 0–309–04169–4.

46. Little-Known Asian Animals with a Promising Economic Future. 1983, 133pp. Describes banteng, madura, mithan, yak, kouprey, babirusa, Javan warty pig, and other obscure but possibly globally useful wild and domesticated animals that are indigenous to Asia. ISBN 0–309–04170–8.

Health

49. Opportunities for the Control of Dracunculiasis. 1983, 65pp. Dracunculiasis is a parasitic disease that temporarily disables many people in remote, rural areas in Africa, India, and the Middle East. Contains the findings and recommendations of distinguished scientists who were brought together to discuss dracunculiasis as an international health problem. ISBN 0–309–04172–4.

55. Manpower Needs and Career Opportunities in the Field Aspects of Vector Biology. 1983, 53pp. Recommends ways to develop and train the manpower necessary to ensure that experts will be available in the future to understand the complex ecological relationships of vectors with human hosts and pathogens that cause such diseases as malaria, dengue fever, filariasis, and schistosomiasis. ISBN 0–309–04252–6.

60. U.S. Capacity to Address Tropical Infectious Diseases. 1987, 225pp. Addresses U.S. manpower and institutional capabilities in both the public and private sectors to address tropical infectious disease problems. ISBN 0–309–04259–3.

Resource Management

50. Environmental Change in the West African Sahel. 1984, 96pp. Identifies measures to help restore critical ecological processes and thereby increase sustainable production in dryland farming, irrigated agriculture, forestry and fuelwood, and animal husbandry. Provides baseline information for the formulation of environmentally sound projects. ISBN 0–309–04173–2.

51. Agroforestry in the West African Sahel. 1984, 86pp. Provides development planners with information regarding traditional agroforestry systems—their relevance to the modern Sahel, their design, social and institutional considerations, problems encountered in the practice of agroforestry, and criteria for the selection of appropriate plant species to be used. ISBN 0–309–04174–0.

General

65. Science and Technology for Development: Prospects Entering the Twenty-First Century. 1988. 79pp. This report commemorates the twenty-fifth anniversary of the U.S. Agency for International Development. The symposium on which this report is based provided an excellent opportunity to describe and assess the contributions of science and technology to the development of Third World countries and to focus attention on what science and technology are likely to accomplish in the decades to come.

Forthcoming Books from BOSTID

The Improvement of Tropical and Subtropical Rangelands. 1989. This report characterizes tropical and subtropical rangelands, describes social adaptations to these rangelands, discusses the impact of socioeconomic and political change upon the management of range resources, and explores culturally and ecologically sound approaches to rangeland rehabilitation. Selected case studies are included.

68. Microlivestock: Little-Known Small Animals with a Promising Economic Future. 1989, approx. 300pp. Discusses the promise of small breeds and species of livestock for Third World villages. Identifies more than 40 species, including miniature breeds of cattle, sheep, goats, and pigs; eight types of poultry; rabbits; guinea pigs and other rodents; dwarf deer and antelope; iguanas; and bees. ISBN 0-309-04265-8.

Traditional Fermented Foods. (1990)

For More Information

To receive more information about BOSTID reports and programs, please fill in the attached coupon and mail it to:

Board on Science and Technology for International Development
Publications and Information Services (HA–476E)
Office of International Affairs
National Research Council
2101 Constitution Avenue, N.W.
Washington, D.C. 20418 USA

Your comments about the value of these reports are also welcome.

Name _____
Title _____
Institution _____
Street Address _____

City _____
Country _____Postal Code _____

67

Name _____
Title _____
Institution _____
Street Address _____

City _____
Country _____Postal Code _____

67